Disaster Risk Reduction

Methods, Approaches and Practices

Series Editor

Rajib Shaw, Keio University, Shonan Fujisawa Campus, Fujisawa, Japan

Disaster risk reduction is a process that leads to the safety of communities and nations. After the 2005 World Conference on Disaster Reduction, held in Kobe, Japan, the Hyogo Framework for Action (HFA) was adopted as a framework for risk reduction. The academic research and higher education in disaster risk reduction has made, and continues to make, a gradual shift from pure basic research to applied, implementation-oriented research. More emphasis is being given to multi-stakeholder collaboration and multi-disciplinary research. Emerging university networks in Asia, Europe, Africa, and the Americas have urged process-oriented research in the disaster risk reduction field. With this in mind, this new series will promote the output of action research on disaster risk reduction, which will be useful for a wide range of stakeholders including academicians, professionals, practitioners, and students and researchers in related fields. The series will focus on emerging needs in the risk reduction field, starting from climate change adaptation, urban ecosystem, coastal risk reduction, education for sustainable development, community-based practices, risk communication, and human security, among other areas. Through academic review, this series will encourage young researchers and practitioners to analyze field practices and link them to theory and policies with logic, data, and evidence. In this way, the series will emphasize evidence-based risk reduction methods, approaches, and practices.

All submitted proposals are carefully reviewed by the series editor.

Publication Ethics and Publication Malpractice Statement: https://www.springernature.com/gp/policies/book-publishing-policies.

Ethical, Conflict-of-Interest, and Editorial Policy Statement: https://www.springer.com/us/authors-editors/journal-author/journal-author-helpdesk/before-you-start/before-you-start/1330#c14214.

Lalit Kumar Dashora · Mohammed Firoz C. ·
Rajib Shaw
Editors

Urban Sustainability

Navigating Environmental Transition Challenges in Global South

Springer

Editors
Lalit Kumar Dashora
School of Global Studies
Thammasat University
Bangkok, Thailand

Mohammed Firoz C.
Department of Architecture and Planning
National Institute of Technology Calicut
Kozhikode, Kerala, India

Rajib Shaw
Graduate School of Media and Governance
Keio University
Fujisawa, Kanagawa, Japan

ISSN 2196-4106 ISSN 2196-4114 (electronic)
Disaster Risk Reduction
ISBN 978-981-96-4889-4 ISBN 978-981-96-4890-0 (eBook)
https://doi.org/10.1007/978-981-96-4890-0

This Springer imprint is published by the registered company Springer Nature Singapore Pte Ltd.
The registered company address is: 152 Beach Road, #21-01/04 Gateway East, Singapore 189721, Singapore

Preface

As urban areas and urban population surge worldwide, the Global South is at the forefront of unprecedented environmental and socio-economic challenges. Rapid urbanization in these regions brings about transformative pressures on natural resources, infrastructure, and societal wellbeing, exacerbating pre-existing vulnerabilities and creating new demands on urban systems. The book *Urban Sustainability: Navigating Environmental Transition Challenges in Global South* seeks to illuminate these challenges while also highlighting pathways toward resilient and sustainable urban futures.

The Global South, comprising a diverse range of low- and middle-income countries, faces a unique set of circumstances in its pursuit of sustainable development. Urban centers in these regions must grapple with the dual imperatives of addressing environmental degradation and meeting the socio-economic needs of fast-growing populations. Issues such as inadequate waste management, air and water pollution, poor housing quality, and limited access to green spaces intensify as cities expand. Furthermore, these areas often face political, economic, and infrastructural limitations that make adopting and implementing sustainable practices all the more challenging.

This book aims to provide a comprehensive exploration of urban sustainability from a multi-disciplinary perspective, drawing insights from environmental science, urban planning, social policy, and economics. Through detailed case studies and comparative analyses, it examines innovative solutions, policies, and community-based approaches that are emerging in the Global South. By looking at successful examples and lessons learned from cities across Asia and the Pacific, it provides a framework for understanding the paths toward sustainable urbanization that are both inclusive and contextually appropriate.

We hope that urban sustainability serves as a valuable resource for students, researchers, policymakers, and practitioners engaged in urban planning, development, and environmental policy. The journey toward sustainable urban transitions in the Global South is fraught with challenges but also filled with opportunities for transformative change. In an era where global environmental crises demand urgent action, this book underscores the essential role of urban sustainability in shaping a more equitable and resilient future for all.

We will be very happy if the readers find this book useful for their work.

Bangkok, Thailand — Lalit Kumar Dashora
Kozhikode, India — Mohammed Firoz C.
Fujisawa, Japan — Rajib Shaw

About This Book

Urban Sustainability: Navigating Environmental Transition Challenges in Global South delves into the pressing need for sustainable urban development in rapidly growing urban areas across the Global South. This comprehensive book explores the complex environmental, social, and economic challenges faced by urban areas in Asia and the Pacific as they confront rapid industrialization, climate change, and resource depletion.

Written for scholars, policymakers, urban planners, and students, the book examines the unique challenges and opportunities present in urban areas of developing countries in Global South. It highlights case studies from small to large size urban areas, showcasing how local governments and communities are innovating to risk reduction, build resilience, and improve quality of life for their residents. The book focuses on key aspects of sustainable urban development, including climate change adaptation and resilience, disaster risk reduction, public health, waste and water management, mobility and transportation, environment management.

One of the book's central themes is the need for inclusive and context-sensitive strategies that address the intersectionality of vulnerability, social equity, and environmental management. It emphasizes the role of participatory governance, community-led initiatives, and local knowledge in crafting effective solutions. With insights from experts in environmental science, urban studies, and social policy, the book offers a holistic approach to tackling these challenges, advocating for transformative practices that bridge the gap between urbanization and sustainability.

Through an engaging and accessible narrative, urban sustainability encourages readers to reimagine the future of urban living in the Global South, where sustainable growth is not just a possibility but a necessity. It is a vital resource for anyone interested in the evolving dynamics of urban resilience and sustainability in a rapidly changing world.

Contents

Editors and Contributors

About the Editors

Lalit Kumar Dashora is an Urban Risk and Resilience Specialist, currently based at the School of Global Studies, Thammasat University, Bangkok, Thailand. Prior to joining SGS at TU, he worked with Asian Development Bank (ADB) in Dushanbe, Tajikistan, and Asian Disaster Preparedness Center (ADPC) in Bangkok, Thailand. He holds a master's degree in Geo-Information Science for Earth Observation, Environmental Modeling and Management from the Faculty of Geo-Information Science and Earth Observation (formerly known as ITC), University of Twente, The Netherlands, and in Urban and Regional Planning from the Faculty of Planning, CEPT University, India. His research interests focus on risk-informed, climate-compatible urban planning and development in developing countries, Small Island Developing States (SIDS), and other climate-fragile regions in Asia and the Pacific. He is also Adviser for the Warning Research Center (WRC) of University College of London (UCL).

Mohammed Firoz C. currently serves as an Associate Professor and the Head of the Department of Architecture and Planning, at the National Institute of Technology (NIT), Calicut, India. He had earlier served as a visiting faculty at the Asian Institute of Technology Bangkok, Thailand, and at the School of Architecture, Architecture Association (AA) London. He is also currently holding the position of research fellow at the Shinawatra University, Thailand. His fields of interest include research on metropolitan and regional planning, sustainable design and development of cities, creative economy for cities, peri-urban studies, environmental management, and climate adaptation. He is actively associated as an adviser and consultant to several organizations besides being in several research committees, editorial boards of reputed journals, etc.

Rajib Shaw is a professor in Graduate School of Media and Governance of Keio University, Japan. He is co-founder of a Delhi-based social entrepreneur startup, Resilience Innovation Knowledge Academy (RIKA), and founder and chair of the board of SEEDS Asia, a Japanese NGO. He is the Co-chair of the United Nations Office for Disaster Risk Reduction (UNDRR) Asia Pacific Science Technology Advisory Group (AP-STAG), and CLA for IPCC's 6th Assessment Report. He has 79 books and more than 450 research papers in the fields of environment, disaster management and climate change. He is the editor in chief of Progress in Disaster Science. He is the recipient of "Pravasi Bharatiya Samman Award (PBSA)" in 2021 for his contribution in education sector. PBSA is the highest honor conferred on overseas Indian and person of Indian origin from the President of India. He also received United Nations Sasakawa Award in 2022 for his lifetime contributions in the field of disaster risk reduction. More about his work can be found in: www.rajibshaw.org.

Contributors

Mumtaz Alam Fiji National University, Suva, Fiji

K.**Anupama** Indian Institute of Technology Kharagpur, Kharagpur, India

Jagath S. Ashok National Institute of Technology Calicut, Kozhikode, Kerala, India

Aswajith K. Babu School of Planning and Architecture Vijayawada, Vijayawada, India

Suchandra Bardhan Jadavpur University, Kolkata, India

Sanjukkta Bhaduri School of Planning and Architecture, New Delhi, India

P.**Bimal** National Institute of Technology Calicut, Kozhikode, Kerala, India

Apoorva Chandra Department of Architecture and Planning, National Institute of Technology Calicut, Kozhikode, India

Devika Chandran National Institute of Technology Calicut, Kozhikode, Kerala, India

K.**Chithra** National Institute of Technology Calicut, Kozhikode, Kerala, India

M. Banu Chitra School of Planning and Architecture Vijayawada, Vijayawada, India

Faiz Ahmed Chundeli School of Planning and Architecture Vijayawada, Vijayawada, Andhra Pradesh, India

Lalit Kumar Dashora School of Global Studies, Thammasat University, Bangkok, Thailand

M. Mavin De Silva Extreme Energy-Density Research Institute, Nagaoka University of Technology, Nagaoka, Niigata, Japan;
Department of Transport Management and Logistics Engineering, Faculty of Engineering, University of Moratuwa, Katubedda, Sri Lanka

Swati Dutta Amity University Punjab, Mohali, India

Navoda Fernando Department of Transport Management and Logistics Engineering, Faculty of Engineering, University of Moratuwa, Katubedda, Sri Lanka

Mohammed Firoz C. Department of Architecture and Planning, National Institute of Technology Calicut, Kozhikode, India

Kunal Avishek Gounder Fiji National University, Suva, Fiji

Janani Mani School of Planning and Architecture Vijayawada, Vijayawada, Andhra Pradesh, India

Taruni Joshi School of Planning and Architecture Vijayawada, Vijayawada, Andhra Pradesh, India

A. K.**Kalidas** National Institute of Technology Calicut, Kozhikode, Kerala, India

Selva Köroğlu Kurt Istanbul Aydin University, Istanbul, Turkey

B.**Krithik** Department of Architecture and Planning, National Institute of Technology Calicut, Kozhikode, India

Amal S. Kumarage Department of Transport Management and Logistics Engineering, Faculty of Engineering, University of Moratuwa, Katubedda, Sri Lanka

Tadachika Nakayama Extreme Energy-Density Research Institute, Nagaoka University of Technology, Nagaoka, Niigata, Japan

Rama Umesh Pandey School of Planning and Architecture, Bhopal, Madhya Pradesh, India

Rachana Patil Department of Architecture and Planning, Visvesvaraya National Institute of Technology, Nagpur, India

Saikat Kumar Paul Indian Institute of Technology Kharagpur, Kharagpur, India

Varun Prasad National Institute of Technology Calicut, Kozhikode, Kerala, India

M.**Bharath Raghavan** National Institute of Technology Calicut, Kozhikode, Kerala, India

Adinarayanane Ramamurthy School of Planning and Architecture Vijayawada, Vijayawada, Andhra Pradesh, India

S. R.**Resmi** National Institute of Technology Calicut, Kozhikode, Kerala, India

Kuladeep Kumar Sadevi Department of Architecture and Planning, National Institute of Technology Calicut, Kozhikode, India

Rajib Shaw Graduate School of Media and Governance, Keio University, Fujisawa, Kanagawa, Japan

Sanskriti Shukla School of Planning and Architecture, Bhopal, Madhya Pradesh, India

T. Sivakumar Department of Transport Management and Logistics Engineering, Faculty of Engineering, University of Moratuwa, Katubedda, Sri Lanka

A. G. T. Sugathapala Department of Mechanical Engineering, Faculty of Engineering, University of Moratuwa, Katubedda, Sri Lanka

Meenal Surawar Department of Architecture and Planning, Visvesvaraya National Institute of Technology, Nagpur, India

Agnes Mariam Xavier School of Planning and Architecture Vijayawada, Vijayawada, Andhra Pradesh, India

Gökçen Firdevs Yücel Caymaz Istanbul Aydin University, Istanbul, Turkey

Chapter 1
Understanding Urbanization, Urban Sustainability, and Environmental Transition Challenges in Global South

Lalit Kumar Dashora, Mohammed Firoz C., and Rajib Shaw

Abstract Today, urban areas are expanding at unprecedented rate. Currently, 56 percent of the global population resides in urban areas and it is expected that nearly 70 percent of the global population to reside in urban areas by 2050. Urbanization in Global South has increased much more rapidly in past half century and beyond than in the Global North and global urbanization is becoming increasingly characterized by urbanization in the Global South. The rapid urbanization presents serious challenges such as urban sprawl, poverty, higher unemployment rates, increased living costs, housing affordability issues, insufficient investment, limited financial and governance capacities, rising inequality and crime, and environmental degradation. Climate change is further complicating and escalating these issues, by making urban areas more vulnerable to emerging extreme weather events, such as heatwaves, droughts, floods, and tropical cyclones. Urban areas in the Global South facing environmental transition challenges that demand new paradigm of urban sustainability, a model that harmoniously balances economic development with ecological resilience in changing climate. The chapter explores the environmental transition challenges in urban area of Global South and calls for innovative multi-level approaches to sustainable urban development.

Keywords Urbanization · Sustainability · Climate change · Global south · Environment transition

L. K. Dashora (✉)
School of Global Studies, Thammasat University, Bangkok, Thailand
e-mail: dashora.lalitkumar@gmail.com

M. Firoz C.
National Institute of Technology Calicut, Kozhikode, India
e-mail: firoz@nitc.ac.in

R. Shaw
Keio University, Fujisawa, Kanagawa, Japan
e-mail: shaw@sfc.keio.ac.jp

L. K. Dashora et al. (eds.), *Urban Sustainability*, Disaster Risk Reduction,
https://doi.org/10.1007/978-981-96-4890-0_1

1.1 Understanding Urbanization in Global South

Urban areas are expanding at unprecedented rate across the globe with 56 percent of the global population residing in such areas, and it is expected that nearly 70 percent by 2050 (World Bank n.d.; Rogers 2020). According to Kundu and Pandey (2020), the global urban population has increased from 0.75 billion in 1950 to 4.22 billion in 2018, while it is projected that by mid-century, nearly 70 percent of the world's population will reside in urban areas. Urban areas are characterized by their complex web of high population density, along with infrastructures and economic activities, generally called cities or towns. It is often marked with different land use such as residential, institutional, recreational, and commercial activities with advanced infrastructure, and diversified industries and livelihood options (Bertinelli and Strobl 2007; McGranahan and Satterthwaite 2014; Nuissl and Siedentop 2021). The term "urbanization" itself refers to the conversion process of rural areas into urban areas due to population growth and economic activities (Bertinelli and Strobl 2007; McGranahan and Satterthwaite 2014; Tacoli et al. 2015).

Urbanization in the Global South has increased much more rapidly in the past half century and beyond than in the Global North (Myers 2021; McGranahan and Satterthwaite 2014). It is also strikingly evident from the trends and spatial pattern of urbanization that urban growth has begun to shift from the Global North to the Global South, especially in Asia and Africa (Kundu and Pandey 2020). According to Myers (2021), global urbanization is becoming increasingly characterized by urbanization in the Global South. In contrast, many urban areas in the Global North, especially in Europe and North America, are experiencing population decline. Today, urban areas in Asia and Africa are growing faster compared to any other part of the world. By future projections, urban areas in Asia and Africa will continue to lead in growth rate between 2015 and 2050 (Kundu and Pandey 2020). In this rapid urbanization process, many large and mega urban areas are growing in Asia and Africa, followed by small and medium-sized urban areas.

Urban areas provide economic growth and diverse livelihood opportunities. Generating over 80% of the world's GDP, cities have the potential to support sustainable growth through enhanced productivity and innovation, provided urbanization is effectively managed (World Bank n.d.). In contrast, however, rapid urban growth presents serious challenges such as urban sprawl, poverty, higher unemployment rates, increased living costs, housing affordability issues, insufficient investment, limited financial and governance capacities, rising inequality and crime, and environmental degradation (Zhang 2016; Ahluwalia 2016; Mokoele and Sebola 2018; Theodorou 2022; Childers et al. 2014). These challenges pose very complex sustainability issues in urban areas of Global South. The combined effects of urbanization, and environmental transition challenges such as climate change, pollution, environmental degradation, and resource depletion bring unusual risks to urban areas.

In the path toward becoming hubs of growth and opportunities, urban areas in the Global South also have pressing environmental transition challenges that demand a new paradigm of urban sustainability, a model that harmoniously balances economic development with ecological resilience in changing climate.

1.2 Understanding Urban Sustainability

Different urban sustainability research and studies provide useful information to enhance both the science and practice of maintaining sustainable urban systems in changing climate (Hiremath et al. 2013; Cohen 2017; Fu and Zhang 2017; Wachsmuth et al. 2016; Zhang and Li 2018; Zeng et al. 2022; Merino-Saum et al. 2020; Michael et al. 2014; Michalina et al. 2021; Li et al. 2018; Phillis et al. 2017; Romero-Lankao et al. 2016; Russo and Cirella 2020; Sharifi 2021; Shen et al. 2011; Soma et al. 2018; Spiliotopoulou and Roseland 2020; Topal et al. 2021; Turcu 2012a, b; Verma and Raghubanshi 2018; Vojnovic 2014). Currently, understanding sustainable urban development is a high priority to cope with environmental transition challenges in the rapidly urbanized world. With rapid urbanization in Global South, achieving sustainability in urban areas has become a major focus (Kaur and Garg 2019; Kundu and Pandey 2020). Sustainable urban development is crucial to cope with environmental transition challenges and for safeguarding the natural environment and the well-being of individuals and society as a whole, as haphazard urbanization can severely affect surrounding ecosystems (Kaur and Garg 2019). In some instances, such practices have also negatively impacted local precipitation, temperature, and the quality of air, water, and soil (Kaur and Garg 2019). Realizing this goal involves not only envisioning sustainable urban environments and implementing strategies but also continuously measuring progress toward sustainable urban development (Cohen 2017). Urban sustainability is essential for the greater regional sustainability at national, regional, and global levels (Huang et al. 2015).

Although urban sustainability practices in Global South are well-documented, the environmental transition challenges, and their effects on urban sustainability have not been fully explored at different scales (Bai et al. 2010; Fitzgerald et al. 2012; Paul 2014; Childers et al. 2014; Buzási and Jäger 2020). Urbanization, environmental transition challenges, and resource deficit are just a few of the many challenges urban areas face in determining for larger sustainability. However, policies and spatial development plans aimed at meeting specific sustainability goals often interact and conflict, creating a complex, multi-objective spatial optimization challenge for policy and decision-makers (Caparros-Midwood et al. 2015).

Urban sustainability is all about making urban areas that can fulfill present requirements of its residents without compromising the future. It is very important that these urban areas be designed and developed in a livable, environmentally friendly, and futuristic manner, as more and more people migrate to urban areas in search of better quality of life. In a nutshell, urban sustainability tries to establish a balance among social, economic, and environmental factors to support both the growth of urban areas and the well-being of its residents.

A sustainable urban area means accessibility to safe, affordable housing and transportation options, competence in waste and water management services, and accessibility to green spaces. In addition, it provides economic and livelihood opportunities available to all through a strong local economy with least environmental degradation. Urban areas that adopt sustainability, work on reduction of air and water pollution, proper waste management, and preservation of natural areas.

In achieving urban sustainability, urban areas use options such as green infrastructure, energy-efficient building as well as the use of renewable sources of energy such as solar and wind power. This is in context to designing buildings and transportation systems that consume less energies as well as produce a minimum amount of carbon emissions. Sustainable urban areas also promote walking, cycling, along with the above-stated options.

1.3 Understanding Environmental Transition Challenges

It is evident from previous sections that urban areas are growing rapidly, especially in Global South countries. This growth is putting additional pressure on existing infrastructure, resources, and ecosystem. In most countries in the Global South, the existing pattern of urban growth happens faster than the actual building of necessary services such as clean water, energy, sanitation, waste disposal, and public transportation. The consequence of such growth is environmental degradation: air and water pollution, loss of green spaces, and mounting waste become increasingly complex challenges for urban planners and policymakers.

There are many complex challenges in the path to sustainable urban development. The environmental transition challenges for urban areas are a somewhat new feature in the twenty-first century, largely associated with the interaction between climate change, rapid urbanization, population growth, and resource constraints. Climate change is indeed basically at the core of the interaction that affects ecosystems, communities, and economies across the Global South. The changes in climate will shift weather patterns and increase emerging extreme events such as heat waves, droughts, floods, and typhoons. Urbanization and industrialization further aggravated the existing situation, which added to the increase in emissions, besides land degradation, loss of biodiversity, shortage of water, and pollution.

Climate change is further complicating and escalating these issues, by making urban areas more vulnerable to emerging extreme weather events, such as heatwaves, droughts, floods, and tropical cyclones. High temperature increases the Urban Heat Island (UHI) effect in most concrete-jammed urban areas, trapping hot air inside and increasing the demand for energy-consuming cooling, which further worsens air quality. Flooding due to a lack of adequate drainage infrastructure or intrusion into natural waterways presents health hazards, economic loss, and the displacement of urban residents. All these challenges with climate, besides the balance of urban growth, do call for innovative multi-level approaches to sustainable urban development.

Urban areas are also facing the challenges of resource scarcity, due to their consumption of large quantities of energy, water, and other resources, which is beyond what the environment can supply sustainably. Affordable housing, food security, and clean energy in high-density urban areas create a need for integrating sustainable methods into the urban planning approaches. Urban stakeholders such as local governments, private sectors, and communities should coordinate to address the challenges related to the transition of the urban environment with a customized approach. Policies focusing urban areas need to redesign for making societies resilient as well as sustainable by raising awareness, promoting green building, and undertaking environmental protection initiatives with communities. Furthermore, resilient cities, smart city approaches and big data-driven solutions can guide, improve resource efficiency and manage environmental transition challenges in urban areas. The environmental transition challenges in urban areas are complex and need for sustainable urban development, which can meet 21st-century needs without putting additional burden on the environment and quality of life for future generations.

1.4 Purpose and Structure of the Book

The book titled "**Urban Sustainability: Navigating Environmental Transition Challenges in Global South**" is a novel investigation into the dynamic interface of urbanization, environmental transition challenges, and sustainability in the rapidly changing urban areas of the Global South. With rigorous evidence-based research and foresight, the book is written and edited to unpack the different dimensions and shades of complexities with environmental transition challenges in urban areas generally and those of the Global South in particular. The book presents the understanding of how urban areas in the Global South are stressed to address the challenging task of sustainability amidst rapid urbanization and population growth, economic transformations, and environmental uncertainties associated with climate change and its entering extremes.

The book tries to bring out the unique set of evidence shaping the sustainability discourse within these urban areas through a critical examination of case studies, policy frameworks, and innovative solutions. Benefits from the proposed book shall not be confined to academia but accrue to the policymakers, urban planners, environmentalists, and citizens at large. The book provides evidence-based case studies for making relevant policies and strategies that guarantee cities can enjoy their development processes sustainably, thus addressing a number of challenges that include resource management, climate resilience, and social equity. In that respect, the book on urban sustainability is indeed a timely and highly essential guide as urban areas of the Global South navigate through the challenges of the twenty-first century in structuring resilient, inclusive, and ecologically conscious cities.

There are sixteen chapters in this book. Following the Introduction in this Chapter, Chap. 2 presents the nexus between heat island and air pollution toward sustainable environment of industrial regions in Raipur, Chhattisgarh, India. This chapter explores the dual challenges of heat islands and air pollution in Raipur and its surrounding industrial regions. Chapter 3 presents the Impact of sea level rise and unpredicted precipitation events on the Indian Western Coastal Region. This chapter examines the threat posed to coastal areas due to sea level rise and unpredicted precipitation with a primary focus on the Western coastal region of India. Chapter 4 presents the Climate Resilience in Urban Centers in Fiji: Strategies for the Global South. This chapter looks at ways to make their urban centers more resilient to these challenges. Chapter 5 presents the Influence of Urban Form on Urban Heat Island and its Planning Implications in Chennai Metropolitan Area, Tamil Nadu, India. The chapter aims to examine spatiotemporal changes in urban climate conditions to identify Urban Heat Island (UHI) hotspots within the study region. Chapter 6 presents the Integrated Simulation Modeling and Geospatial Analysis for Assessing Flood and Urban Heat Vulnerability. This chapter focuses on Thrissur, a rapidly urbanizing region in Kerala, India, presenting an integrated analysis framework for assessing flood and urban heat vulnerabilities and evaluating future climate change scenarios. Chapter 7 presents the Assessing the Environmental Vulnerability of Coastal City, Kochi, Kerala State, India. The study aims to analyze the city's distinct features and environmental factors, including existing pollution levels, shoreline change using the Digital Shoreline Analysis System (DSAS) tool, and the physical susceptibility of the area to environmental hazards. Chapter 8 presents the Exploring the role of Socio-cultural Factors in Disaster Risk Reduction—A Case of Indigenous Koli communities of Mumbai. The chapter explores the heightened vulnerability of Mumbai's indigenous Koli fishing communities to climate change-induced disasters, accentuated by rapid urbanization and their historical ties to the region's ecosystem. Chapter 9 presents the Exploring the risk and socio-economic vulnerability of riverbank erosion hazard along the Upper Tapi River, India. This chapter introduces a thorough geospatial methodology for evaluating riverine erosion hazards and vulnerability in the Upper Tapi River Basin. Chapter 10 presents the Nature-Based Solutions for Mitigating Zoonotic Disease Spillover Risk in Urban Areas. The chapter focuses on understanding the possible intervention points for nature-based solutions in the interaction pathway of urban stressors and risk factors that contribute to zoonotic spillovers, based on an upstream approach. Chapter 11 presents the Re-Envisioning of India's Silicon Valley as A Water-Sensitive City: Bengaluru. This chapter aims to analyze Bengaluru's current water management practices by addressing gaps in the city and proposing strategic recommendations to transform it into a water-sensitive city. Chapter 12 presents the Transitioning Sri Lanka's Urban Transport: A System Dynamics Approach to Electrification and Decarbonization. This chapter outlines the pathways Sri Lanka should take to achieve its electric vehicle transition targets. Chapter 13 presents the Urban Environment Studies that can help in designing Environmentally Conducive Neighborhoods, with case studies from three Indian cities. The chapter identifies eight neighborhoods each representing a particular residential pattern from the three case study cities of Amritsar, Chandigarh, and Gurugram and presents a detailed study of the relationship between the different physical density variables (denoting the pattern of development) and indicators of environmental quality. It shows that all the measures of density (people density, building density, and spatial density) have significant impact on environmental quality. Chapter 14 presents the Landfill Site Selection Directives: A Landscape-Centric Perspective in the Context of Kerala. This chapter focuses on the landfill site selection directives, in the context of Kerala, India, with a landscape-centric approach. Chapter 15 presents the Analyzing Urban Space Design Principles for the Homeless: Examples of Gezi, Anıt, and Sıracevizler Parks in Istanbul, Turkey. The chapter aims to investigate the outdoor design requirements for the homeless, who are increasing daily as an important social problem worldwide.

References

Ahluwalia IJ (2016) Challenges of urbanisation in India. Contemporary issues in development economics. Palgrave Macmillan UK, London, pp 163–177

Bai X, Roberts B, Chen J (2010) Urban sustainability experiments in Asia: patterns and pathways. Environ Sci Policy 13(4):312–325.

Bertinelli L, Strobl E (2007) Urbanisation, urban concentration and economic development. Urban Stud 44(13):2499–2510

Buzási A, Jäger BS (2020) District-scale assessment of urban sustainability. Sustain Cities Soc 62:102388

Caparros-Midwood D, Barr S, Dawson R (2015) Optimised spatial planning to meet long term urban sustainability objectives. Comput Environ Urban Syst 54:154–164

Childers DL, Pickett ST, Grove JM, Ogden L, Whitmer A (2014) Advancing urban sustainability theory and action: challenges and opportunities. Landsc Urban Plan 125:320–328

Cohen M (2017) A systematic review of urban sustainability assessment literature. Sustainability 9(11):2048

Fitzgerald BG, O'Doherty T, Moles R, O'Regan B (2012) A quantitative method for the evaluation of policies to enhance urban sustainability. Elsevier Ecol Indic 18

Fu Y, Zhang X (2017) Trajectory of urban sustainability concepts: a 35-year bibliometric analysis. Cities 60:113–123

Hiremath RB, Balachandra P, Kumar B, Bansode SS, Murali J (2013) Indicator-based urban sustainability—a review. Energy Sustain Dev 17(6):555–563

Huang L, Wu J, Yan L (2015) Defining and measuring urban sustainability: a review of indicators. Landscape Ecol 30:1175–1193

Kaur H, Garg P (2019) Urban sustainability assessment tools: a review. J Clean Prod 210:146–158

Kundu D, Pandey AK (2020) World urbanisation: trends and patterns. In: Kundu D, Sietchiping R, Kinyanjui M (eds) Developing national urban policies. Springer, Singapore. https://doi.org/10.1007/978-981-15-3738-7_2

Li N, Chan D, Mao Q, Hsu K, Fu Z (2018) Urban sustainability education: challenges and pedagogical experiments. Habitat Int 71:70–80

McGranahan G, Satterthwaite D (2014) Urbanisation concepts and trends, vol 220. International Institute for Environment and Development

Merino-Saum A, Halla P, Superti V, Boesch A, Binder CR (2020) Indicators for urban sustainability: key lessons from a systematic analysis of 67 measurement initiatives. Ecol Ind 119:106879

Michael FL, Noor ZZ, Figueroa MJ (2014) Review of urban sustainability indicators assessment–case study between Asian countries. Habitat Int 44:491–500

Michalina D, Mederly P, Diefenbacher H, Held B (2021) Sustainable urban development: a review of urban sustainability indicator frameworks. Sustainability 13(16):9348

Mokoele J, Sebola M (2018) Unplanned urbanisation in South African cities: the emergence of urban environmental problems. Bus Manage Rev 9(3):574–584

Myers G (2021) Urbanisation in the global south. In: Shackleton CM, Cilliers SS, Davoren E, du Toit MJ (eds) Urban ecology in the global south. Cities and Nature. Springer, Cham. https://doi.org/10.1007/978-3-030-67650-6_2
Nuissl H, Siedentop S (2021) Urbanisation and land use change. In: Sustainable land management in a European context: a co-design approach, pp 75–99
Phillis YA, Kouikoglou VS, Verdugo C (2017) Urban sustainability assessment and ranking of cities. Comput Environ Urban Syst 64:254–265
Rogers D (2020) Understanding urbanism. In: Rogers D, Keane A, Alizadeh T, Nelson J (eds) Understanding urbanism. Palgrave Macmillan, Singapore. https://doi.org/10.1007/978-981-15-4386-9_1
Romero-Lankao P, Gnatz DM, Wilhelmi O, Hayden M (2016) Urban sustainability and resilience: from theory to practice. Sustainability 8(12):1224
Russo A, Cirella GT (2020) Urban sustainability: integrating ecology in city design and planning. In: Sustainable human–nature relations: environmental scholarship, economic evaluation, urban strategies, pp 187–204
Sharifi A (2021) Urban sustainability assessment: an overview and bibliometric analysis. Ecol Ind 121:107102
Shen LY, Ochoa JJ, Shah MN, Zhang X (2011) The application of urban sustainability indicators–a comparison between various practices. Habitat Int 35(1):17–29
Soma K, Dijkshoorn-Dekker MWC, Polman NBP (2018) Stakeholder contributions through transitions towards urban sustainability. Sustain Cities Soc 37:438–450
Spiliotopoulou M, Roseland M (2020) Urban sustainability: from theory influences to practical agendas. Sustainability 12(18):7245
Tacoli C, McGranahan G, Satterthwaite D (2015) Urbanisation, rural-urban migration and urban poverty. Human Settlements Group, International Institute for Environment and Development, London, UK
Theodorou P (2022) The effects of urbanisation on ecological interactions. Curr Opin Insect Sci 52:100922
Topal HF, Hunt DV, Rogers CD (2021) Exploring urban sustainability understanding and behaviour: a systematic review towards a conceptual framework. Sustainability 13(3):1139
Turcu C (2012a) Local experiences of urban sustainability: researching housing market renewal interventions in three English neighbourhoods. Prog Plan 78(3):101–150
Turcu C (2012b) Re-thinking sustainability indicators: local perspectives of urban sustainability. J Environ Plann Manag 56(5) (Taylor and Francis)
Verma P, Raghubanshi AS (2018) Urban sustainability indicators: challenges and opportunities. Ecol Ind 93:282–291
Vojnovic I (2014) Urban sustainability: research, politics, policy and practice. Cities 41:S30–S44
Wachsmuth D, Cohen DA, Angelo H (2016) Expand the frontiers of urban sustainability. Nature 536(7617):391–393
World Bank (n.d.) Urban development. https://www.worldbank.org/en/topic/urbandevelopment/overview
Zeng X, Yu Y, Yang S, Lv Y, Sarker MNI (2022) Urban resilience for urban sustainability: concepts, dimensions, and perspectives. Sustainability 14(5):2481
Zhang XQ (2016) The trends, promises and challenges of urbanisation in the world. Habitat Int 54:241–252
Zhang X, Li H (2018) Urban resilience and urban sustainability: what we know and what do not know? Cities 72:141–148

Chapter 2
Assessing the Nexus Between Heat Island and Air Pollution Toward Sustainable Environment of Industrial Regions in Raipur, Chhattisgarh, India

Adinarayanane Ramamurthy, Taruni Joshi, and Faiz Ahmed Chundeli

Abstract This chapter explores the dual challenges of heat islands and air pollution in Raipur and its surrounding industrial regions. The research examines the physical, social, and environmental conditions of industrial zones, with a focus on hazards and their spatial impacts. A heat island map was generated using key control parameters such as building density, land surface temperature, vegetation cover, water bodies, wind patterns, land use, and humidity. Similarly, an air pollution map was created, incorporating pollutants such as PM10, PM2.5, CO, CO_2, NOx, and SO_2, alongside air quality parameters. The air quality index was quantified to identify and analyze the pollution hotspots, termed as "pollution islands," within the study area. The study utilizes scientific tools, including ArcGIS and the Kriging method, to analyze the relationship between heat islands and pollution islands. Through the identification of hotspot areas in affected villages and an assessment of socio-environmental vulnerabilities, the research identifies the zones most at risk. Further, a comprehensive environmental risk assessment, highlighting the high-risk areas. Based on these findings, the study proposes a conceptual framework for mitigating the impacts of environmental risks and suggests planning strategies aimed at promoting sustainability within the region.

Keywords Urbanization · Heat island · Air pollution · Hazard mapping · Environmental risk · Sustainable framework

---

A. Ramamurthy (✉) · T. Joshi · F. A. Chundeli
School of Planning and Architecture Vijayawada, Vijayawada, Andhra Pradesh, India
e-mail: adinarayanane@spav.ac.in

T. Joshi
e-mail: taruni.joshi16@gmail.com

F. A. Chundeli
e-mail: faizahmed@spav.ac.in

L. K. Dashora et al. (eds.), *Urban Sustainability*, Disaster Risk Reduction,
https://doi.org/10.1007/978-981-96-4890-0_2

2.1 Introduction

Industrial regions serve as engines of economic growth; however, their relentless pursuit of progress often incurs significant environmental costs. This burden is not only borne by the surrounding ecosystems but also affects the health and well-being of the communities dependent on these industries. (Majumder 2010; Middel et al. 2020) This research investigates the critical issue of the interconnected challenges of heat islands and air pollution, focusing specifically on Raipur, Chhattisgarh, India. Raipur, a city undergoing rapid industrialization, exemplifies the complex interplay between these two phenomena. The dense concentration of factories and infrastructure, coupled with a scarcity of green spaces, contributes to the formation of a microclimate characterized by significantly higher temperatures compared to surrounding areas (Ramachandra and Shruthi 2007; Pelaez-Santi 2024). This heat island effect exacerbates the already severe air pollution caused by industrial emissions (Moro 2020; Miah et al. 2016). These pollutants not only degrade air quality but also trap heat, creating a vicious cycle that intensifies both issues (Kumar 2023; Yang et al. 2020).

The combined effects of heat islands and air pollution pose significant threats to environmental integrity, social well-being, and public health in Raipur. Elevated temperatures facilitate the formation of harmful ground-level ozone, further compromising air quality (Kulkarni and Patil 2015). The adverse effects of heat stress and pollutants contribute to respiratory problems, heat-related illnesses, and long-term environmental degradation (Lukyanets 2021). This research aims to address these pressing challenges by proposing a sustainable development framework tailored specifically for Raipur. Through a comprehensive analysis of the factors driving heat islands and air pollution, as well as their spatial interactions, the study will identify co-mitigation strategies and recommend effective planning interventions (Ionescu et al. 2011; Hwang et al. 2011). Ultimately, the goal is to pave the way for sustainable development practices that prioritize environmental health, social well-being, and economic prosperity for Raipur and other industrial regions facing similar challenges.

Urban Dynamics and Environmental Sustainability

Rapid urbanization presents a double-edged sword for cities: it drives economic growth while simultaneously posing significant environmental challenges. This phenomenon is particularly pronounced in developing countries like India, where cities such as Raipur struggle to balance progress with environmental sustainability. Urban sprawl often replaces natural landscapes with buildings and infrastructure, leading to the formation of "Urban Heat Islands" (UHIs) characterized by elevated temperatures (Shreevastava 2019; Jaganmohan et al. 2012; Garg and Chand 2008; Chan and Yao 2008) (Fig. 2.1).

These increased temperatures adversely affect the environment and public health by reducing soil moisture and exacerbating respiratory issues due to trapped pollutants. Moreover, Raipur's industrial activities further compound the problem,

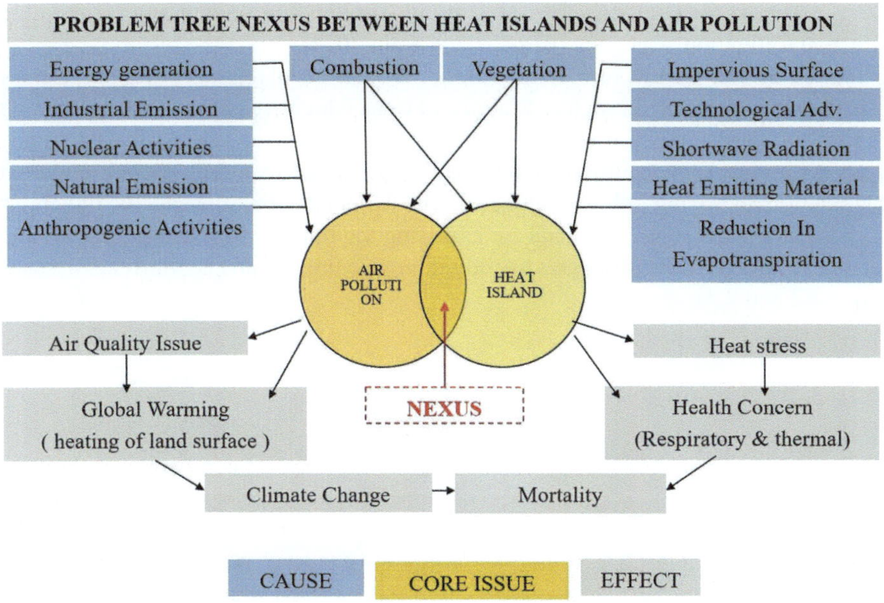

Fig. 2.1 Identification of problems of air pollution and heat island. *Source* Generated by the authors, 2024

releasing pollutants that not only worsen the UHI effect but also deteriorate air quality. Understanding the connection between land-use changes and rising temperatures is essential for addressing these challenges (Gogoi 2019). To tackle these issues, sustainable urban planning becomes crucial. Promoting green spaces, such as parks and urban forests, can help mitigate the UHI effect. Additionally, incorporating reflective materials in infrastructure design can significantly reduce heat absorption. Implementing stricter regulations on industrial emissions, alongside promoting cleaner energy sources and enhancing public transportation, are vital steps toward achieving cleaner air and a more sustainable urban environment (Halecki 2023).

2.2 Study Site

Chhattisgarh, located in central India, is rich in natural resources such as coal, iron ore, and limestone, which have significantly fueled its industrial growth. Raipur, the capital city of Chhattisgarh, serves as a major industrial hub, housing numerous factories and extensive infrastructure (Chhattisgarh 2024). However, the rapid industrialization and urbanization in this region have led to serious environmental challenges, notably the formation of heat islands and elevated levels of air pollution. These issues not only threaten the local environment but also pose risks to public health and overall quality of life for the residents of Raipur. Understanding the dynamics of this study site is essential for assessing the implications of industrial growth and formulating effective mitigation strategies (Birhane and Kant 2021).

The study area for this research focuses on the industrial regions of Raipur, Chhattisgarh, particularly areas such as Siltara and Urla, which are characterized by a high concentration of industrial activities. While these regions have experienced significant economic growth, this progress has come at the expense of environmental degradation and rising public health concerns. Understanding the specific dynamics of these industrial zones is crucial for assessing the impacts of industrialization and for developing effective strategies to mitigate environmental and health-related issues (Figs. 2.2 and 2.3).

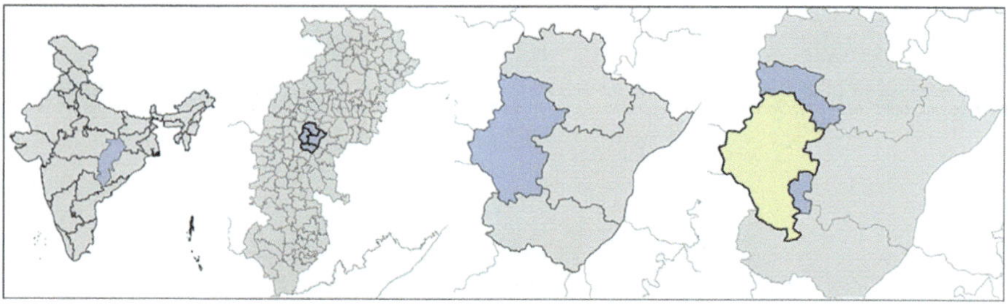

Fig. 2.2 Study area location. *Source* Generated by the authors, 2024

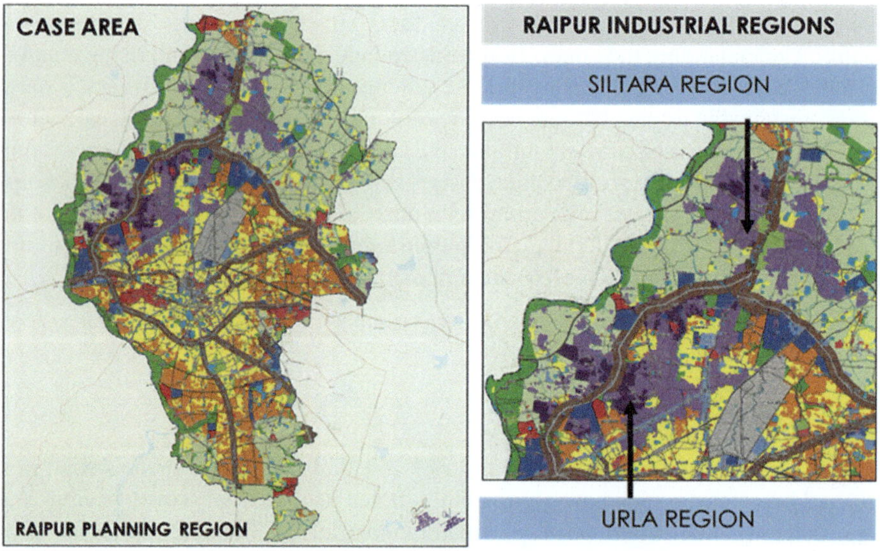

Fig. 2.3 Industrial region. *Source* Town and planning department, 2024

The Raipur Planning Region is located in the east-central part of Chhattisgarh, India, and has a population of approximately 1,027,244 people, resulting in a population density of 6,598 people per square kilometer. Raipur, the largest city in this region, also accounts for the same population figure. The region is bordered by Madhya Pradesh to the north, Odisha to the east, and Andhra Pradesh to the south. It is home to a diverse range of industries, including steel, cement, and power generation, making it a significant industrial hub. Additionally, the Raipur Planning Region plays an important role in agriculture and forestry, contributing to the overall economic landscape of Chhattisgarh.

2.3 Data Collection and Analysis

Identifying urban environmental issues necessitates the use of various data sets. To pinpoint heat island hotspots, authors employ satellite imagery to analyze factors such as land use, vegetation density, land surface temperature, and water presence. Additionally, air temperature, humidity, and wind data are considered to provide a comprehensive understanding of the microclimate. For assessing air pollution hotspots, data on specific pollutants, including PM2.5 and SO_2, is crucial (Salamanca et al. 2014; Aboelenein et al. 2020). Social vulnerability assessments require the collection of socioeconomic data, access to healthcare facilities, and the availability of green spaces (Asimakopoulos et al. 2012). This multifaceted data approach is essential for targeting interventions aimed at fostering a healthier urban environment (Table 2.1).

2.4 Heat and Pollution Hazard Analysis

2.4.1 Heat Island Hotspots

Heat islands are analyzed using a combination of thermal load and dynamic potential analysis. Thermal load focuses on the parameters contributing to heat accumulation, utilizing data such as land surface temperature (LST), air temperature, the normalized difference built-up index (NDBI), and land use/land cover (LULC) to generate a thermal load map (Mohamed et al. 2011; Cheng et al. 2021). In contrast, dynamic potential analysis emphasizes the parameters that can help mitigate heat. This analysis considers factors such as vegetation cover (measured by the normalized difference vegetation index, NDVI), water presence (assessed using the normalized difference water index, NDWI), prevailing wind patterns, and humidity levels to create a dynamic potential map (Ali et al. 2022). This dual approach allows for a comprehensive understanding of the heat island effect and informs strategies for its mitigation (Gartland 2008).

Table 2.1 Identified parameters

Parameters for identifying heat island hotspots	
Normalized difference built-up index	
Band 5 and band 7	LANDSAT 8
Normalized difference vegetation index	
Band 4 and band 5	LANDSAT 8
Normalized difference water index	
Band 3 and band 5	LANDSAT 8
Land surface temperature	
Band 10	LANDSAT 8
Climate data	
Air temperature	India Meteorological Department Met Centre Raipur & Directorate of Town and Country Planning, Chhattisgarh
Relative humidity	
Prevailing wind	
Parameters for identifying air pollution hotspots	
PM10	Chhattisgarh Environment Conservation Board
PM2.5	
CO	
CO_2	
NOx	
SO_2	
Parameters for vulnerability	
Population	Census Data 2011
Population density	
No. of people having age 06 or below	
Female population	
Illiterate population	
Literate population	
Working population	
Marginal worker	
Disabled population	
Primary and secondary Health care units	
Maternity and child care center	

Source Compiled by author, 2024

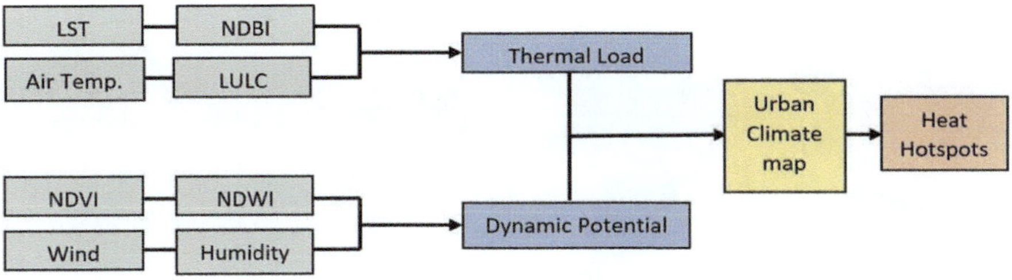

Fig. 2.4 Methodology for identification of heat island hotspots. *Source* Generated by the authors, 2024

By integrating the specified parameters and utilizing the weighted overlay tool in ArcGIS, authors generate thermal load and dynamic potential maps. These maps are then combined using the weighted overlay method to create a comprehensive urban climate map. Heat hotspots are identified by applying the following formula, UHI = Ts−Tm /SD (Figs. 2.4, 2.5 and 2.6).

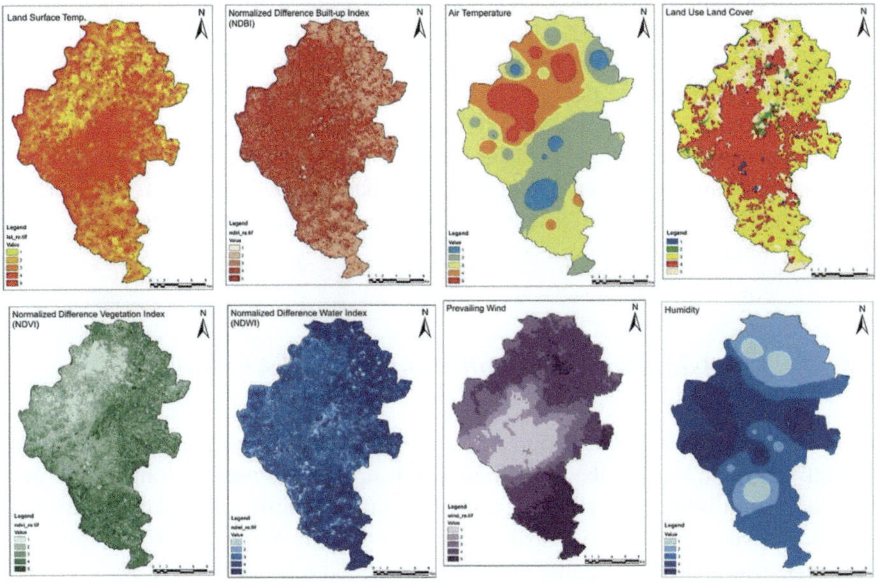

Fig. 2.5 **a** Land surface temperature, **b** normalized difference built-up index, **c** air temperature, **d** land use/land cover, **e** normalized difference vegetation index, **f** normalized difference water index, **g** prevailing wind, **h** humidity. *Source* Generated by the authors, 2024

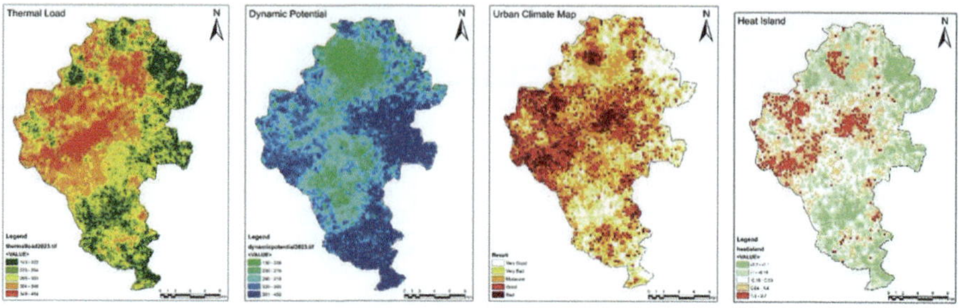

Fig. 2.6 **a** Thermal load map, **b** dynamic potential map, **c** urban climate map, **d** heat island hotspot map. *Source* Generated by the authors, 2024

2.4.2 Air Pollution Island Hotspots

Air pollution hotspots are characterized by high concentrations of pollutants, including PM10, PM2.5, CO, CO2, SO2, and NO2. An analysis of the air quality index (AQI) indicates that the Urla and Siltara Industrial Regions near Raipur exhibit unhealthy pollution levels that exceed regulatory safety limits. Urla is particularly affected, highlighting an urgent need for targeted emission controls. Surrounding villages, such as Sarora and Sondora, also experience unhealthy air quality, underscoring the extensive impact of industrial emissions on local communities. In contrast, Raipur's main municipal area shows a moderate AQI, suggesting that effective urban planning and stringent industrial regulations have been successful in reducing pollution levels. This analysis emphasizes the necessity for comprehensive air quality management strategies and targeted interventions, especially in the highly polluted industrial areas (Fig. 2.7).

2.4.3 Heat Island and Air Pollution Hazard Analysis

By analyzing areas with overlapping heat and air pollution, we identified 64 environmental "hotspots" within the 108 villages of Raipur's planning region. This targeted approach highlights the villages facing the most significant environmental challenges. These hotspots are distributed across Raipur Municipal Corporation (27 villages), Birgaon Municipality (6 villages), and the broader planning region (30 villages). They suffer from air pollution driven by factors such as coal-fired power plants and industrial facilities, which emit pollutants including particulate matter, nitrogen oxides, sulfur dioxide, and carbon monoxide (Figs. 2.8 and 2.9).

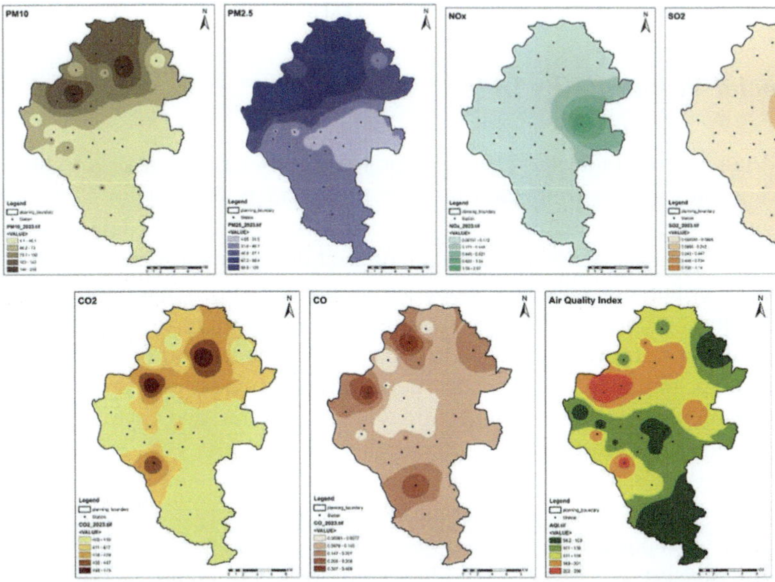

Fig. 2.7 **a** PM10, **b** PM2.5, **c** NOx, **d** SO_2, **e** CO_2, **f** CO, **g** air quality hotspots. *Source* Generated by the authors, 2024

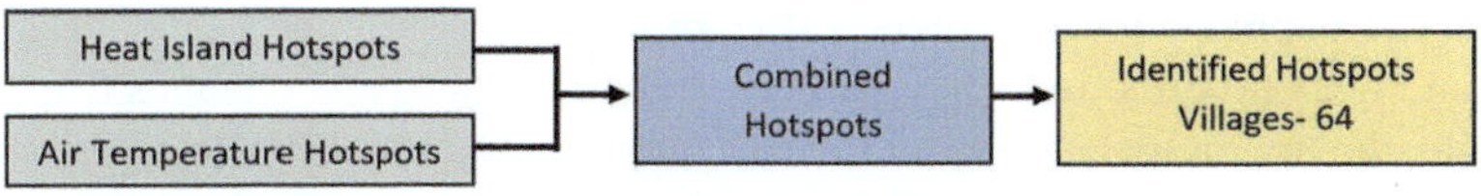

Fig. 2.8 Methodology for heat island and air pollution hotspots

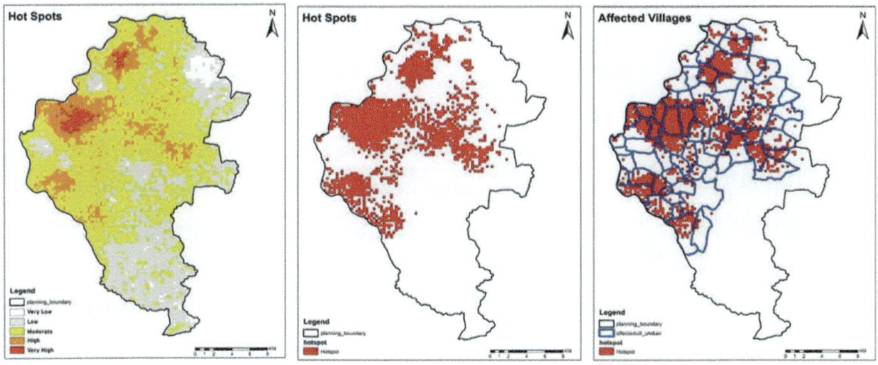

Fig. 2.9 Combined heat and air pollution, **b** Nexus hotspots, **c** affected villages. *Source* Generated by the authors, 2024

2.5 Socio-environment Vulnerability Analysis

2.5.1 Social Vulnerability Index

The Social Vulnerability Index (SVI) is a comprehensive tool used to assess communities' susceptibility to environmental hazards. It considers three main elements: sensitivity, exposure, and adaptive capacity. Sensitivity examines demographic factors that may increase vulnerability, including child population, female population, illiterate population, disabled population, and working population. Exposure focuses on environmental factors such as land surface temperature, population density, and air temperature. Adaptive capacity evaluates resources and infrastructure that help communities cope, including literate population, primary and community health centers, maternal and child welfare facilities, and vegetation. The SVI is calculated using the formula: Vulnerability = Sensitivity + Exposure − Adaptive Capacity. This nuanced approach recognizes that vulnerability is not just about exposure to hazards, but also about the inherent characteristics of a population and their ability to respond and recover (Figs. 2.10 and 2.11).

Sensitivity: Sensitivity refers to the vulnerability of a community or population to negative impacts from hazards such as air pollution and the heat island effect. **Exposure**: Exposure denotes the level of contact or interaction that a community or population has with hazards like air pollution and the heat island effect. **Adaptive Capacity**: Adaptive capacity refers to the ability of a community or population to adjust and respond effectively to environmental challenges and changes (Fig. 2.12).

While the municipal corporation and municipalities exhibit a high adaptive capacity, this is not the case for villages with industrial areas. Communities characterized by high sensitivity and low adaptive capacity are more vulnerable to environmental hazards. Villages like Siltara face increased vulnerability due to social factors

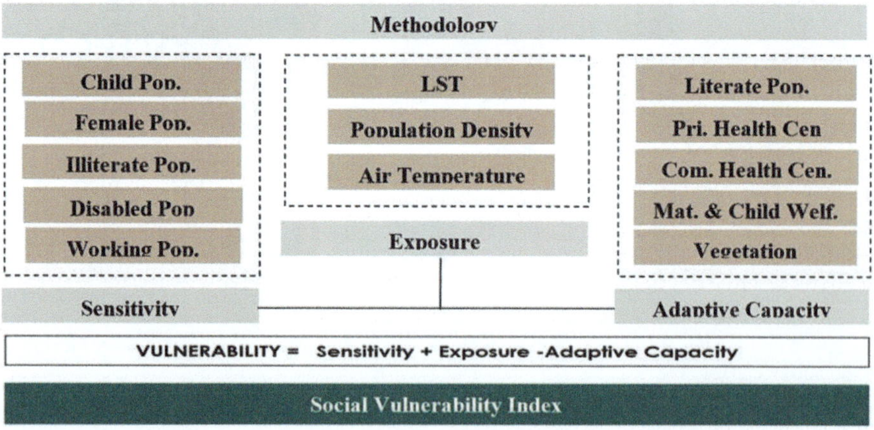

Fig. 2.10 Methodology for social vulnerability index. *Source* Generated by the authors, 2024

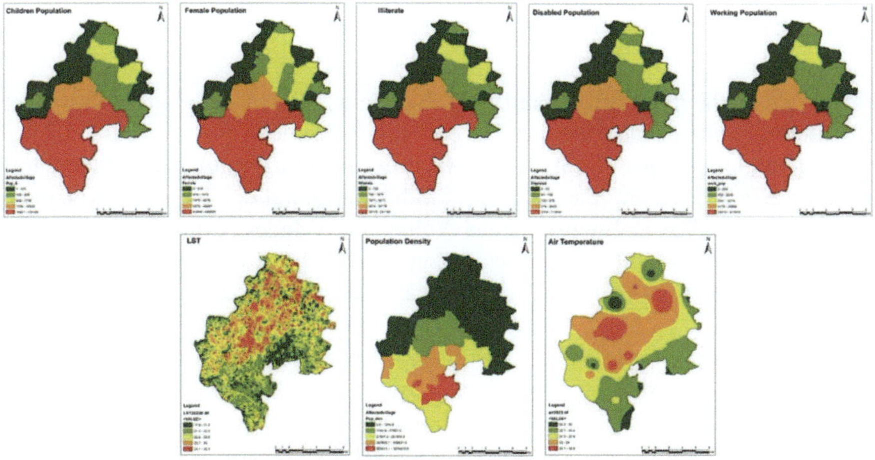

Fig. 2.11 **a** Children 0–6 years population, **b** female population, **c** illiterate population, **d** disabled population, **e** working population, **f** land surface temperature, **g** population density, **h** air temperature. *Source* Generated by the authors, 2024

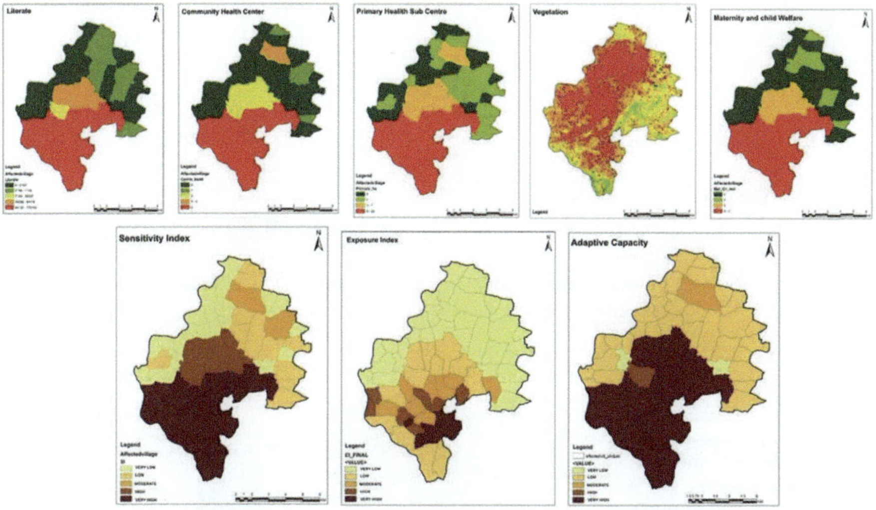

Fig. 2.12 **a** Literate, **b** community health center, **c** primary health subcenter, **d** vegetation, **e** maternity and child welfare, **f** sensitivity index, **g** exposure index, **h** adaptive capacity. *Source* Generated by the authors, 2024

such as poverty, limited resources, and inadequate infrastructure. As a result, these communities are likely to experience greater negative impacts from environmental challenges.

2.5.2 *Environmental Vulnerability Index*

Areas such as the municipalities of Urla, Borjhara, and Sarora, along with villages like Siltara, Sondara, and Mandhar, are particularly prone to environmental damage. This vulnerability arises from various factors, including challenging topography, limited green spaces, and the impacts of industrial pollution. Topography x Industrial Location x AQI = Environmental Vulnerability Index. Combining social and environmental vulnerabilities enables a comprehensive assessment of overall vulnerability, facilitating the identification of areas that are most at risk (Figs. 2.13 and 2.14).

Both social and environmental factors assist in identifying areas at high risk of issues. The regions exhibiting high overall vulnerability include Raipur Khas, Urla, Borjhara, Acholi, Siltara, Mandhar, and Sondara.

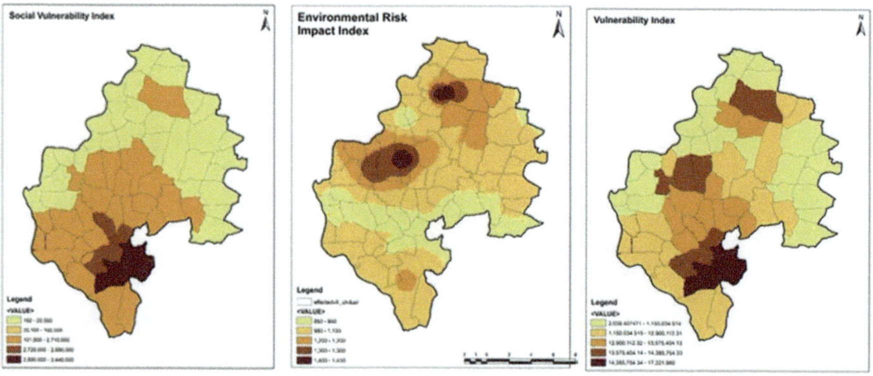

Fig. 2.13 **a** Social vulnerability index, **b** Environmental vulnerability Index, **c** Socio-environmental vulnerability index. *Source* Generated by the authors, 2024

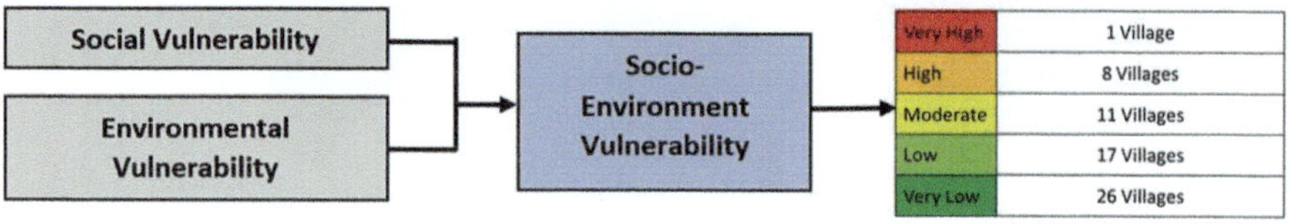

Fig. 2.14 Number of socio-environment vulnerable villages. *Source* Generated by the authors, 2024

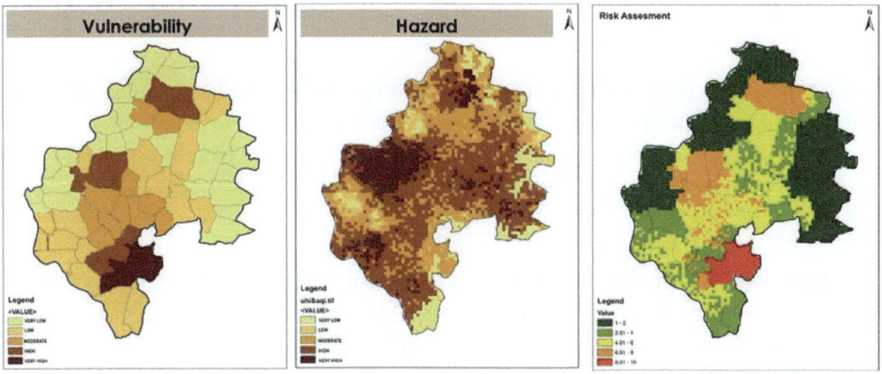

Fig. 2.15 **a** Vulnerability, **b** hazard, **c** risk assessment. *Source* Generated by the authors, 2024

2.6 Environmental Risk Assessment

Risk is defined as the potential for adverse consequences affecting human or ecological systems, acknowledging the diverse values and objectives associated with these systems. The United Nations Office for Disaster Risk Reduction (UNDRR) emphasizes that risk encompasses not only existing dangers (hazards) but also our vulnerability to them. In essence, risk is the product of hazard and vulnerability, i.e., Risk = Vulnerability × Hazard.

Risk assessment integrates vulnerability and hazard analysis to evaluate overall risk. Raipur Khas, characterized by its dense population, is highly vulnerable, resulting in significant risk. The Urla Industrial Region, which hosts heavily polluting industries in Urla, Borjhara, and Acholi, represents a high-hazard zone, contributing to elevated risk levels. Meanwhile, the Siltara Industrial Region, despite exhibiting lower pollution hazards, faces high risk due to its heightened vulnerability to other factors (Fig. 2.15).

2.7 Conceptual Planning Framework to Mitigate Environmental Risk Through Urban Acupuncture

Our cities face a multitude of environmental threats, ranging from climate change and pollution to dwindling resources. These interconnected challenges demand innovative solutions. Urban acupuncture emerges as a powerful framework to address these issues and guide cities toward a more sustainable future. This strategic approach utilizes targeted interventions in key urban areas, triggering positive ripple effects that can transform the entire city.

Why?

A Multi-Pronged Approach: Mitigating Heat Islands and Air Pollution for Thriving Cities

Our cities face a dual threat: rising temperatures due to the heat island effect and deteriorating air quality. To combat this, we need a comprehensive framework that simultaneously addresses both issues, leading to numerous benefits for our health, environment, and social well-being.

The Goals:

1. **Decreased Land Surface Temperature and Improved Air Quality**: This is the core objective. By implementing strategies such as promoting green spaces, using reflective materials for buildings, and regulating traffic, we can lower urban temperatures and reduce air pollutants.
2. **Reduced Greenhouse Gas Emissions**: Lowering energy consumption through cooler buildings and cleaner transportation choices leads to a significant reduction in greenhouse gas emissions, mitigating climate change.
3. **Water and Energy Recycling**: Sustainable practices like rainwater harvesting and energy-efficient infrastructure can create a more closed-loop system, reducing pressure on precious resources.
4. **Social Interaction and Improved Health**: A cooler, cleaner city encourages outdoor activities, fostering social interaction and improving physical and mental health. This translates to a more vibrant and resilient community.
5. **Protection of Vulnerable Groups**: The elderly, children, and individuals with respiratory illnesses are most vulnerable to the impacts of heat and air pollution. The framework prioritizes their well-being through targeted measures.
6. **Citizen Engagement and Sense of Belonging**: Involving citizens in planning and implementation creates a sense of ownership and belonging. Educational initiatives can raise awareness about environmental issues and empower citizens to make informed choices.
7. **Employment and Energy Savings**: The framework fosters the creation of green jobs in areas like renewable energy and sustainable construction. Energy savings through improved infrastructure further benefit both residents and the economy.

Who?

The Key Players: Working Together for Cleaner, Cooler Cities

Building a framework to combat heat islands and air pollution necessitates a collaborative effort from diverse stakeholders:

- **Government Organizations**: They play a crucial role in setting policies and regulations, including urban planning that prioritizes green spaces, incentives for energy-efficient buildings and transportation, and regulations to curb industrial pollution. Additionally, government agencies can fund research and development of innovative solutions.

- **Citizens**: The active participation of citizens is vital. Individuals can adopt practices such as using public transport, conserving energy, and planting trees in their own spaces. Engaging in community gardens, clean-up initiatives, and advocating for sustainable policies further strengthens the impact.
- **Local Volunteers**: Community volunteers are essential for on-the-ground action. They can contribute by planting trees, organizing neighborhood clean-up events, and raising awareness about environmental issues.
- **Manufacturers and Industry Owners**: Industries have a significant role to play. Investing in cleaner production processes, adopting renewable energy sources, and developing eco-friendly materials are crucial steps.

By working together, these groups can create a powerful force for change. Governments can provide the framework, citizens can drive demand for sustainable practices, volunteers can mobilize the community, and industries can develop and implement cleaner technologies. Through this collaborative effort, we can build healthier, more resilient cities for the future.

Where?

While the previously discussed framework can be applied at various scales, let's focus on how it can be implemented in Raipur and the affected villages of Urla, Siltara, and Sondora:

Regional Level (Raipur):

- **Government Organizations**: The Raipur Development Authority (RDA) can create a comprehensive plan incorporating green spaces, sustainable building codes, and regulations to curb industrial emissions. They can collaborate with the Chhattisgarh Environment Conservation Board to monitor air quality and enforce regulations.
- **Citizens**: Raipur residents can form environmental groups to advocate for green policies and organize awareness campaigns. Initiatives such as carpooling, promoting cycling, and adopting energy-efficient practices at home can significantly contribute.
- **Local Businesses and Industries**: Raipur's industries can invest in pollution control technologies and explore cleaner production processes. Partnerships with research institutions can foster innovation in sustainable practices.

Affected Village Level (Urla, Siltara, Sondra)

Municipal Corporation: Extend services to these villages through waste management and tree plantation programs.

Villagers: Encourage residents to plant trees and revive traditional practices like rainwater harvesting.

Local NGOs: Collaborate with villagers to promote organic farming and eco-tourism initiatives.

By implementing this framework at both regional and village levels, Raipur can effectively address heat islands and air pollution.

How?

Building a Greener, Cooler Raipur

The framework can be operationalized through:

- **Urban Planning**: Revise land-use policies to promote compact, walkable communities with green spaces.
- **Infrastructure**: Invest in green infrastructure, such as permeable pavements and green roofs, to enhance cooling and manage stormwater.
- **Transportation**: Develop efficient public transport and cycling paths to reduce emissions and improve air quality.
- **Community Engagement**: Educate residents about sustainable practices and involve them in decision-making.
- **Monitoring**: Establish a framework to assess the effectiveness of strategies for continuous improvement.

These measures will enhance urban resilience and improve the quality of life for Raipur's residents (Fig. 2.16).

Smart Technologies:

- **Air Quality Monitoring Network:** Install a real-time air quality monitoring network across Raipur to identify pollution hotspots and guide targeted interventions.
- **Smart Traffic Management:** Implement smart traffic management systems to reduce congestion and cut down on vehicle emissions.

Sustainable Climate Design:

- **Water Spray Systems:** Strategically install water spray systems in high-traffic areas to cool the air through evaporation.
- **Green Infrastructure (Green Roofs and Walls):** Incentivize the construction of green roofs and walls on buildings to insulate them, reduce energy consumption, and provide natural cooling (Fig. 2.17).

Water Infrastructure:

- **Permeable Pavements:** Use permeable pavements that allow rainwater to infiltrate the ground, replenishing groundwater reserves and reducing surface runoff.
- **Retention Ponds:** Create retention ponds to collect rainwater and other non-potable uses, reducing pressure on freshwater resources.
- **Fountains:** Incorporate water features like fountains in parks and public spaces to create a cooling effect through evaporation.

Building Design and Materials:

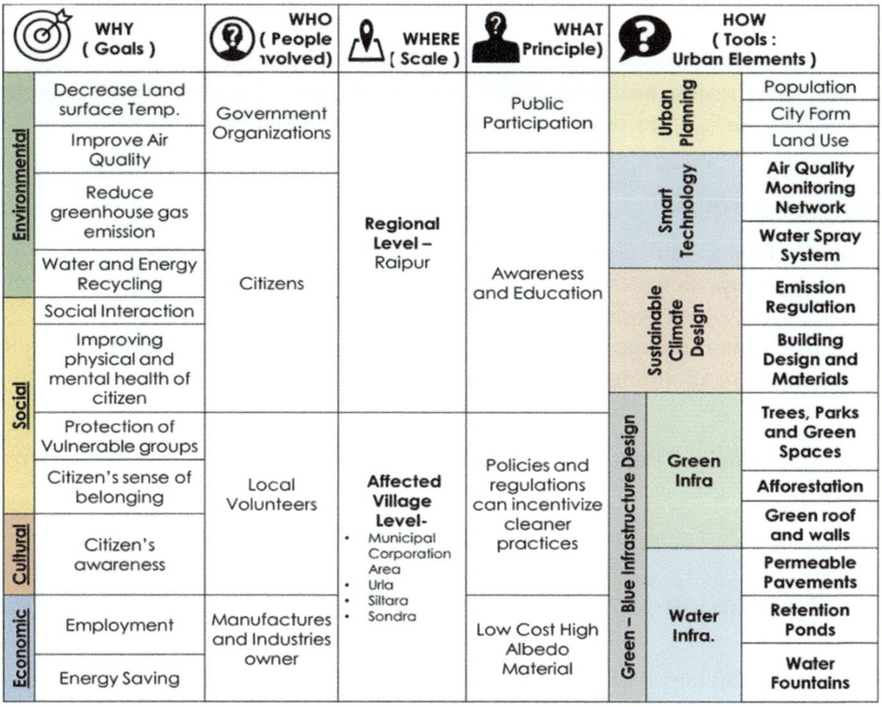

Fig. 2.16 5WH framework to mitigate environmental risk. *Source* Generated by the authors, 2024

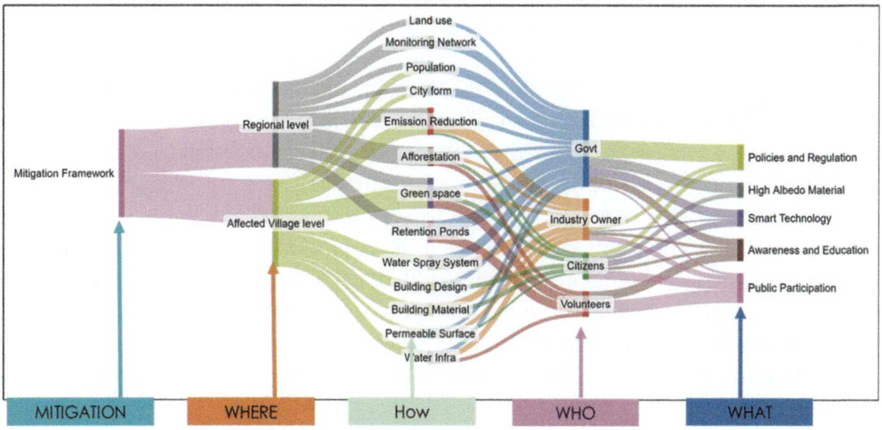

Fig. 2.17 Conceptual mitigation planning framework. *Source* Generated by the authors, 2024

- **Emission Regulations:** Enact stricter emission regulations for industries and encourage them to adopt cleaner technologies.
- **Building Design and Materials:** Promote the use of reflective materials for building surfaces to deflect sunlight and heat. Implement building codes that encourage energy-efficient construction and passive cooling techniques.

Green Infrastructure:

- **Trees, Parks, and Green Spaces:** Increase tree cover throughout Raipur by planting trees along streets, in parks, and on vacant lots. Create larger parks and green spaces to provide shade, cool the air, and improve air quality.
- **Afforestation:** Initiate afforestation programs around Raipur to create green belts that act as natural lungs for the city, filtering pollutants and providing a cooling effect.

By strategically utilizing these tools and encouraging citizen participation, Raipur can transform itself into a cooler, cleaner, and more sustainable city for all.

2.8 Conclusion

This paper examines the interconnected issues of heat islands and air pollution in Raipur's industrial zones. It uses advanced spatial analysis to identify high-risk areas and proposes a multifaceted approach for mitigation. Key recommendations include implementing green infrastructure and nature-based solutions. The research advocates for creating green spaces, green roofs, and permeable surfaces to combat urban heat and improve air quality. Community engagement and stakeholder collaboration are emphasized as crucial elements. The study promotes involving local volunteers, municipal corporations, and affected villages in environmental protection initiatives. Raising public awareness about the dangers of heat islands and air pollution is highlighted as a vital component. Educating citizens can empower them to make informed choices and advocate for stricter environmental regulations. The paper also stresses the importance of data-driven decision-making. It recommends using advanced mapping techniques and social vulnerability assessments to guide resource allocation and prioritize interventions. The approach balances top-down policies with bottom-up community participation. This comprehensive strategy aims to create a cooler, cleaner, and more resilient Raipur, offering insights applicable to industrial regions worldwide.

References

Aboelenein S, El-Shorbagy W, El-Massry A, Ali M (2020) Quantifying environmental and social vulnerability: role of urban heat island and air quality—a case study of Camden, NJ. Environ Pollut 259:113878

Ali A, Pasha GR, Farooq S, Khan A, Shah KH (2022) Identification of air pollution hotspots using remote sensing and GIS approaches: a case study of Islamabad, Pakistan. Sensors 22(13):4627
Asimakopoulos DN, Bais AF, Amiridis V, Zerefos CS (2012) Spatial and temporal variations of air pollution over large cities by use of satellite observations. Atmos Res 109:129–139
Birhane D, Kant P (2021) Analyzing vulnerability of climate change and developing adaptation strategies for Chhattisgarh, India using GIS and remote sensing techniques. Climate 9(10):145
Chan EYY, Yao X (2008) Air pollution in mega cities in China. Atmos Environ 42(1):1–42
Cheng X, Xu N, Liu H, Zhao Y, Tang J, Fan X, Yang R (2021) Synergistic control of air pollution and heat waves in the urban built environment: challenges and opportunities. J Clean Prod 280:124571
Garg SK, Chand I (2008) Satellite remote sensing aids in air pollution monitoring over Delhi city. Environ Monit Assess 140(1–3):393–403
Gartland L (2008) Heat islands: understanding and mitigating heat in urban areas. Routledge.
Gogoi PP (2019) Land use and land cover change. Sci Rep 9:8859
Halecki W (2023) Assessment and planning of green spaces in urban parks: a review. Sustain Cities Soc 88
Hwang RL, Lin TP, Tsai YI (2011) Relationship between urban heat island intensities and anthropic heat emissions with the level of urbanization. Int J Climatol 31(2):324–334
Ionescu C, Malache A, Stanescu CR, Chisalita D (2011) Estimation of urban heat island intensity by using remote sensing. In: IEEE international symposium on applied computational intelligence and informatics (SACI). IEEE, pp 345–349
Jaganmohan M, Surampalli RY, Zhang TC, Amirtharajah A (2012) Urban sustainability indicators. American Society of Civil Engineers
Kulkarni S, Patil RS (2015) Assessment of ambient air quality in an industrial area, Maharashtra, India. Asian J Atmos Environ 9(1):43–50
Kumar M (2023). AQI India. https://www.aqi.in/blog/en-in/urban-heat-island-effect-causes-and-cities-rising-temperatures/
Lukyanets EG (2021) Combined effect of hot weather and outdoor air pollution on respiratory health: literature review. Atmosphere 2021(12):790
Majumder R (2010) Environmental costs of industrialisation: a study of durgapur region in West Bengal. MPRA Paper 35054, University Library of Munich, Germany
Miah MD, Rahman MM, Hasan MK (2016) Air quality index of Asa-dam industrial area, Ilorin, Kwara state. Egypt J Remote Sens Space Sci 19(1):117–123
Middel A, Selover N, Hagen B, Chhetri N (2020) Impact of urban heat island mitigation strategies on heat-related morbidity in Phoenix, Arizona, under a changing climate. Urban Clim 32:100617
Mohamed AR, Murad AWA, Aziz HA (2011) On the relationship between land surface temperature and air temperature in urban environments. Urban Clim 1:38–48
Moro D (2020) Environmental effects of industrialization. In: Scientific inquiry into various conditions of human life in different parts of the world
Peláez-Santi MA (2024) Environmental cost and environmental management in non-primary. Int J Relig 5(8):829–838
Ramachandra TV, Shruthi BV (2007) Temporal variations in surface urban heat island across cities: study of mid-sized Indian cities from TRMM-TMI imagery. J Geophys Res Atmos 112(D20)
Salamanca F, Georgescu M, Mahalov A, Moustaoui M, Wang M (2014) Anthropic heating of the urban environment due to air conditioning. J Geophys Res Atmos 119(10):5949–5965
Shreevastava AB (2019) Paradoxical impact of sprawling intra-urban heat islets: reducing mean surface temperatures while enhancing local extremes. Sci Rep Nat Res 9:19681
Yang YI, Cheng FY, Sharma A, Chan EY, Chen LC, Chao CYH (2020) Impact of urbanization and spatial patterns on urban heat island. Int J Environ Res Public Health 17(12):4401

Chapter 3
Impact of Sea Level Rise and Unpredicted Precipitation Events on the Indian Western Coastal Region

Meenal Surawar and Rachana Patil

Abstract The rapid growth of urban areas is causing significant changes in the environment. Consequently, this urbanization is resulting in various environmental impacts such as increased greenhouse gas emissions, anthropogenic heat, modifications in land surfaces. These changes are disrupting the water budget and energy balance, resulting in a number of consequences of climate change. These consequences include more frequent heat waves due to the Urban Heat Island effect, flash floods due to increased intensity and frequency of unpredictable precipitation events, economic and environmental losses due to more frequent and intense unforeseen events, rising sea levels due to global warming, storm surges, etc. Unplanned urban areas are especially susceptible to these hazards. Therefore, it is crucial to accurately measure and evaluate these changes to enable the implementation of suitable measures and policies to reduce the potential impacts on vulnerable communities and ecosystems. According to the analysis by the Intergovernmental Panel on Climate Change, coastal regions are particularly at risk due to these alterations. This chapter examines the threat posed to coastal areas due to sea level rise and unpredicted precipitation with a primary focus on the Western coastal region of India. This region is distinguished by its unique cultural, ecological, and economic significance. Stretching along the Arabian Sea for approximately 1600 km, it encompasses a diverse array of ecosystems, including mangroves, estuaries, and coral reefs. Additionally, it is home to major cities, crucial ports, and numerous fishing communities, rendering it pivotal from both an ecological and economic standpoint. Hence, comprehending its vulnerability to rising sea levels is paramount for regional planning and sustainable development. This chapter delves into a thorough analysis of the proportion of areas at risk due to increasing sea levels and unpredicted precipitation events. It investigates sea level rise and precipitation predictions made by the Intergovernmental Panel on Climate Change (IPCC), considering various scenarios based

M. Surawar (✉) · R. Patil
Department of Architecture and Planning, Visvesvaraya National Institute of Technology, Nagpur, India
e-mail: meenalsurawar@arc.vnit.ac.in

R. Patil
e-mail: er.rachanapatil@gmail.com

L. K. Dashora et al. (eds.), *Urban Sustainability*, Disaster Risk Reduction,
https://doi.org/10.1007/978-981-96-4890-0_3

on Shared Socioeconomic Pathways (SSPs) and different levels of global warming, specifically within the Indian Western coastal region. This data forms the basis for quantifying the potential extent of sea level rise and its implications for the Western coastal region. The analysis indicates that approximately 15% of the Western coastal region will experience impacts from rising sea levels by the year 2100. Furthermore, the analysis incorporates a range of potential scenarios for future development and global climate change, as outlined in the Shared Socioeconomic Pathways, presenting a spectrum of potential outcomes. Recognizing the sensitivity of coastal ecosystems, this chapter emphasizes the importance of studying and implementing appropriate policies. It explores nature-based solutions to mitigate the effects of sea level rise and unpredicted precipitation events, concluding with a focus on policies suitable for coastal areas.

Keywords Sea level rise · Unpredicted precipitation events · Indian Western coastal region · Nature-based solution

3.1 Introduction

The speed of urbanization is rapidly increasing, prompting the need for urban areas to expand in tandem to accommodate this growth (Sridhar 2007). Despite the benefits urbanization brings, it also brings along several disadvantages such as congestion, the rise of slums, traffic issues, and various weather and climate-related risks (Andersson et al. 2021; Lu et al. 2021; Agonafir et al. 2023; Côté et al. 2023).

Throughout the last century, there has been clear evidence of the phenomenon of global warming, and its continuation is expected as long as greenhouse gas emissions persist. The adverse effects of this warming on ecosystems, societies, and the global economy are largely attributed to shifts in extreme weather patterns and climate change (Allan et al. 2023). The risks associated with climate change have escalated worldwide due to rising temperatures, leading to significant damages for societies, economies, and natural environments (Thomas and López 2015).

Particularly in densely populated nations like India, climate change has exposed more people to losses (Xu et al. 2022). In 2015, the Paris Agreement set the ambitious target of limiting global warming to 2.0 °C, or ideally 1.5 °C, above preindustrial levels. However, even a 1.5 °C limit poses substantial challenges for the global population (Mitchell et al. 2016; Schleussner et al. 2016; King and Karoly 2017; Aerenson et al. 2018; Lin et al. 2018; Jones et al. 2018). Given the far-reaching impact of global warming across vast regions, it is crucial to develop a comprehensive understanding of these developments.

According to the IPCC report, coastal regions face heightened vulnerability due to climate change as a result of greater exposure to storm surges triggered by rising sea levels and erosion along coastlines during intense storms, endangering both coastal communities and ecosystems. Coastal regions, characterized as the interface between the land and sea, encompass merely 4% of terrestrial areas and 11% of marine waters. Nevertheless, they are home to approximately one-third of the global population (Clark et al. 2017). Many of the world's largest urban centers and megacities are situated along coastlines (Von Glasow et al. 2013), such as Tokyo, New York, Newark, and Shanghai, with populations of 37.2, 20.4, and 20.2 million residents, respectively, (Blackburn et al. 2019). Projections indicate a further rise in coastal population, reaching 625 million by 2030 in low-lying areas (Neumann et al. 2015). Concerningly, significant portions of vital coastal ecosystems have already been lost, including 50% of salt marshes, 35% of mangroves, 30% of coral reefs, and 29% of seagrass meadows (Barbier et al. 2011). The restoration and degradation of these natural habitats are recognized as major drivers of change in coastal regions.

Coastal urban centers face heightened impacts due to their high population density and susceptibility to rising sea levels, storm surges, and tropical cyclones, as indicated by recent studies projecting more severe scenarios under the conditions of climate change (Revi 2008; Kantamaneni et al. 2019). Given these trends, the vulnerability to rising sea levels, compounded by unchecked development in densely populated cities and heavy reliance on coastal livelihoods, is a significant concern (Faccini et al. 2018; Nithila Devi et al. 2019; Abass et al. 2020; Castells-Quintana et al. 2022; Ingle and Chattopadhyay 2022). Therefore, understanding the impact of sea level rise on coastal regions and implementing suitable sustainable measures is imperative.

India, being the world's most populous country, also contains an extensive coastline. The challenges confronting its coastal areas are significant. Ranking 18th globally in terms of coastline length, India's total coastal stretch measures 7517 km, including the Andaman and Nicobar Islands in the Bay of Bengal and the Lakshadweep Islands in the Arabian Sea (Dhiman et al. 2019). In this chapter, the western coastal region of India is focused (Fig. 3.1).

The IPCC has forecasted data for precipitation patterns and sea level rise for various Shared Socioeconomic Pathways scenarios (SSPs) spanning until the year 2100. Study of such long-term impact is crucial for adaptation planning. Given the extensive and unpredictable nature of these phenomena, it is recommended to employ a multi-model approach. The Climate Change Knowledge Portal (CCKP) offers such a multi-model ensemble (MME) of climatic data based on the Coupled Model Intercomparison Project, presently in its 6th phase (CMIP6) (Lan et al. 2021). For precipitation, the data projected by the IPCC is grouped into 20-year intervals, starting from 1900 and extending to 2100, for precipitation. For sea level rise, the data is provided for specific years at decade intervals, covering the period from 2020 to 2150. The data predicted by NASA is taken as reference for sea level rise (https://sealevel.nasa.gov/). To get long-term view, interval of 2080–2099 is taken under consideration for off season precipitation and data of 2100 is taken for sea level rise.

This chapter utilizes the latest set of Shared Socioeconomic Pathways scenarios (SSPs) introduced by CMIP6, which is a part of the World Climate Research Program (WCRP), to analyze extreme precipitation events and the effects of sea level rise in the Western coastal region (Harris et al. 2020) (Table 3.1). These specific scenarios are based on socioeconomic development trajectories and atmospheric greenhouse gas concentration pathways (Siabi et al. 2023).

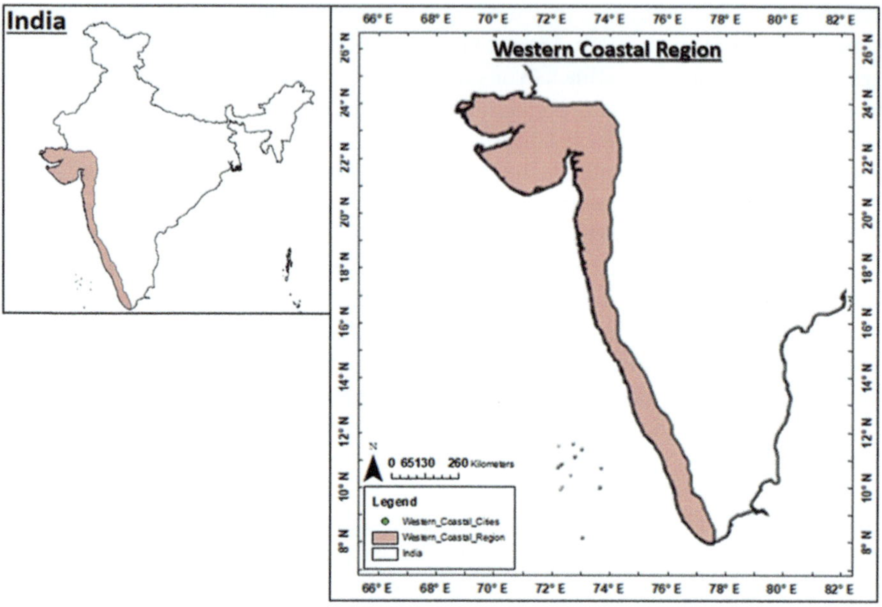

Fig. 3.1 Study area

Table 3.1 Shared Socioeconomic Pathways Scenario

Sr. No	SSP	Scenario	GHG emission	Radiative forcing level in 2100 (W/m^2)
1	SSP 1–1.9	Sustainability—low challenges to mitigation and adaptation	Very low GHG emissions: CO_2 emissions cut to net zero around 2050	1.9
2	SSP 1–2.6	Middle of the road—medium challenges to mitigation and adaptation	Low GHG emissions: CO_2 emissions cut to net zero around 2075	2.6
3	SSP 2–4.5	Regional rivalry—high challenges to mitigation and adaptation	Intermediate GHG emissions: CO_2 emissions around current levels until 2050, then falling but not reaching net zero by 2100	4.5
4	SSP 4–7.0	Inequality—low challenges to mitigation, high challenges to adaptation	High GHG emissions: CO_2 emissions double by 2100	7
5	SSP 5–8.5	Fossil-fueled development—high challenges to mitigation, low challenges to adaptation	Very high GHG emissions: CO_2 emissions triple by 2075	8.5

This model offers a robust foundation for addressing unforeseeable events. Precise prediction of such events is necessary in terms of spatial as well as temporal extent in order to lay groundwork for future planning, facilitating the implementation of suitable solutions.

3.2 Area Affected by Unpredicted Precipitation

IPCC data categorizes precipitation into intervals spanning 2020–2039, 2040–2059, 2060–2079, and 2080–2099. For analysis, the indicator Rx5, representing the average largest 5-day cumulative precipitation in millimeters, is selected. To provide a temporal overview, two intervals are chosen: 2020–2039 and 2080–2099. Within these periods, a distinct trend emerges in both summer and winter seasons, indicating a rise in heavy rainfall events, as indicated by Rx5 values, particularly in southern regions (Fig. 3.2). These events denote intense and off season precipitation over a specified timeframe. Across both seasons, Scenario 5–8.5 consistently exhibits the highest percentage of Rx5 events compared to other scenarios.

To quantify the spatial diversity, the indicator is segmented into uniform class intervals. These intervals are determined by subtracting the lowest and highest indicator values derived from the analysis and then dividing by the number of classes, which is taken as 5. This method enables the quantification in the percentage of the area affected by extreme events across different scenarios. Table 3.2 gives the percentage area under each class.

In summer for the period 2020–2039, the Rx5 values exhibit an interesting distribution (Fig. 3.2A). For Scenario 5–8.5 (Fig. 3.2A.e) second-highest class interval, ranging from 122 to 155 mm, covers a significant portion of the area when compared

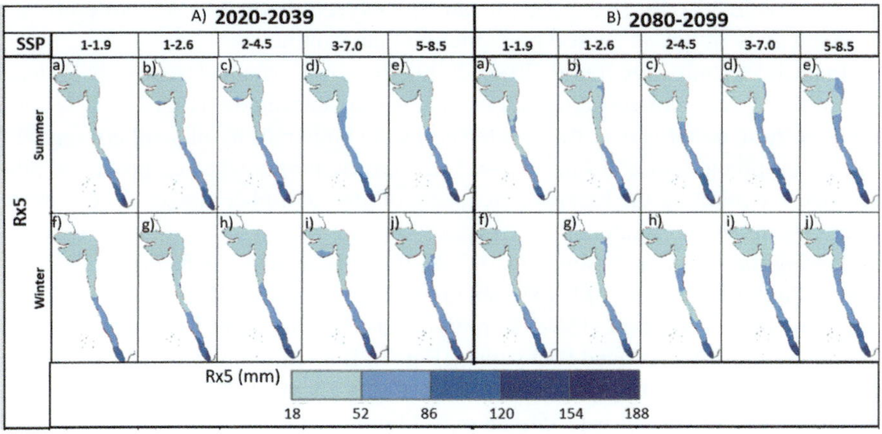

Fig. 3.2 Spatial distribution of Rx5 (Patil & Surawar, 2025)

Table 3.2 Percentage area covered by Rx5 events

Rx5	Summer					Winter				
(mm)	119	126	245	370	585	119	126	245	370	585
2020–2039										
23–56	82.99	74.7	81.36	68.28	61.41	80.77	79.56	78.16	67.72	53.41
56–89	11.75	14.89	12.13	19.72	26.75	13.85	12.59	14.63	20.28	30.75
89–122	3.87	8.84	3.26	8.07	7.76	3.62	5.81	4.26	7.07	9.76
122–155	1.39	1.57	2.66	3.02	**3.09**	1.76	2.04	2.16	3.02	**4.29**
155–188	0	0	0.59	0.91	**0.99**	0	0	0.79	**1.91**	1.79
2080–2099										
23–56	80.44	69.23	68.1	58.03	52.39	81.44	69.37	70.5	54.75	46.87
56–89	12.62	22.08	16.54	29.37	30.6	11.62	20.08	15.54	32.45	36
89–122	5.85	7.05	10.57	6.09	9.35	6.85	9.05	9.57	6.39	7.4
122–155	1.09	1.64	3.13	4	**5.09**	1.09	1.5	3.13	4.21	**6.12**
155–188	0	0	1.66	2.51	3.57	0	0	1.26	2.2	**3.61**

to other SSP scenarios, accounting for 3.09% of the total area (Table 3.2). Meanwhile, the highest-class interval, ranging from 155 to 188 mm, covers 0.99% of the area. Notably, Scenario 1–1.9 displays lower Rx5 values in comparison to all other scenarios (Fig. 3.2A.a).

The winter season during this period follows a similar pattern, where Rx5 events continue to be concentrated in the southern regions (Fig. 3.2B), but here Scenario 3–7.0 claims the highest percentage of Rx5 higher class event, i.e., 1.91% of class 155–188 mm (Table 3.2). In scenario 5–8.5, the second-highest class interval (122–155 mm) covers a larger area than in summer, accounting for 4.29% of the total area. Scenario 1–1.9 continues to display lower percentage area exposed to higher value classes (Table 3.2). From the data, it can be analyzed that winter are more prone to high intensity of the off season precipitation than summer.

In the later study time frame (2080–2099), there is a sustained trend of heightened Rx5 values, predominantly concentrated in the southern regions (Fig. 3.2). The continuity of these patterns over time underscores the vulnerability of southern region to off season precipitation. Scenario 5–8.5 remains dominant when compared to other scenarios, exhibiting the highest percentage of area in higher classes of Rx5 events in both summer and winter.

During the summer season (Fig. 3.2.B), under Scenario 5–8.5 (Fig. 3.2B.e), the second-highest interval (122–155 mm) now encompasses 5.09% of the total area, while the highest interval (155–188 mm) covers 3.57% (Table 3.2). Scenario 1–1.9 continues to exhibit reduced Rx5 values (Fig. 3.2B.a), aligning with trends observed in 2020–2039. Interestingly, Scenarios 119 (Fig. 3.2B.a) and 126 (Fig. 3.2B.b) show no area in the highest class (155–188 mm) (Table 3.2), indicating reduced exposure to extreme events.

In contrast, the winter season witnesses a more pronounced increase in Rx5 values compared to summer, aligning with the pattern observed in the preceding period (Fig. 3.2B). The second-highest interval (122–155 mm) under Scenario 5–8.5 covers a larger area, representing 6.12% of the total area, while the highest interval (155–188 mm) covers 3.61% (Table 3.2). Notably, these values exceed those observed in the corresponding summer season. Similarly, Scenario 1–1.9 exhibits lower Rx5 values (Fig. 3.2).

From this analysis, it is understood that Southern region of study area is prone to off season precipitation. Winter are more prone to high intensity of the off season precipitation than summer. These results align with previous research indicating that in coastal areas, the interaction between sea and land breezes modifies convective waves, causing increased off season precipitation during winter compared to summer (Patil and Surawar 2023). Implementing policies in line with Scenario 1–1.9 which is based on sustainable development could lead to an improvement in conditions, while continued development as per Scenario 5–8.5 which is based on fossil-fueled development could exacerbate the situation.

3.3 Area Affected by Sea Level Rise

The analysis for areas affected by sea level rise is based on the sea level prediction provided by the IPCC Sixth Assessment Report published by NASA (https://sealevel.nasa.gov/). This prediction varies across different locations due to regional-scale factors. Various additional processes influence local sea level changes in relation to the Global Mean Sea Level (GMSL), including vertical land motion, ocean circulation and density changes, and gravitational, rotational, and deformational effects resulting from the redistribution of water and ice mass between land and the ocean. These processes create a spatial pattern that generally increases sea level rise in low latitudes while decreasing it in high latitudes.

Data regarding sea level rise projections for the year 2100 under different Shared Socioeconomic Pathway (SSP) scenarios is available for the station mentioned in (Fig. 3.3a). To estimate the approximate sea level rise, the averages of sea level data at the specified location are utilized. Result obtained from the analysis indicates that nearly 15% of the area will be affected by sea level rise if no preventive measures are taken (Fig. 3.3b).

From Table 3.3, it can be observed that sea level rise for scenario SSP 1–1.9 is less when compared to other scenario. This again emphasis the importance of implementing the policies inlined with scenario SSP 1–1.9. It is observed that sea level rise at Bhaunagar is more when compared to other station (Table 3.3).

Table 3.3 provides a comparison of sea level rise projections across different scenarios, with a notable observation regarding the SSP 1–1.9 scenario. This specific scenario indicates a lower projected sea level rise compared to the other scenarios listed. This finding suggests that if policies aligned with the SSP 1–1.9 scenario,

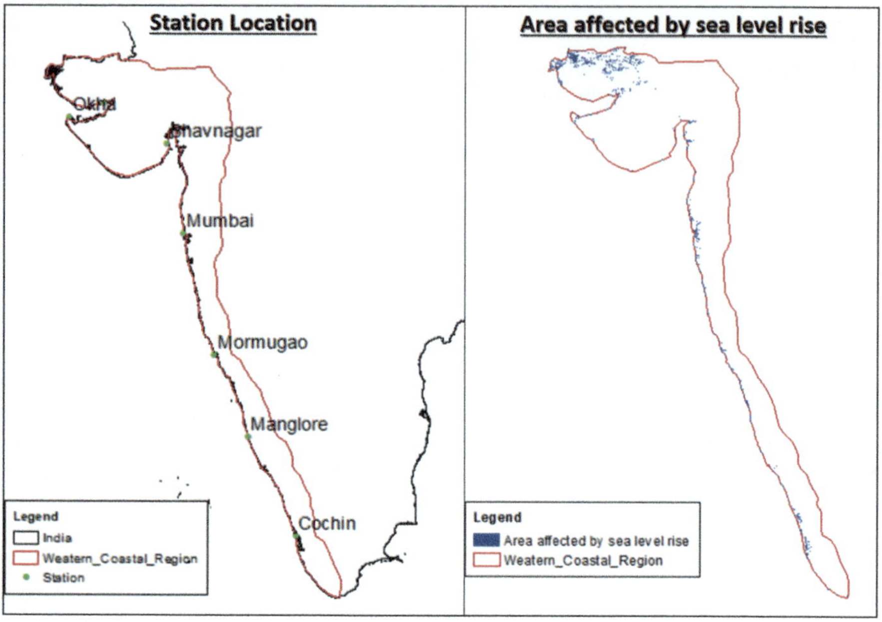

Fig. 3.3 **a** Station location **b** Area affected by sea level rise

Table 3.3 Sea level rise in m by 2100 for different SSP scenario

Station	Scenario				
	SSP 1–1.9	SSP 1–2.6	SSP 2–4.5	SSP 3–7.0	SSP 5–8.5
Kandala	0.28	0.34	0.46	0.57	0.7
Okha	0.3	0.36	0.48	0.6	0.72
Bhaunagar	0.53	0.59	0.71	0.82	0.95
Mumbai	0.28	0.34	0.46	0.58	0.7
Mormugao	0.32	0.38	0.51	0.63	0.75
Manglore	0.33	0.39	0.52	0.64	0.75
Cochin	0.4	0.42	0.58	0.68	0.82
Average	**0.35**	**0.40**	**0.53**	**0.65**	**0.77**

which emphasizes sustainable development, are effectively implemented, the potential impact of sea level rise by the year 2100 could be mitigated more effectively than under other scenarios.

Furthermore, a closer examination of sea level rise at specific locations reveals variability. For instance, at the Bhaunagar station, the projected sea level rise is notably higher compared to other stations. This variation emphasizes the localized nature of sea level rise, influenced by factors such as geographical features, ocean currents, and local land subsidence. Understanding these local disparities is crucial for effective adaptation and mitigation strategies in coastal areas.

3.4 Impacts of Off Season Precipitation and Sea Level Rise

Addressing off season precipitation and sea level rise in coastal regions is imperative due to the escalating impacts of climate change. These phenomena contribute to coastal erosion, flooding, and inundation, exacerbating the vulnerability of coastal communities and ecosystems. Moreover, they facilitate the intrusion of tidal waters into estuaries and river systems, leading to the contamination of freshwater reservoirs and agricultural lands. The consequences extend to the degradation of nesting beaches, displacement of populations, and loss of vital habitats like mangroves and seagrasses, which support diverse and economically valuable species.

Furthermore, the effects of climate change compound existing pressures on coastal areas, including urbanization, overfishing, tourism, and infrastructure development. These anthropogenic and natural stressors, coupled with rising sea levels and unpredictable precipitation patterns, intensify the challenges faced by coastal governance and management worldwide. Delaying action only exacerbates these challenges, underscoring the urgent need for proactive measures to safeguard coastal regions and their inhabitants.

Coastal regions, characterized by their fragility and diverse ecosystems, necessitate mitigation strategies implemented with caution. Traditional approaches to mitigating coastal flooding typically involved the use of hard-engineered structures, often referred to as "gray" infrastructure. Gray infrastructure, known for its durability, ease of implementation, and immediate effectiveness, is employed as a rapid response to coastal flooding and erosion incidents. Primarily focused on safeguarding human lives and minimizing economic losses, these structures were frequently constructed in densely populated urban areas without sufficient consideration for their integration into the natural landscape or their long-term effects on coastal ecosystem functions. However, the deployment of gray infrastructure has resulted in adverse environmental consequences, including alterations in sediment dynamics and disruptions to local ecology. Moreover, these structures necessitate ongoing maintenance throughout their operational lifespan. Despite their immediate benefits, gray infrastructures are not optimal for mitigating the impacts of climate change and other anthropogenic disturbances on coastal ecosystems.

In recent decades, there has been a growing trend toward employing nature-based solutions (NBS) to mitigate coastal floods. NBS involve combining measures to protect, restore, and sustainably manage ecosystems, while also addressing socioeconomic considerations to ensure human well-being and preserve biodiversity. These solutions harness the natural capacity of coastal ecosystems to serve as barriers, attenuating and dissipating the impact of storm surges and high-water levels. They are increasingly recognized as a viable and more efficient alternative to traditional coastal protection methods, as they not only offer protection but also provide ecosystem services and additional co-benefits. NBS techniques are particularly crucial for developing countries, where limited financial and technological resources constrain their ability to pursue engineering-based adaptations. In many of these countries, adaptation efforts may take a backseat to more immediate concerns such as poverty alleviation and meeting basic needs. Therefore, NBS present a practical and accessible approach to coastal protection in such contexts.

3.5 Nature-Based Solution

Like many ecological concepts, there are various approaches to defining and classifying nature-based solutions (NBS) for mitigating coastal floods. However, a widely accepted categorization, as outlined by several authors and adopted by Gómez Martín et al. divides NBS for coastal flood mitigation into three types: NBS type 1, characterized by minimal human interventions; NBS type 2, involving moderate human interventions; and NBS type 3, focusing on the establishment of new ecosystems and hybrid solutions.

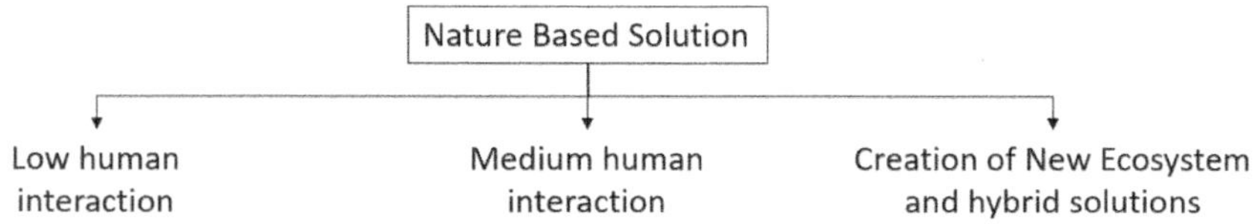

3.5.1 NBS Type 1: Low Human Intervention NBS

These mitigation strategies prioritize leveraging natural processes and the functions of ecosystems, requiring minimal human intervention beyond initial efforts for restoration or enhancement. They capitalize on the inherent capabilities of natural habitats and processes to provide protection along coastlines, including mechanisms such as wave attenuation, erosion prevention, and the creation of habitats. Importantly, these approaches are recognized for their sustainability and cost-effectiveness over extended periods. By working with nature rather than against it, they often yield benefits that endure over time while minimizing adverse impacts on the environment. Following are some examples of NBS Type 1.

(a) **Mangrove Restoration**: This involves planting mangrove seedlings in areas where mangrove forests have been degraded or lost. Mangroves serve as natural buffers against storm surges and erosion, contributing significantly to coastal protection (Gerona-Daga and Salmo 2022; Kinya et al. 2023; Sam et al. 2023).
(b) **Salt Marsh Creation:** The establishment or restoration of salt marsh habitats can effectively reduce wave energy and stabilize sediments along coastlines.

These marshes serve as natural barriers against erosion, offering protection to coastal areas (Baptist et al. 2021; Jordan and Fröhle, 2022; Marin-Diaz et al. 2023).

(c) **Beach Nourishment:** This technique involves adding sand to eroded beaches, which helps widen the shoreline and dissipate wave energy. In addition to providing coastal protection, it enhances the recreational and aesthetic values of the beach environment (Speybroeck et al. 2006; Chen et al. 2022; Kindeberg et al. 2023; Motta Zanin et al. 2024).

(d) **Oyster Reef Restoration:** Rebuilding oyster reefs is another strategy that enhances coastal resilience. Oyster reefs reduce wave energy, stabilize sediments, and provide habitat for various marine species, thereby contributing to the overall health and stability of coastal ecosystems (Chowdhury et al. 2019; Howie and Bishop 2021; Hynes et al. 2022; Perricone et al. 2023).

These nature-based approaches not only address immediate coastal protection needs but also promote the restoration and preservation of valuable ecosystems, ensuring their long-term health and functionality.

3.5.2 NBS Type 2: Medium Human Intervention NBS

These nature-based approaches not only address immediate coastal protection needs but also promote the restoration and preservation of valuable ecosystems, ensuring their long-term health and functionality. This approach to coastal flood mitigation involves a degree of human intervention in natural processes, aimed at balancing the needs of human development with the preservation of natural ecosystems. It acknowledges the importance of maintaining a harmonious relationship between coastal infrastructure and the surrounding environment. While it may require periodic maintenance to ensure its effectiveness, this method emphasizes the utilization of natural elements alongside engineered solutions. Following are few examples of NBS type 2.

(a) **Managed Retreat:** This strategy entails relocating or redesigning coastal infrastructure away from vulnerable areas. By allowing natural coastal processes to operate more freely, such as beach migration and dune formation, it enhances long-term protection against sea level rise and erosion. Managed retreat may involve relocating buildings, roads, and other structures further inland to safer locations (O'Donnell 2022; Schuerch et al. 2022; Setter et al. 2023).

(b) **Dune Construction and Management:** Building and maintaining sand dunes along coastlines is a proactive measure to absorb wave energy, prevent erosion, and stabilize shorelines. This process often involves planting and managing vegetation within the dune system to enhance its effectiveness. Dunes act as natural barriers against storm surges and provide crucial habitat for coastal wildlife (Costa et al. 2023; Johnston et al. 2023; Della Bella et al. 2024).

(c) **Living Shorelines:** This approach focuses on creating or restoring natural shoreline habitats, such as marshes, mangroves, or oyster reefs, while integrating engineered structures like sills or breakwaters strategically. By combining natural features with human-made interventions, living shorelines provide effective coastal protection while promoting biodiversity and ecosystem health. They also offer additional benefits such as carbon sequestration, water filtration, and recreational opportunities (Currin 2018; O'Donnell 2017; Smith et al. 2020).

In essence, these strategies emphasize the importance of working in harmony with nature to mitigate coastal flood risks. By leveraging the inherent resilience of natural ecosystems, coastal communities can build a more sustainable future while safeguarding both human lives and the environment.

3.5.3 NBS Type 3: High Human Intervention NBS

High human intervention nature-based solutions (NBS) involve the use of engineered structures or interventions that mimic natural processes to mitigate coastal risks. While they may not rely solely on natural processes, they still aim to address coastal protection needs while considering ecological aspects.

(a) **Seawalls:** Though constructed by humans, seawalls aim to provide immediate protection against erosion and storm surges. They are designed to mimic the function of natural coastal features, such as cliffs or rock formations, by absorbing and redirecting wave energy. However, their construction can disrupt natural coastal processes and lead to ecological impacts such as habitat loss (Development Bank 2023).
(b) **Groins:** These structures are engineered to trap sand and stabilize beaches. While they involve human intervention, their purpose is to mimic the function of natural features like barrier islands or spits. Groins attempt to manage sediment transport in a way that mimics natural processes, but they can also cause erosion on adjacent beaches (Setyandito et al. 2020; Torres-Freyermuth et al. 2020).
(c) **Breakwaters:** Similar to seawalls, breakwaters aim to reduce wave energy and protect coastal areas from erosion. They are typically placed offshore and are designed to mimic the function of natural reefs or shoals. However, their construction can alter wave patterns and sediment transport, impacting adjacent ecosystems (Pinto and Neves 2004; Rubinato et al. 2020).
(d) **Artificial Reefs:** These structures, such as sunken ships or specially designed reef modules, are deployed to mitigate waves, stabilize sediments, and provide habitat for marine life. While they are human-made, they serve to mimic the function of natural reefs and promote biodiversity. However, they may require ongoing maintenance to ensure their effectiveness (Silva et al. 2016).

While these high human intervention NBS involve significant engineering and construction, they are categorized as nature-based solutions because they aim to replicate or complement natural coastal processes. However, it's important to consider the potential ecological impacts and sustainability of these interventions, as they may disrupt natural ecosystems if not carefully planned and managed.

Each of these methods offers different levels of human intervention in coastal protection, with varying degrees of effectiveness, cost, and environmental impact. The selection of an appropriate approach depends on factors such as local conditions, project goals, and stakeholder preferences.

References

Abass K, Buor D, Afriyie K, Dumedah G, Segbefi AY, Guodaar L, Garsonu EK, Adu-Gyamfi S, Forkuor D, Ofosu A, Mohammed A, Gyasi RM (2020) Urban sprawl and green space depletion: implications for flood incidence in Kumasi, Ghana. Int J Disaster Risk Reduct 51. https://doi.org/10.1016/j.ijdrr.2020.101915

Aerenson T, Tebaldi C, Sanderson B, Lamarque JF (2018) Changes in a suite of indicators of extreme temperature and precipitation under 1.5 and 2 degrees warming. Environ Res Lett 13(3). https://doi.org/10.1088/1748-9326/aaafd6

Agonafir C, Lakhankar T, Khanbilvardi R, Krakauer N, Radell D, Devineni N (2023) A review of recent advances in urban flood research. Water Secur 19. Elsevier B.V. https://doi.org/10.1016/j.wasec.2023.100141

Allan R, Arias P, Berger S, Canadell J (2023) Summary for policymakers. In: Climate change 2021—the physical science basis, pp 3–32. Cambridge University Press. https://doi.org/10.1017/9781009157896.001

Andersson E, Borgström S, Haase D, Langemeyer J, Wolff M, McPhearson T (2021) Urban resilience thinking in practice: ensuring flows of benefit from green and blue infrastructure. Ecol Soc 26(4). https://doi.org/10.5751/ES-12691-260439

Baptist MJ, Dankers P, Cleveringa J, Sittoni L, Willemsen PWJM, van Puijenbroek MEB, de Vries BML, Leuven JRFW, Coumou L, Kramer H, Elschot K (2021) Salt marsh construction as a nature-based solution in an estuarine social-ecological system. Nat Based Solutions 1:100005. https://doi.org/10.1016/j.nbsj.2021.100005

Barbier EB, Hacker SD, Kennedy C, Koch EW, Stier AC, Silliman BR (2011) The value of estuarine and coastal ecosystem services. Ecol Monogr 81(2):169–193. https://doi.org/10.1890/10-1510.1

Blackburn S, Pelling M, Marques C (2019) Megacities and the coast: global context and scope for transformation. In: Coasts and estuaries: the future, pp 661–669. Elsevier. https://doi.org/10.1016/B978-0-12-814003-1.00038-1

Castells-Quintana D, del Pilar Lopez-Uribe M, McDermott TKJ (2022) Population displacement and urban conflict: global evidence from more than 3300 flood events. J Dev Econ 158. https://doi.org/10.1016/j.jdeveco.2022.102922

Chen WL, Muller P, Grabowski RC, Dodd N (2022) Green nourishment: an innovative nature-based solution for coastal erosion. Front Mar Sci 8. https://doi.org/10.3389/fmars.2021.814589

Chowdhury MSN, Walles B, Sharifuzzaman S, Shahadat Hossain M, Ysebaert T, Smaal AC (2019) Oyster breakwater reefs promote adjacent mudflat stability and salt marsh growth in a monsoon dominated subtropical coast. Sci Rep 9(1). https://doi.org/10.1038/s41598-019-44925-6

Clark CW, Rice AN, Ponirakis DW, Dugan, Erbe PJ, Reichmuth C, Cunningham C, Lucke K, Dooling K, Nedelec R, Campbell SL, Radford J, Simpson AN, Merchant SD (2017) Marine acoustic ecologies and acoustic habitats: concepts, metrics, and realities. J Acoust Soc Am 27. Springer-Verlag

Costa GP, Musumeci RE, Cáceres I (2023) Coastal dunes: a nature-based solution to face coastal hazards:615–626. https://doi.org/10.1142/9789811275135_0058

Côté J-N, Germain M, Levac E, Lavigne E (2023) Vulnerability assessment of heat waves within a risk framework using artificial intelligence. Sci Total Environ 169355. https://doi.org/10.1016/j.scitotenv.2023.169355

Currin CA (2018) Living shorelines for coastal resilience. In: Coastal wetlands: an integrated ecosystem approach, pp. 1023–1053. Elsevier. https://doi.org/10.1016/B978-0-444-63893-9.00030-7

Della Bella A, Del Vecchio S, Fantinato E, Buffa G (2024) Coastal dune restoration: a checklist approach to site selection. Land 13(2). https://doi.org/10.3390/land13020135

Development Bank A (2023) Building-coastal resilience through nature-based and integrated solutions: nature-based seawall conceptual design review. www.o2marine.com.au

Dhiman R, VishnuRadhan R, Eldho TI, Inamdar A (2019) Flood risk and adaptation in Indian coastal cities: recent scenarios. Appl Water Sci 9(1). Springer Verlag. https://doi.org/10.1007/s13201-018-0881-9

Faccini F, Luino F, Paliaga G, Sacchini A, Turconi L, de Jong C (2018) Role of rainfall intensity and urban sprawl in the 2014 flash flood in Genoa City, Bisagno catchment (Liguria, Italy). Appl Geogr 98:224–241. https://doi.org/10.1016/j.apgeog.2018.07.022

Gerona-Daga MEB, Salmo SG (2022) A systematic review of mangrove restoration studies in Southeast Asia: challenges and opportunities for the United Nation's decade on ecosystem restoration. Front Mar Sci 9. Frontiers Media S.A. https://doi.org/10.3389/fmars.2022.987737

Gómez Martín E, Máñez Costa M, Schwerdtner Máñez K (2020) An operationalized classification of nature based solutions for water-related hazards: from theory to practice. Ecol Econ 167. https://doi.org/10.1016/j.ecolecon.2019.106460

Harris I, Osborn TJ, Jones P, Lister D (2020) Version 4 of the CRU TS monthly high-resolution gridded multivariate climate dataset. Sci Data 7(1). https://doi.org/10.1038/s41597-020-0453-3

How Nature-Based Solutions in Coastal Areas can Help Address Global Crises Frontiers of Change in collaboration with (n.d.)

Howie AH, Bishop MJ (2021) Contemporary oyster reef restoration: responding to a changing world. Front Ecol Evol 9. Frontiers Media S.A. https://doi.org/10.3389/fevo.2021.689915

Hynes S, Burger R, Tudella J, Norton D, Chen W (2022) Estimating the costs and benefits of protecting a coastal amenity from climate change-related hazards: nature based solutions via oyster reef restoration versus grey infrastructure. Ecol Econ 194. https://doi.org/10.1016/j.ecolecon.2022.107349

Ian H, Osborn TJ, Jones P, Lister DH (2021) Climate change knowledge portal. Retrieved from The World Bank Group 2021. Climate Change Knowledge Portal. The World Bank, Washington DC. https://climateknowledgeportal.worldbank.org

Ingle K, Chattopadhyay S (2022) A Place-based approach to assess the vulnerability of communities to urban floods: case of Nagpur, India. Int J Disaster Risk Reduct 75. https://doi.org/10.1016/j.ijdrr.2022.102933

Johnston KK, Dugan JE, Hubbard DM, Emery KA, Grubbs MW (2023) Using dune restoration on an urban beach as a coastal resilience approach. Front Mar Sci 10. https://doi.org/10.3389/fmars.2023.1187488

Jones B, Tebaldi C, O'Neill BC, Oleson K, Gao J (2018) Avoiding population exposure to heat-related extremes: demographic change vs climate change. Clim Change 146(3–4):423–437. https://doi.org/10.1007/s10584-017-2133-7

Jordan P, Fröhle P (2022) Bridging the gap between coastal engineering and nature conservation?: a review of coastal ecosystems as nature-based solutions for coastal protection. J Coast Conserv 26(2). Springer Science and Business Media B.V. https://doi.org/10.1007/s11852-021-00848-x

Kantamaneni K, Rani NNVS, Rice L, Sur K, Thayaparan M, Kulatunga U, Rege R, Yenneti K, Campos LC (2019) A systematic review of coastal vulnerability assessment studies along Andhra Pradesh, India: a critical evaluation of data gathering, risk levels and mitigation strategies. Water (Switzerland) 11(2). https://doi.org/10.3390/w11020393

Kindeberg T, Almström B, Skoog M, Olsson PA, Hollander J (2023) Toward a multifunctional nature-based coastal defense: a review of the interaction between beach nourishment and ecological restoration. Nordic J Botany (1). https://doi.org/10.1111/njb.03751

King AD, Karoly DJ (2017) Climate extremes in Europe at 1.5 and 2 degrees of global warming. Environ Res Lett 12(11). https://doi.org/10.1088/1748-9326/aa8e2c

Kinya G, Kairo JG, Nyoike RN, Gachoki JN, Githinji BK, Githaiga MN (2023) Nature-based solutions: restoring mangroves for shoreline protection and other ecosystem services at Gazi Bay, Kenya. https://doi.org/10.20944/preprints202312.0099.v1

Lin L, Wang Z, Xu Y, Zhang X, Zhang H, Dong W (2018) Additional intensification of seasonal heat and flooding extreme over China in a 2 °C warmer world compared to 1.5 °C. Earth's Future 6(7):968–978. https://doi.org/10.1029/2018EF000862

Lu J, Li B, Li H, Al-Barakani A (2021) Expansion of city scale, traffic modes, traffic congestion, and air pollution. Cities 108. https://doi.org/10.1016/j.cities.2020.102974

Marin-Diaz B, van der Wal D, Kaptein L, Martinez-Garcia P, Lashley CH, de Jong K, Nieuwenhuis JW, Govers LL, Olff H, Bouma TJ (2023) Using salt marshes for coastal protection: effective but hard to get where needed most. J Appl Ecol 60(7):1286–1301. https://doi.org/10.1111/1365-2664.14413

Mitchell D, James R, Forster PM, Betts RA, Shiogama H, Allen M (2016) Realizing the impacts of a 1.5 °C warmer world. Nat Clim Change 6(8):735–737. Nature Publishing Group. https://doi.org/10.1038/nclimate3055

Motta Zanin G, Muwafu SP, Máñez Costa M (2024) Nature-based solutions for coastal risk management in the Mediterranean basin: a literature review. J Environ Manag 356. Academic Press. https://doi.org/10.1016/j.jenvman.2024.120667

Neumann B, Vafeidis AT, Zimmermann J, Nicholls RJ (2015) Future coastal population growth and exposure to sea-level rise and coastal flooding—a global assessment. PLoS One 10(3). https://doi.org/10.1371/journal.pone.0118571

Nithila Devi N, Sridharan B, Kuiry SN (2019) Impact of urban sprawl on future flooding in Chennai city, India. J Hydrol 574:486–496. https://doi.org/10.1016/j.jhydrol.2019.04.041

O'Donnell JED (2017) Living shorelines: a review of literature relevant to new England coasts. J Coast Res 33(2):435–451. Coastal Education Research Foundation Inc. https://doi.org/10.2112/JCOASTRES-D-15-00184.1

O'Donnell T (2022) Managed retreat and planned retreat: a systematic literature review. Phil Trans R Soc B Biol Sci 377(1854). Royal Society Publishing. https://doi.org/10.1098/rstb.2021.0129

Patil R, Surawar M (2023) Impact of urban heat island on formation of precipitation in Indian western coastal cities. J Contemp Urban Aff 7(2):37–55. https://doi.org/10.25034/ijcua.2023.v7n2-3

Patil R, Surawar M (2025) An approach for analyzing unpredicted heat and precipitation events using spatiotemporal big data: a case study of Indian Western coastal cities. Urban Clim 61:102414. https://doi.org/10.1016/j.uclim.2025.102414

Perricone V, Mutalipassi M, Mele A, Buono M, Vicinanza D, & Contestabile P (2023) Nature-based and bioinspired solutions for coastal protection: An overview among key ecosystems and a promising pathway for new functional and sustainable designs. ICES J Mar Sci 80(5):1218–1239. Oxford University Press. https://doi.org/10.1093/icesjms/fsad080

Pinto FT, Neves ACV (2004) Environmental aspects of using detached breakwaters for coastal protection purposes. Manag Environ Qual Int J 15(1):62–71. Emerald Group Publishing Ltd. https://doi.org/10.1108/14777830410513621

Revi A (2008) Climate change risk: an adaptation and mitigation agenda for Indian cities. Environ Urban 20(1):207–229. https://doi.org/10.1177/0956247808089157

Rubinato M, Heyworth J, Hart J (2020) Protecting coastlines from flooding in a changing climate: a preliminary experimental study to investigate a sustainable approach. Water (Switzerland) 12(9). https://doi.org/10.3390/w12092471

Sam K, Zabbey N, Gbaa ND, Ezurike JC, Okoro CM (2023) Towards a framework for mangrove restoration and conservation in Nigeria. Reg Stud Mar Sci 66. Elsevier B.V. https://doi.org/10.1016/j.rsma.2023.103154

Schleussner CF, Lissner TK, Fischer EM, Wohland J, Perrette M, Golly A, Rogelj J, Childers K, Schewe J, Frieler K, Mengel M, Hare W, Schaeffer M (2016) Differential climate impacts for policy-relevant limits to global warming: the case of 1.5 °C and 2 °C. Earth Syst Dyn 7(2):327–351. https://doi.org/10.5194/esd-7-327-2016

Schuerch M, Mossman HL, Moore HE, Christie E, Kiesel J (2022) Invited perspectives: managed realignment as a solution to mitigate coastal flood risks—optimizing success through knowledge co-production. Nat Hazard 22(9):2879–2890. https://doi.org/10.5194/nhess-22-2879-2022

Setter RO, Han RX, Tavares KD, Newfield C, Terry A, Roberson IM, Tarui N, Coffman M (2023) Managing retreat for sandy beach areas under sea level rise. Sci Rep 13(1). https://doi.org/10.1038/s41598-023-38939-4

Setyandito O, Purnama AC, Yuwono N, Juliastuti J, Wijayanti Y (2020) Shoreline change with groin coastal protection structure at north Java beach. ComTech: Comput Math Eng Appl 11(1):19–28. https://doi.org/10.21512/comtech.v11i1.6022

Siabi EK, Awafo EA, Kabo-bah AT, Derkyi NSA, Akpoti K, Mortey EM, Yazdanie M (2023) Assessment of shared socioeconomic pathway (SSP) climate scenarios and its impacts on the Greater Accra region. Urban Clim 49. https://doi.org/10.1016/j.uclim.2023.101432

Silva R, Mendoza E, Mariño-Tapia I, Martínez ML, Escalante E (2016) An artificial reef improves coastal protection and provides a base for coral recovery. J Coastal Res 1(75):467–471. https://doi.org/10.2112/SI75-094.1

Smith CS, Rudd ME, Gittman RK, Melvin EC, Patterson VS, Renzi JJ, Wellman EH, Silliman BR (2020) Coming to terms with living shorelines: a scoping review of novel restoration strategies for shoreline protection. Front Mar Sci 7:1–14. Frontiers Media S.A. https://doi.org/10.3389/fmars.2020.00434

Speybroeck J, Bonte D, Courtens W, Gheskiere T, Grootaert P, Maelfait JP, Mathys M, Provoost S, Sabbe K, Stienen EWM, Van Lancker V, Vincx M, Degraer S (2006) Beach nourishment: an ecologically sound coastal defence alternative? a review. Aquat Conserv Mar Freshwater Ecosyst 16(4):419–435. https://doi.org/10.1002/aqc.733

Sridhar KS (2007) Density gradients and their determinants: evidence from India. Reg Sci Urban Econ 37(3):314–344. https://doi.org/10.1016/j.regsciurbeco.2006.11.001

Thomas V, López R (2015) Global increase in climate-related disasters. www.adb.org

Torres-Freyermuth A, Hofman A, Clemente Tuz-Pech J, Medellín G, Roos PC (2020) Design and performance of permeable groins on a low-energy natural beach. J Mar Sci Eng 8(4). https://doi.org/10.3390/JMSE8040283

Von Glasow R, Jickells TD, Baklanov A, Carmichael GR, Church TM, Gallardo L, Hughes C, Kanakidou M, Liss PS, Mee L, Raine R, Ramachandran P, Ramesh R, Sundseth K, Tsunogai U, Uematsu M, Zhu T (2013) Megacities and large urban agglomerations in the coastal zone: interactions between atmosphere, land, and marine ecosystems. Ambio 42(1):13–28. https://doi.org/10.1007/s13280-012-0343-9

Xu H, Chen H, Wang H (2022) Increased populations will be exposed to the dangerous precipitation extremes across China in the future. Front Earth Sci 10. https://doi.org/10.3389/feart.2022.963042

Chapter 4
Climate Resilience in Urban Centres in Fiji: Strategies for the Global South

Mumtaz Alam and Kunal Avishek Gounder

Abstract Fiji and other nations in the Global South are experiencing fast urbanisation and the effects of climate change. This chapter looks at ways to make their urban centres more resilient to these challenges. Issues such as urban heat islands and severe weather occurrences are examined, and strategies tailored to individual regions are discussed, such as ecosystem-based solutions, community-based adaptation, and climate-integrated urban design. Sustainable urban development strategies and frameworks of Fiji are discussed, including the National Climate Change Policy and the Green Growth Framework. In this chapter, we'll look at how cities like Suva tackle urban carbon emissions by implementing green infrastructure, renewable energy, and better transit options. Although there has been some improvement, there are still obstacles to overcome during implementation, such as inefficient administration, lack of resources, and outdated technology. Amongst the suggestions are the creation of well-defined policies, the optimisation of administrative procedures, and the correct distribution of funds for climate adaptation. It is critical and doable to construct climate-resilient metropolitan centres in the Global South; nevertheless, this takes continuous dedication, creativity, and coordination amongst several stakeholders, as Fiji's experience shows. The lessons learnt could be very useful for other developing nations dealing with climate change and urbanisation.

Keywords Climate resilience · Strategic approaches · Risk assessment · Sustainable development · Global south

4.1 Introduction

Generally, increasing the density of cities is challenging because half of the global population resides in metropolitan regions. Consequently, the remaining space is slowly eroded without respect for environmental considerations, resulting in areas

M. Alam (✉) · K. A. Gounder
Fiji National University, Suva, Fiji
e-mail: mumtaz.alam@fnu.ac.fj

L. K. Dashora et al. (eds.), *Urban Sustainability*, Disaster Risk Reduction,
https://doi.org/10.1007/978-981-96-4890-0_4

with more impermeable surfaces and increasingly difficult circumstances. Furthermore, the interconnectedness of operational components inside urban areas plays a crucial role in their ability to withstand and adapt to shifting surroundings (Serre and Heinzlef 2018). There is a general agreement to enhance the resilience of cities and integrate sustainable development initiatives with urban resilience. However, scientists have not reached a consensus on the whole definition of resilience. Various resilient patterns are observed either within individual cities or across multiple cities, resulting in higher expenses for implementing robust methods due to the disparities that cause uneven distribution of urban services (Bresch and Spiegel 2011). In works about cities and climate change, the concept of resilience is starting to make more and more headlines. Common phrases like 'climate resilient', 'climate-proofing', and 'the resilient city' highlight the importance of urban systems, urban constituents, and cities' capacity to recover from climate-related stressors and shocks swiftly (Leichenko 2011). The capacity of a city to endure and adapt to various shocks and pressures is known as urban resilience, which includes a lot more than only the effects of climate change. We can classify the many sources that went into developing this complex idea into four broad categories: The first area deals with urban ecological resilience, which involves studying the relationship between man-made and naturally occurring systems; the second with urban hazards and disaster risk reduction, which includes both man-made and natural disasters; the third with the resilience of urban and regional economies, which involves studying the ability to withstand economic shocks and fluctuating market conditions; the fourth with the promotion of resilience through urban governance and institutions, which involves thinking about how leadership, policies, and community involvement can make cities stronger in general. These interrelated viewpoints offer a thorough framework for comprehending and resolving the diverse problems encountered by contemporary urban areas (Leichenko 2011). Hulme (2009) found that the concept of resilience is very malleable, much like the flexibility of climate change. Building resilience is often mistaken for making one less susceptible to harm or more able to adjust to new circumstances. If we want the word 'resilience' to stay relevant, we must keep asking ourselves how we apply the concept to cities (Boyd et al. 2008).

Fiji, like numerous other countries in the Global South, is seeing a swift process of urbanisation as a growing number of individuals are transitioning from rural to urban regions. Nevertheless, the process of urbanisation also brings heightened susceptibility to the impacts of climate change. Poor urban development design and insufficient consideration of climate vulnerability components increase the likelihood of extreme weather events and other climate-related risks causing destruction. Ensuring climate resilience in urban centres in the Global South is crucial for achieving sustainable development goal 11: Sustainable Cities and Communities of the United Nations since it safeguards lives and livelihoods and promotes long-term sustainability. Urban centres in the Global South face an escalating risk as extreme weather events become more frequent and intense, a trend that has seen a large rise. Fiji, a developing nation in the Global South, is susceptible to severe occurrences due to its geographical position, climate-sensitive industries, and other socio-economic challenges. This chapter focuses on enhancing climate resilience in urban areas of Fiji, with a particular emphasis on sustainability and adaptation strategies for the Global South.

Fiji's metropolitan areas face many climate-related challenges, including escalating sea levels, heightened tropical cyclones, frequent flooding, and urban heat problems. The primary economic hubs of Fiji, namely Suva and Nadi, are particularly susceptible to risks, mostly due to their coastal location. Climate vulnerabilities significantly threaten the overall stability of the economy, livelihoods, and infrastructure. Growing cities in developing countries, such as those in the Global South, are adversely affected by fast urbanisation and increasing climatic risks. This could lead to higher urban vulnerability and the disruption of the complex support systems that urban centres depend on. The South Pacific region, including Fiji, is currently under extreme stress due to several factors, including fast urbanisation, the unplanned expansion of informal settlements near urban areas, isolation from other regions, inadequate infrastructure, a lack of institutional capacity and the current and past effects of natural disasters, which all work together to reduce the resilience of the region (McEvoy et al. 2020).

Countries need a strategic adaptation framework to protect the urban infrastructure and development against extreme weather events. And if adaptation involves determining what to keep, improve, and reject to shape the future, resilience provides the context necessary to justify these choices. The goals of catastrophic risk reduction and climate change adaptation are constrained by resilience narratives, which also provide legitimacy to the people who develop and possess particular types of knowledge (Borie et al. 2019). Resilience is not a neutral tool; policymakers should focus on how resilience is developed and reviewed. Since resilience is highly accepted, it is necessary to be transparent about the presumptions underlying various viewpoints (Alexander 2013; Brown 2014).

4.1.1 Climate Change and Natural Disasters in Fiji

Fiji is amongst the Pacific Island Countries (PIC) experiencing adverse climate change effects. Despite being amongst the nations with the lowest carbon emissions—0.006% of all emissions worldwide—Fiji suffers severe consequences, including financial strains (Syngellakis et al. 2018). As with other SIDs, Fiji is presently seeing rising sea levels, temperature increases, coastline erosion, and a rise in the frequency and severity of natural disasters. Flooding is Fiji's most frequent natural catastrophe, caused by increased precipitation by inefficient mitigating measures. The Fijian government responded to climate change by enacting the 'National Climate Change Policy' in 2012, which served as a roadmap for project development and execution throughout the nation (McNamara and Des Combes 2015). As the President of COP23, Fiji actively participates in advocating on behalf of the Pacific region against climate change on a global scale (Corbett et al. 2018). However, Fiji confronts several obstacles to economic growth and climate change mitigation efforts since it is a SID with a limited budget and resources. 830 villages that need to relocate due to climate change have been recognised by the Fijian government (McMichael and Katonivualiku 2020). Relocating 48 villages urgently is necessary (McMichael and Katonivualiku 2020). Due to its tropical location, Fiji usually encounters strong tropical cyclones between November and April. The Pacific Ocean plays a significant role in the development of tropical cyclones by supplying a significant amount of warm ocean water that piques their interest. Periodic droughts are a by-product of Fiji's tropical climate. The South Pacific Convergence Zone is primarily the location of El Nino events, which are defined as the warming of ocean water or surface. As a result, Fiji has hot, dry weather from December to February. EL Nino also contributed to the worst drought that Fiji has ever seen, which occurred from 1997 to 1998 (Agrawala et al. 2003). Trenberth and Hoar (1997) go on to say that an increase in El Nino weather patterns is a result of or a link to global warming, which is a direct result of greenhouse gas emissions. The tropical east Pacific Ocean is seeing an increase in climate connected to El Nino, according to Trenberth and Hoar's climate model. Using the climate model from the Coupled Model Intercomparison Project (CMIP) experiments, Chand et al. (2017) further integrates the predictions of changes in El Nino-Southern Oscillation (ENSO) driven Tropical Cyclones towards the end of the twenty-first century and increases the frequency of tropical cyclones by approximately 40% in future El Niño events compared to present-climate El Niño events—and decreases in future La Niña events—around a group of small island nations (e.g., Fiji, Vanuatu, and Marshall Islands) (Chand et al. 2020). Their discoveries concerning how to address climate change and prepare for its effects have significant implications for small, susceptible island governments in the South Pacific.

4.1.2 Problem Statement

Fiji urban areas are increasingly vulnerable to escalating dangers due to the rising frequency and intensity of extreme weather occurrences. The main economic centres of Fiji, namely Suva and Nadi, are highly vulnerable because of their proximity to the coast. The economy, livelihoods, and infrastructure are greatly at risk due to climate vulnerability. Developing cities such as Fiji are negatively impacted by fast urbanisation and rising climate hazards, resulting in heightened urban vulnerability and the disruption of vital support systems that urban centres rely on.

4.1.3 Research Gaps

Despite the increasing recognition of climate resilience requirements in urban settings, there are still some areas of study that have not been well addressed:

- There is a lack of research on climate resilience solutions that are specifically designed for tiny island developing governments such as Fiji.
- There is a lack of enough study on the use of traditional knowledge in contemporary urban planning to enhance climate resilience.
- Insufficient evaluations of the efficacy of current climate policies and frameworks in Fiji's urban setting.
- Inadequate comprehension of the economic consequences of climate resilience measures in urban areas of Fiji.

4.1.4 Objectives

The primary aims of this paper are:

- To examine the present obstacles encountered by urban centres in Fiji because of climate change and growing urbanisation.
- To assess the efficacy of current climate resilience methods in Fiji's urban setting.
- To investigate novel strategies for improving the ability of Fijian urban areas to withstand and recover from climate-related challenges, with particular emphasis on utilising ecosystem-based and community-based adaptations.
- To evaluate the consequences of resilient urban development on Fiji's socio-economic landscape.
- To identify obstacles to the implementation of climate resilience measures and suggest strategies for overcoming these difficulties.

The chapter seeks to provide valuable insights into urban climate resilience strategies that can be implemented in Fiji as well as other small island developing states and countries in the Global South that are confronted with similar challenges. This chapter aims to contribute to the existing knowledge in this field by addressing the objectives and the identified research gaps.

4.2 Materials and Methods

This chapter employs a blend of qualitative and quantitative approaches to examine climate resilience strategies for metropolitan areas in developing nations, with a specific emphasis on Fiji. The research method integrates a literature review, examines case studies, and develops a conceptual framework. An extensive analysis was conducted on the existing literature about the impacts of climate change, the vulnerability of urban areas, and strategies to improve resilience. The study specifically concentrated on Small Island Developing States (SIDS) like Fiji. This review comprehensively analysed scholarly publications, policy documents, and reports from international organisations to gain a deep understanding of the challenges and potential solutions. The selection of Fiji as a case study was made to illustrate the specific climate-related challenges faced by urban areas in the Global South. The data on Fiji's climate risks, urbanisation patterns, and ongoing adaption activities was collected from authoritative government sources, scientific research, and global assessments. Analysed data was used to identify noteworthy issues and prospective chances for intervention. A conceptual framework was developed to provide strategic approaches for enhancing climate resilience in urban settings. This was achieved by conducting a comprehensive literature review and meticulous analysis of case studies. This framework, seen in Fig. 4.1, integrates contextual aspects, theoretical views, and focused techniques across urban planning, infrastructure development, and ecosystem-based adaptation. The framework also emphasises the importance of engaging stakeholders and continuously monitoring the implementation process. The chapter's organisation adheres to this conceptual framework, initially examining the difficulties encountered by metropolitan regions, subsequently investigating region-specific strategic methodologies, and ultimately expounding on the significance of infrastructure in climate resilience. The analysis incorporates examples and case studies from Fiji and other countries in the Global South to demonstrate the practical implementation of the stated solutions. This methodology enables a thorough analysis of climate resilience in urban areas in developing countries, offering useful insights for researchers, policymakers, and urban planners involved in sustainable and resilient urban development.

4.2.1 Conceptual Framework

The authors have presented a conceptual framework for Climate Resilience in Urban Centres in Fiji, as seen in Fig. 4.1. The format is designed as a flowchart comprising numerous hierarchical layers. The main topic is divided into two main categories: 'Contextual Factors' (which includes geographic, socio-economic, cultural, governance and policies, and resources) and 'Theoretical lenses' (which includes sustainable development, urban planning, disaster risk reduction, and climate change adaptation). These contribute to 'Strategies', which are categorised into three domains: 'Urban Planning and Design', 'Infrastructure Development', and 'Ecosystem-based Adaptation'. The methods culminate in 'Stakeholder Engagement', which is then followed by 'Implementation and Monitoring', ultimately leading to the establishment of 'Climate-Resilient Urban Centres'. The chart employs distinct chromatic rectangles to symbolise each level and notion, resulting in a visually structured depiction of the framework.

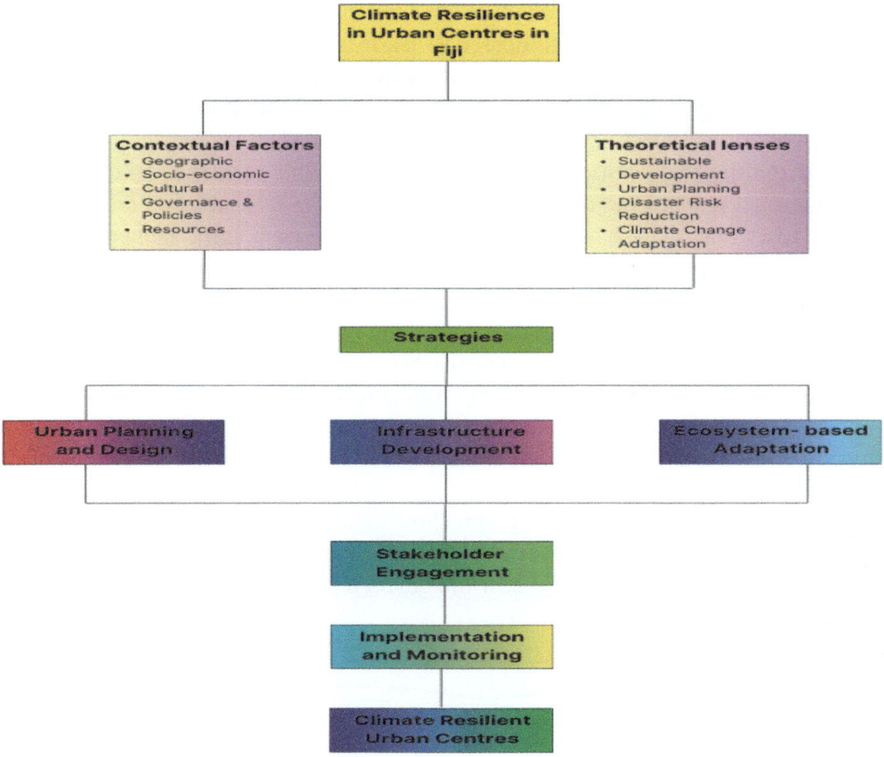

Fig. 4.1 Conceptual framework (*Source* Authors)

4.3 Results and Discussion

4.3.1 Challenges Faced by Urban Centres

Fiji is one of the most urbanised island states in the Pacific, with rapid urbanisation posing severe challenges. Approximately 55.9 per cent of Fiji's population resides in urban areas, an increase of 50.7 per cent in 2007 (Fiji Bureau of Statistics 2018). Rapid urbanisation with poor planning has led to many socio-economic problems in urban Fiji, such as unemployment, poverty, squatter settlements, and limited access to services (Phillips and Keen 2016). The current research trend shows that poorly planned urbanisations lead to many other challenges that are not climate resilient. When discussing urbanisation, the development of infrastructures and buildings is a compulsory component. Therefore, large volumes of natural resources, especially land and water, are used by the growing cities, which harms biodiversity, freshwater supplies, and habitat loss (Bera et al. 2023). In addition to being problematic, the increasing urban population and industrialization have caused an increase in temperature by 5–11 degrees Celsius compared to nearby regions (Kalhor and Mahdisoltani 2015). This has a long-term effect on human health and the environment.

The percentage of people living in urban areas in different Pacific Island nations and regions from 1950 to 2050 is depicted in Fig. 4.2. The percentage of the population living in urban areas is shown on the y-axis, which can take values between zero and one hundred per cent. The years 1950–2050 are displayed on the x-axis. Several lines, one for each nation or area, make up the graph. Important points to note are that Nauru's population resides in metropolitan areas. The urbanisation rates are projected to reach high percentages by 2050 in Guam, the Northern Mariana Islands, and American Samoa. The rate of urbanisation in Fiji is somewhere in the middle, rising slowly but surely. Vanuatu and the Solomon Islands' urbanisation rates are relatively low and amongst the lowest on the chart. Over time, the global average, represented by a thick black line, rises. In addition to a section for Pacific SIDS, it has trend lines for more expensive regions such as Polynesia, Micronesia, and Melanesia. According to the projections, most regions would likely continue their urbanisation patterns until the year 2050. The graph shows that the urbanisation patterns of the Pacific Island nations are very different from one another. Some nations have very high urban population percentages, while others have much lower ones.

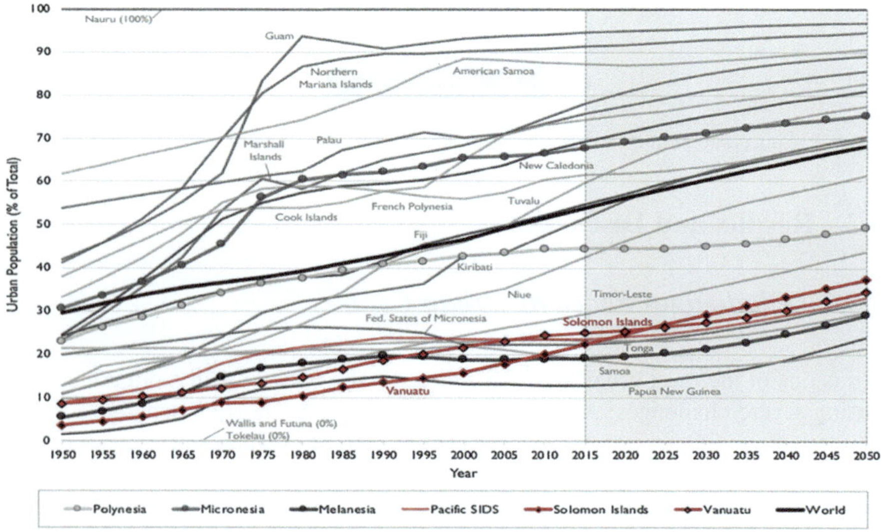

Fig. 4.2 Historical and projected urbanisation by Pacific SIDS sub-regions and countries (*Source* Trundle 2020)

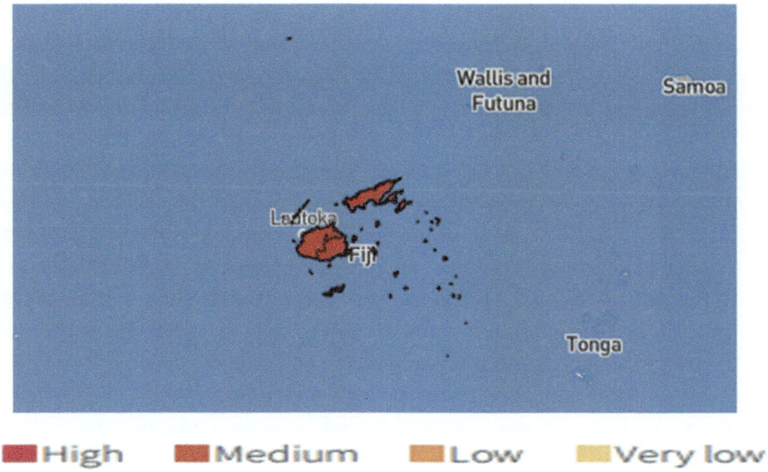

Fig. 4.3 Fiji heat threat (*Source* ThinkHazard 2020)

4.3.1.1 Urban Heat Island

Urban heat island is a challenge due to urbanisation. The number of natural environments is depleting, and grey infrastructure is increasing; hence, due to more concrete buildings, traffic and lack of vegetation, urban areas are often warmer than their rural counterparts, aggravating health problems and high energy consumption (Bera et al. 2023). Asian Global Change Research states that urban heat has two main effects: more energy used for air conditioning and fans to cool down, affecting the residents' health (Arifwidodo et al. n.d.). Ramsay et al. (2023) also added that the primary cause of urban heat islands is the change in land cover, which urbanisation has brought about, affecting the morphology and surface energy balance. These widespread changes in land cover result in increased heat radiation absorption and retention in urban centres, raising surrounding and surface temperatures compared to nearby non-urban areas (Ramsey et al. 2023). According to the ThinkHazard web-based tool, Fiji's heat threat is categorised as medium (Fig. 4.3). This indicates a more than 25% probability that over the next five years, there will be at least one instance of extended exposure to intense heat that causes heat stress (ThinkHazard Report: Fiji 2020).

4.3.1.2 Extreme Weather Events

Urban centres of Fiji are threatened by flooding, tropical cyclones, and other extreme weather events caused by climate change. These harsh circumstances not only cause new problems for the urban centres but also create pre-existing infrastructure problems such as disposal of waste and traffic congestion. Nearly all the rise in economic losses may be attributed to changes in socio-economic factors, including affluence, population expansions and increased developments in vulnerable locations. When hazards occur, like floods, combined with the exposure and susceptibility of communities and human settlements, they become disasters, resulting to socio-economic impacts (Mendoza and Schwarza 2019). Extreme weather conditions may cause infrastructure damage and power outages, interrupting human activity and urban services. Over the past 60 years, there has been a notable increase in the damages that natural disasters inflict upon human society (Chen et al. 2020). These damages have many socio-economic problems, such as the closure of businesses, unemployment, an outbreak of diseases, the decline in tourism activities and an increase in the cost of rebuilding. Fiji being a country mostly relying on service trade, suffers a lot; the closure of the airport, tourism activities, and exports could cripple the financial status as service trade contributes to 70% of Gross Domestic Product (GDP) (Radinibaravi 2024). Gard and Veitayaki (2017) further add that human-induced climate change hurts Pacific Island countries, especially in the densely populated urban and economic centres on the coastal areas, making them even more susceptible to extreme weather and increasing sea levels. Figure 4.4 shows the surface air temperature and precipitation trend in Western Fiji, which is most prone to extreme weather events like flooding and drought. Urban centres in Western Fiji contribute the highest amount of GDP due to tourism and manufacturing industries.

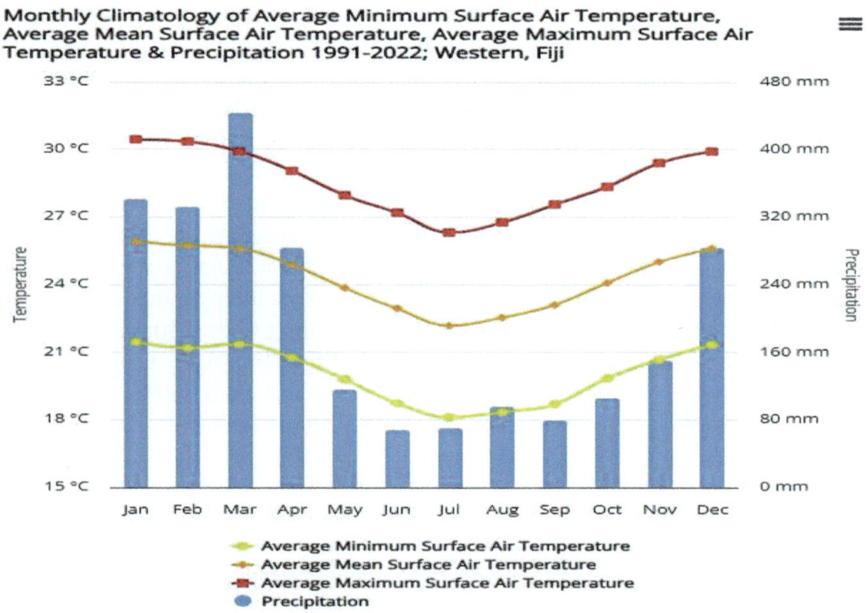

Fig. 4.4 Air temperature (*Source* World Bank Group)

4.3.2 Strategies for Climate Resilience

4.3.2.1 Climate-Integrated Urban Planning and Development

Planning urban development is essential to creating climate resilience. Zoning laws that prohibit building in high-risk areas, such as floodplains and coastal zones, and encourage green space that may absorb floodwaters and lessen heat islands are a few examples of sustainable urban land use planning. Carter et al. (2015) supports the idea that planning and design must consider the possible effects of the changing climate in more depth in order to be robust to these changes. One of the fundamental methods of contemporary planning is integrated planning, which makes sense given the complexity of spatial phenomena (Milojević 2018). Currently, sustainable development is predicated on the economy, society, and environment developing balanced while considering geographical and current linkages. Integrating spatial-physical and social elements of urban space development is also incorporated. Integrated planning is recognised as the means of attaining a resilient and sustainable urban environment and space, and this recognition is reflected in formulating policies for sustainable urban development and establishing institutional frameworks at all levels. A well-planned, resilient urban development will have less destruction caused by extreme weather events and be more climate friendly. In Fiji, the previous administration structured many policies and frameworks that align and guide Fiji's action towards becoming climate resilient. Policies and frameworks are key elements to guide the country in development and decision-making. Currently, there are many policies Fiji has adopted to ensure the nation is on the track to achieve the sustainable development goals (SDG) and become climate resilient due to vulnerability.

Table 4.1 outlines the primary climate and development policies implemented in Fiji. The following policies and frameworks are included: National Climate Change Policy (2018–2030), Green Growth Framework (2014), National Adaptation Plan (2018), Climate Vulnerability Assessment (2017), National Disaster Risk Reduction Policy (2018–2030), Fiji Low Emission Development Strategy, and Fiji's 5 and 20-Year National Development Plans. Each policy is succinctly outlined, highlighting its primary goals in relation to climate resilience, urban growth, and sustainability.

In 2017, the Ministry of Economy and Ministry of Local Government signed a Memorandum of Understanding (MoU) with Singapore Cooperation Enterprise (SCE) on behalf of the Government of Fiji (GoF) to collaborate on the exchange of Singapore's integrated urban solutions via urban master planning. For the next fifty years, the project aims to develop a comprehensive plan for Fiji's key metropolitan centres and infrastructure. A Strategic Master Plan Study of Viti Levu Island and a Conceptual Master Plan of Greater Suva, Nadi, and Lautoka are the project's two main outputs, broken down into six sections (Advisory Plan Rural Growth Centres 2024).

Table 4.1 Fijian government climate policies and descriptions

Policies and framework	Description
National climate change policy (NCCP) (2018–2030)	The main objective of NCCP is to change the role of the private sector, increase climate funding, and make connections between environmental protection, risk reduction, emission reduction, mitigation and adaptation. Policy strongly emphasises mitigation and adaptation strategies for urban planning and development
Green growth framework for Fiji (2014)	This framework focuses on sustainable development with techniques to increase urban resilience. Urban energy efficiency, clean transportation and green infrastructure are a few pillars of the framework
National adaptation plan (NAP) (2018)	NAP focuses on many projects, such as food, water, health, conservation, and infrastructure. Under infrastructure, it focuses on how building resilient infrastructures in urban areas improves services such as water and sanitation
Climate vulnerability assessment: making Fiji climate resilient (2017)	The Fiji Government conducted an assessment with the World Bank's support to assess the climate vulnerabilities and recommend Fiji's national development priorities for the next 5, 10, and 20 years. The assessment also identifies aging urban infrastructures and climate-vulnerable urban infrastructures
National disaster risk reduction policy (2018–2030)	The policy focuses on disaster risk reduction and adaptation. It also implements building codes and standards to ensure that new infrastructure developments are disaster resilient
Fiji low emission development strategy	The overall aim is to decarbonise and reduce Greenhouse Gases. It also promotes renewable energy, sustainable transport and energy-efficient buildings. This can reduce urban heat island issues
Fiji's 5 years and 20 years national development plan	This plan sets a vision for many projects, including infrastructure development. Climate resilience and environment are the key components where all development should be climate resilient, such as tourism, infrastructure, agriculture, fisheries, health, and education

Source Author's compilation

4.3.2.2 Community-Based Adaptation (CBA)

Urban CBA is an effective strategy in developing countries to deal with the challenges of rapid urbanisation and increased risk of climate change vulnerabilities. Urban CBA is a challenge as the focus of urban areas is to maximise revenue while least bothering about the environment. Therefore, policies and frameworks must be implemented effectively to achieve the goals. Archer et al. (2014) add that the possibility for more effective risk reduction arises when community-level initiatives are mainstreamed into and supported by city-level planning systems. This also enhances governance and accountability, builds capacity and transfers authority to the community level. CBA is a community-led method built on meaningful engagements and proactive involvement of local people and organisations, which frequently sets it apart from other kinds of adaptation (Remling and Veitayaki 2016). In Fiji, CBA is mainly initiated by Non-Government Organisations (NGOs) and enterprises. In Fiji, where there is a significant risk of natural disasters and climate change, the Adaptation Fund Organisation is working on a project to increase the resilience of informal urban communities. The Greater Suva Urban Area (Lami), Lautoka, Sigatoka, and Nadi are four major urban regions where the project is concentrated. The project's main goal is to enhance the adaptability of Fiji's informal urban settlements—which are extremely susceptible to natural disasters and climate change (Gomes 2021). Entrepreneurs in urban centres are also showing initiatives to deal with climate change in Fiji after implementing policies and frameworks by the government. The tourism industry, such as Fijian resorts, is installing solar power plants and wind energy. Radisson Blu Resort in Nadi adopted the Environment Policy in 1989 (Tourism Fiji n.d). Six Senses on Malolo Island in Fiji installed the Southern Hemisphere's largest off-grid solar and became Fiji's first hundred per cent solar-powered resort (Tourism Fiji n.d). Coca-Cola Fiji erected more than 3,800 solar panels at the Suva manufacturing plant to provide a 40% cleaner renewable energy source (CCAF Solar Launch 2017). The Fiji Government is also working with the local communities on river dredging projects. In Western Fiji, major urban centres like Sigatoka, Nadi, Ba, and Rakiraki are located close to the river, making them vulnerable to flooding due to river overflow. River dredging projects protect the nearby urban areas by increasing the water discharge capacity of the rivers by extracting silt materials that accumulate on rivers over a period. Dredging of the Nadi River protects 875 households, five villages, farms and Nadi town by removing 1.2 million cubic metres of silt (Krishant 2022).

Suva, the capital city of Fiji, is committed to sustainability and green initiatives in the urban Pacific. Suva municipal council has started developing a green and sustainable urban environment, such as a comprehensive examination of the policies, initiatives, and projects that could transform Suva into a sustainable role model for the Global South. To address environmental issues, Suva has witnessed increased cooperation between the public, businesses, international organisations, and NGOs in recent years. These projects are centred around integrating sustainable practices into everyday life and transitioning to more ecologically conscious alternatives in various sectors. Suva's sustainability and green initiatives encompass a diverse array of subjects, ranging from waste management to renewable energy, urban planning, and community engagement, all to exert a substantial impact on environmental conservation and promote sustainable lifestyles. Suva's urban planners employ a green urban development strategy, a crucial element of urban sustainability. This involves strategising urban areas to minimise their negative environmental effects and promote sustainable materials and efficient energy consumption.

To attain its objective of urban sustainability, Suva is also exploring state-of-the-art technology, including rainwater harvesting, ecologically sound garbage disposal systems, and more environmentally friendly forms of transportation. Suva is actively investigating novel approaches, such as rainwater harvesting, sustainable transportation systems, and eco-friendly waste management strategies, to advance its urban sustainability goals. The city is reducing its carbon emissions and promoting environmentally friendly transportation by establishing pedestrian zones and introducing a bicycle-sharing programme (ETIC Hotels 2023). These initiatives aid in reducing traffic congestion, decreasing pollution levels, and encouraging healthier lifestyles amongst the population. The city is implementing bicycle-sharing programmes and pedestrian walkways to reduce emissions and promote the adoption of eco-friendly modes of transportation. These measures bolster the community's endeavours to promote healthy lifestyles, mitigate pollution, and alleviate transportation congestion.

4.3.2.3 Ecosystem-Based Approach (EbA)

Ecosystem-based adaptation (EbA) has gained global recognition as an effective technique for addressing climate change impacts. Both scholars and governments are increasingly intrigued by these strategies because of their capacity to facilitate sustainable transitions. The concept of 'ecosystem-based adaptation' is a contemporary approach that involves utilising biodiversity and ecosystem services to assist individuals adapt to their environment. This theory has gained recognition in recent years (Wamslen 2015). Its objective is to strategically utilise natural services as a protective measure against the impacts presented by climate change. According to a study conducted by Wamsler in 2015, incorporating EbA into urban development and planning requires implementing several methods. These strategies include beginning new initiatives, making changes to organisational structures, amending legislation, and promoting collaboration amongst stakeholders. United Nations Economic and Social Commission for Asia and the Pacific (UNESCAP) (2018) states that EbA offers a chance to enhance awareness, educate young people, and engage communities, all while restoring a traditional connection to the ocean that is being lost in urban areas due to urbanisation. Implementing Ecosystem-based Adaptation (EbA) strategies in urban areas can contribute to the establishment of a sustainable environment by employing effective urban planning and well-managed green and blue spaces. This is crucial for enhancing urban resilience and promoting the overall well-being of urban people. The Secretariat of the Pacific Regional Environment Programme (SPREP) is now implementing a project named Pacific Ecosystem-Based Adaptation to Climate Change (PEBACC) in Fiji. The project's objective is to advocate for methods to address the effects of climate change, with a specific focus on urban greening (SPREP 2021).

The cities of Suva, Lautoka, Labasa, and Savusavu in Fiji are developing sustainable energy and climate action strategies to mitigate the negative impacts of climate change and contribute to its mitigation efforts. The national government is developing a 'Nationally Determined Contribution' climate strategy to fulfil its obligations under the Paris Agreement. They are also engaging in collaboration for this purpose. Their objective is to customise this national policy to suit the local context, enabling cities and municipalities to implement it effectively. The municipal councils of Suva, Lautoka, Labasa, and Savusavu have joined the Global Covenant of Mayors, putting these metropolitan centres in a group that will offer technical assistance for identifying and addressing climate-related hazards unique to their regions (Delegation of the European Union to the Pacific 2023). Some issues being addressed include waste management, coastal erosion, floods, and droughts. The Global Covenant of Mayors offers vital assistance by assessing the effects of climate change locally and assisting cities in putting appropriate adaptation plans into place. In addition, the programme helps urban areas participate in climate change efforts and create futures that are sustainable by facilitating access to international funds. Signatories have the ability to access essential technical help, assistance, and resources, allowing cities to expedite their energy and climate projects (Delegation of the European Union to the Pacific 2023).

4.3.3 Implications of Resilient Urban

Sustainable development relies more on resilient urban areas, especially in the light of climate change. Cities must change as climate-related hazards increase to protect citizens' health and safety and preserve economic stability. Rethinking urban infrastructure to resist harsh weather, increasing sea levels, and temperature shifts is part of this adaptation. Integrating green areas, permeability surfaces, and strong storm water management systems reduces urban heat island impacts and flooding concerns. A city's capacity to sustain electricity during emergencies is improved via distributed renewable energy systems and energy-efficient structures. Redesigning transportation networks will make them less susceptible to disasters and more adaptable. Shamsuddin (2020) argues that for the implementation of policy and strategies, it is crucial to consult with various stakeholders to ensure the success of the implementations. This is because of various income levels. People live in urban centres in the global south, and low-income earners cannot afford expensive resilient strategies, such as installing solar panels on rooftops. The concept of one size fits all must be avoided at the planning stage for resilient urban centres. Especially when the rural-to-urban drift has increased in the global south; thus, there will be financial constraints.

By the use of feedback, the implementation phase can contribute to the understanding of urban resilience. In another way, implementing urban resilience may change how people perceive resilience (Shamsuddin 2020). For instance, storm surge barriers might change how resilience is conceptualised by relating it to physical threats and countermeasures. This is particularly true given how challenging it is to define the precise boundary between policymaking and execution. The outcome of testing the urban resilience strategy must be improved based on the loopholes of the previous project before implementing it in the next urban centres. Policies and frameworks must change over time and adapt the necessary modifications needed for implementation. As Ministry of Commerce Trade Tourism and Transport Permanent Secretary Shaheen Ali highlights, urban centres need to respond to shocks in an environment supported by policy. Fiji needs plans and regulations that are strong, flexible enough to change with the times, and open to fresh interpretations of what it means to live a good life (Salusalu 2022). Trade-offs between the integrated resilience framework's complexity and the viability of its implementation need to be carefully considered, particularly in cities in the global south where resources are few and vulnerable and exposed urban populations are many and expanding (Rözer et al. 2022).

4.3.4 Challenges Faced in the Implementation of the Resilient Urban

Shamsuddin (2020) highlights that forming deeply ingrained administrative systems, which burden administration, hinder processes, and reduce mobility, is a significant component of resilience obstruction. These organisational systems include regulatory protocols, managerial structures, several tiers of decision-making and authorisation, and regulatory supervision. It is crucial to remember that these arrangements can still be effective even in the case of burdensome administrative processes or inflexible managerial hierarchies. In Fiji, the lack of administrative staff is one of the barriers to delays in administrative work, resulting in delays in approval. Finance is a major factor in low-carbon resilience, and green finance is paramount. Notwithstanding, obstacles present at the regional, national, and global levels, such as budgetary concerns, transient investments, and inadequate financial literacy within the private industry (He et al. 2020).

He et al. (2020) further adds that human capital is also essential to governance; a major difficulty might arise from a lack of professionals and leaders with the necessary specialised experience. Enhancing the leadership team created explicitly for low-carbon resilience requires cultivation and education. Given the significance of unique technology, transferability and effectiveness uncertainties might lead to technical issues. In order to overcome these technological obstacles, thorough investigation and empirical investigations are essential. In global south countries like Fiji, various times have been experienced when the government misused the funds for climate-related projects to some other projects. This hinders the climate resilience of developing states as they do not prioritise and utilise the purpose of the funds.

4.3.5 Recommendations

Resilient urban areas are a need in the global south due to the catastrophic impacts of climate change on the urban infrastructure and residents, leading to many socio-economic impacts. Fiji is one of the leading countries in the Global South regarding resiliency against climate change, including resilience in urban areas. To allow effective and efficient resilient, the following actions are needed:

4.3.5.1 Effective Policies and Frameworks

Every country needs to have clear policies and frameworks that can guide the country's actions and planning for the coming years. Establishing policies and frameworks shows the country's commitment to action against climate change. Not only that, but it also sets the investors' confidence that the government is committed to achieving resiliency. International aid providers such as the European Union, World Bank, and other agencies are very particular about the policies and strategies set out by the governments of developing states like the Global South before providing technical and financial assistance.

4.3.5.2 Proper Use of Allocated Resources

Diversion of resources from its allocated project to another is no surprise in developing states. Hence, this must be avoided, as allocating resources to the original project is more important. Especially in the Pacific, the funds received for climate mitigation and adaptation projects are diverted to projects like roads, bridges and public building renovation rather than building sea walls and wave breakers, which are urgently needed. Diverting resources to another project without donor consultation and approval can lead to future trust issues and may result in the discontinuation of aid.

4.3.5.3 Efficient Administration Process

As mentioned in the challenges faced in implementing resilient urban, administration procedures must be simplified to make the process easier and faster. In the Pacific, paperwork takes weeks due to long procedures and limited task force. Therefore, it becomes crucial to simplify the process. It will encourage investors and businesses to plan and invest in strategies that will make them resilient. For instance, in Suva, changing the structure of the building to make it greener and installing renewable energy sources will take months to get all the approvals done. Hence, it discourages others willing to convert grey infrastructure into green infrastructure.

4.3.6 Conclusions

In conclusion, like many other urban areas in the Global South, Fiji's metropolitan centres confront serious difficulties from rapidly increasing urbanisation and climate change. Nonetheless, Fiji has become a frontrunner amongst poor countries in implementing climate resilience plans. This chapter has looked at several important strategies, such as ecosystem-based approaches that use natural systems to build resilience, community-based adaptation initiatives that involve local stakeholders and make use of indigenous knowledge, climate-integrated urban planning and development supported by extensive national policies and frameworks, and creative projects in big cities like Suva that create more sustainable, low-carbon urban environments. Despite improvement, several major challenges remain, such as inefficient administrative processes, a lack of resources, and a need for more technological know-how.

This chapter suggests creating and successfully implementing clear policies and frameworks, ensuring that resources allocated for climate adaptation are allocated and used appropriately, and streamlining administrative procedures to support resilience initiatives to advance further urban climate resilience in Fiji and other Global South countries. Building climate-resilient metropolitan centres in the Global South is vital and feasible, as demonstrated by Fiji's experience. Nevertheless, it takes persistent dedication, creative thinking, and cooperation from all societal levels. Lessons from Fiji may prove important for other developing countries experiencing similar issues as climate impacts worsen.

References

Advisory Plan Rural Growth Centres (2024) Ministry of trade, co-operatives, small and medium enterprises. https://www.mcttt.gov.fj/division/town-and-country-planning/advisory-plans-for-rural-growth-centres/. Accessed 20 June 2024

Agrawala S, Ota T, Risbey J, Hagenstad M, Smith J, van Aalst M, Prasad B et al (2003) Development and climate change in Fiji: focus on coastal mangroves. OECD, Paris, France, pp 1–56

Alexander DE (2013) Resilience and disaster risk reduction: an etymological journey. Nat Hazard 13(11):2707–2716

Archer D, Almansi F, DiGregorio M, Roberts D, Sharma D, Syam D (2014) Moving towards inclusive urban adaptation: approaches to integrating community-based adaptation to climate change at city and national scale. Clim Dev 6(4):345–356

Arifwidodo S, Abdulharis R, Kubota T (n.d.) Understanding urban heat island effect and its implications to climate change adaptation strategies in major southeast Asian cities. Asia-Pacific Network for Global Change Research. https://www.apn-gcr.org/project/understanding-urban-heat-island-effect-and-its-implications-to-climate-change-adaptation-strategies-in-major-southeast-asian-cities/

Bera B, Chinta S, Mahajan DA, Sailaja A, Mahajan R (2023) Urbanization and its impact on environmental sustainability: a comprehensive review. J Harbin Eng Univ 44(8):1310–1318

Borie M, Pelling M, Ziervogel G, Hyams K (2019) Mapping narratives of urban resilience in the global south. Glob Environ Chang 54:203–213

Boyd E, Osbahr H, Ericksen PJ, Tompkins EL, Lemos MC, Miller F (2008) Resilience and 'climatizing' development: examples and policy implications. Development 51(3):390–396

Bresch DN, Spiegel A (2011) Economics of climate adaptation (ECA)–shaping climate-resilient development a framework for decision-making
Brown K (2014) Global environmental change I: a social turn for resilience? Prog Hum Geogr 38(1):107–117
Carter JG, Cavan G, Connelly A, Guy S, Handley J, Kazmierczak A (2015) Climate change and the city: building capacity for urban adaptation. Prog Plan 95:1–66
CCAF Solar Launch (2017) Fiji. https://www.cocacolaep.com/fj/news-and-insight/ccaf-solar-launch/
Chand SS, Tory KJ, Ye H, Walsh KJE (2017) Projected increase in El Niño-driven tropical cyclone frequency in the Pacific. Nat Clim Chang 7:123–127
Chand SS, Dowdy A, Bell S, Tory K (2020) A review of South Pacific tropical cyclones: impacts of natural climate variability and climate change. Clim Change Impacts Pacific:251–273
Chen Z, Gong Z, Yang S, Ma Q, Kan C (2020) Impact of extreme weather events on urban human flow: a perspective from location-based service data. Comput Environ Urban Syst 83:101520
Corbett J, Yi-Chong X, Weller P (2018) Climate change and the active participation of small states in international organisations. The round Table 107(1):103–105
Delegation of the European Union to the Pacific (2023) Nine Pacific cities make green covenant. EEAS. https://www.eeas.europa.eu/delegations/fiji/nine-pacific-cities-make-green-covenant_en. Accessed 21 June 2024
ETIC Hotels (2023) Sustainability and green initiatives in Suva, Fiji. ETIC J. https://www.etichotels.com/journal/sustainability-and-green-initiatives-in-suva-fiji/. Accessed 22 June 2024
Fiji Bureau of Statistics (2018) Fiji bureau of statistics releases 2017 census results. Fiji Bureau of Statistics. https://www.fiji.gov.fj/Media-Centre/News/Fiji-Bureau-of-Statistics-Releases-2017-Census-Res#:~:text=%C2%B7%2055.9%20percent%20of%20Fiji%E2%80%99s%20population%20reside%20in,movement%20of%20people%20from%20rural%20to%20urban%20areas. Accessed 6 June 2024
Gard AR, Veitayaki J (2017) In the wake of Winston–climate change, mobility and resiliency in FIJI. Int J Saf Secur Eng 7(2):157–168
Gomes A (2021) Increasing the resilience of informal urban settlements in Fiji that are highly vulnerable to climate change and disaster risks—adaptation fund. Adaptation Fund. https://www.adaptation-fund.org/project/increasing-resilience-informal-urban-settlements-fiji-highly-vulnerable-climate-change-disaster-risks-2/
He X, Lin M, Chen TL, Lue B, Tseng PC, Cao W, Chiang PC (2020) Implementation plan for low-carbon resilient city towards sustainable development goals: challenges and perspectives. Aerosol Air Qual Res 20(3):444–464
Hulme M (2009) Why we disagree about climate change: understanding controversy, inaction and opportunity. Cambridge University Press
Kalhor K, Mahdisoltani M (2015) Urbanization and its effects on the environment and society along with sustainable development. In: Third international symposium on environmental and water resources engineering, Tehran, Iran, vol 2(3) pp 1–12
Krishant N (2022) Nadi river dredging works commences. FijiVillage. https://www.fijivillage.com/news/Nadi-River-Dredging-works-commences-5xfr84/. Accessed 19 June 2024
Leichenko R (2011) Climate change and urban resilience. Curr Opin Environ Sustain 3(3):164–168. https://doi.org/10.1016/j.cosust.2010.12.014
McEvoy D, Mitchell D, Trundle A (2020) Land tenure and urban climate resilience in the south pacific. Climate Dev 12(1):1–11
McMichael C, Katonivualiku M (2020) Thick temporalities of planned relocation in Fiji. Geoforum 108:286–294
McNamara KE, Des Combes HJ (2015) Planning for community relocations due to climate change in Fiji. Int J Disaster Risk Sci 6:315–319
Mendoza MT, Schwarze R (2019) Sequential disaster forensics: a case study on direct and socio-economic impacts. Sustainability 11(21):5898

Milojević B (2018) Integrated urban planning in theory and practice. In: International conference on contemporary theory and practice in construction/Међународна конференција Савремена теорија и пракса у градитељству (No 13)
Phillips T, Keen M (2016) Sharing the city: urban growth and governance in Suva, Fiji
Radinibaravi M (2024) Significant growth in trade in services. The Fiji Times. https://www.fijitimes.com.fj/significant-growth-in-trade-in-services/
Ramsay EE, Duffy GA, Burge K, Taruc RR, Fleming GM, Faber PA, Chown SL (2023) Spatio-temporal development of the urban heat island in a socioeconomically diverse tropical city. Environ Pollut 316:120443
Remling E, Veitayaki J (2016) Community-based action in Fiji's Gau Island: a model for the Pacific? Int J Clim Change Strat Manag 8(3):375–398
Rözer V, Mehryar S, Surminski S (2022) From managing risk to increasing resilience: a review on the development of urban flood resilience, its assessment and the implications for decision making. Environ Res Lett 17(12):123006
Salusalu J (2022) Partnership develops master plan, for progress of major cities, towns. Fiji Sun. https://www.fijisun.com.fj/2022/11/10/partnership-develops-master-plan-for-progress-of-major-cities-towns/#google_vignette. Accessed 22 June 2024
Secretariat of the Pacific Regional Environment Programme (SPREP) (2021) SPREP pacific ecosystem based adaptation to climate change (PEBACC) project overview (extended version). https://www.fijiclimatechangeportal.gov.fj/videos/sprep-pacific-ecosystem-based-adaptation-to-climate-change-pebacc-project-overview-extended-version/
Serre D, Heinzlef C (2018) Assessing and mapping urban resilience to floods with respect to cascading effects through critical infrastructure networks. Int J Disaster Risk Reduct 30:235–243. https://doi.org/10.1016/j.ijdrr.2018.02.018
Shamsuddin S (2020) Resilience resistance: the challenges and implications of urban resilience implementation. Cities 103:102763
Syngellakis K, Anastasia O, Schulz A, Sagar A, Kang JSC, Johnston P, Mani F et al (2018) Fiji low emission development strategy (2018–2050). Ministry of Economy, Government of the Republic of Fiji, Fiji
ThinkHazard Report: Fiji (2020) thinkhazard.org. https://thinkhazard.org/en/report/83-fiji/EH. Accessed 12 June 2024
Tourism Fiji (n.d.) 6 Ways Fijian resorts practice sustainability favourite. https://www.fiji.travel/things-to-do/sustainability/6-ways-fijian-resorts-goes-sustainable. Accessed 19 June 2024
Trenberth KE, Hoar TJ (1997) El Niño and climate change. Geophys Res Lett 24(23):3057–3060
Trundle A (2020) Resilient cities in a sea of Islands: Informality and climate change in the south pacific. Cities 97:102496
United Nations Economic and Social Commission for Asia and the Pacific (ESCAP) (2018) Knowledge exchange on ocean cities 2018 promoting nature-based solutions in Pacific Island urban settlements. Workshop Report Suva, Fiji
Wamsler C (2015) Mainstreaming ecosystem-based adaptation: transformation toward sustainability in urban governance and planning. Ecol Soc 20(2).

Chapter 5
Influence of Urban Form on Urban Heat Island and Its Planning Implications in Chennai Metropolitan Area, Tamil Nadu, India

Adinarayanane Ramamurthy, Janani Mani, and Faiz Ahmed Chundeli

Abstract According to the World Urbanization Prospects report by the United Nations, the global urban population has grown significantly in recent decades, with 55% of the population living in urban areas in 2018. This figure is expected to rise to 68% by 2050. In the context of Asia, rapid urban expansion and widespread infrastructure development have contributed to the growing challenge of the urban heat island (UHI) effect, along with its associated environmental and social issues. Despite these concerns, many cities are not giving adequate attention to implementing effective heat mitigation strategies, which could exacerbate the long-term impacts of UHI. This phenomenon of rapid urbanization is leading to climate change, forming a great threat to humans and the surrounding ecosystem. Keeping the aforesaid knowledge and understanding of the functional linkages between urban morphology and UHI intensity, Chennai, a Coastal City of Southern India has been chosen for further detailed investigation. This research study delves into exploring intricate relationships through selected regions by using ENVI-Met simulation techniques, in diverse urban fabrics with different land uses under various Local Climate Zones and with an evolved UHI prediction model for the projected year 2043 A.D., by employing Machine-Learning Algorithms—Artificial Neural Network, Cellular Automata modeling. Based on the results an integrated planning framework was developed to recommend planning guidelines and mitigation strategies and to suggest an Area, Zone, Point Approach (AZP Approach) at the Neighborhood level and City level to achieve sustainable urban development of the study region.

A. Ramamurthy (✉) · J. Mani · F. A. Chundeli
School of Planning and Architecture Vijayawada, Vijayawada, Andhra Pradesh, India
e-mail: adinarayanane@spav.ac.in

J. Mani
e-mail: ar.jananimani317@gmail.com

F. A. Chundeli
e-mail: faizahmed@spav.ac.in

L. K. Dashora et al. (eds.), *Urban Sustainability*, Disaster Risk Reduction,
https://doi.org/10.1007/978-981-96-4890-0_5

Keywords Urbanization · UHI · Urban morphology · UHI prediction model · Integrated planning approach · Sustainable development

5.1 Introduction

Urbanization is a multifaceted socio-economic process that transforms the built environment, transitioning rural areas into urban settlements and altering the spatial distribution of populations from rural to urban regions (Cleugh et al. 2011). In India, the urban population is expected to rise significantly, from 470 million in 2020 to 814 million by 2050, accounting for 60% of the total population. This shift toward urban living necessitates a thorough examination of the effects associated with the creation and expansion of urban spaces for sustainable urban development. The interdependent relationship between urbanization and climate change highlights the complex dynamics between human progress and environmental impacts (Jain and Singh 2022). Rapid urbanization contributes to the Urban Heat Island (UHI) effect, intensifying local temperatures and amplifying energy demands for cooling, thus elevating greenhouse gas emissions. Land use changes associated with urban expansion, such as deforestation and altered drainage patterns, further impact local climates and precipitation patterns (Lehnert et al. 2021; Saputra and Setiawan 2019). Conversely, climate change poses challenges to urban areas, heightening the vulnerability of infrastructure and populations to extreme weather events. Adaptation and resilience in urban planning are imperative, necessitating strategies like sustainable infrastructure development, green spaces, and climate-resilient buildings (Demuzere et al. 2020; Wong and Chen 2021). The social dimension is pivotal, as climate change disproportionately affects vulnerable urban populations. As cities grapple with the dual challenges of rapid urbanization and climate change, integrated, sustainable urban planning emerges as a crucial paradigm for fostering resilient, equitable, and Environmentally conscious urban environments (Reba and Seto 2020).

5.2 Conceptual Framework

The research methodology is the fundamental framework and standards that guide the research process. It explains the precise procedures, strategies, and methodologies utilized to gather, analyze, and interpret data to answer the research question. UHI studies involve a multifaceted approach that integrates meteorology, climatology, urban planning, and Environmental science (Sidiqui and Bhat 2022). This section outlines the research methodology employed in investigating the UHI phenomenon, focusing on the key methods and strategies used to gather, tools to analyze data and expected outcomes along with the process of data collection, scientific tools and analytical techniques briefed in Fig. 5.1. The chapter focuses on mitigating

Urban Heat Islands (UHIs) in the Chennai Metropolitan Area, Tamil Nadu, India, by assessing the impact of UHIs and their dependence on urban morphology.

The research aims to examine spatiotemporal changes in urban climate conditions to identify Urban Heat Island (UHI) hotspots within the study region. The intention is to create Heat Stress Maps, Dynamic Potential Maps to generate an Urban Climatic Map, to understand the distribution and severity of UHI impacts. By mapping out areas with higher UHI intensity, researchers can understand the vulnerable sites of heat islands, allowing for focused interventions and mitigation techniques (Arnfield 2003). Understanding these patterns and hotspots is critical for urban planners addressing UHI concerns and working to create more sustainable and resilient urban settings (Ng and Ren 2015).

The second objective investigates the impact of urban morphology at the neighborhood level on the formation and severity of Urban Heat Islands (UHIs). This helps us to better understand the precise characteristics of communities that increase UHI formation by examining how these components contribute to the heat island effect. Identifying these essential elements can assist in designing tailored mitigation efforts to lessen UHI consequences and increase urban climate resilience (Kaloustian and Bou-Zeid 2016).

The third objective aims to create a model that predicts the future intensity of Urban Heat Islands (UHIs) for the year 2043 AD. This entails using modern tools like the Artificial Neural Network and the Methods Of Land Use Change Evaluation (MOLUSCE) Tool to develop a spatial model that can anticipate UHI trends based on urban climatic features and other control variables (Imran and Islam 2021; Di Sabatino and Buccolieri 2020). By using this prediction model, we intend to foresee how UHI intensity will change over time, giving significant information to forecast UHI trends can help guide the adoption of proactive targeted mitigation techniques and sustainable urban development practices in the study region, allowing UHIs to be efficiently managed and reduced (Aria and Cuccurullo 2017).

The final objective aims to develop an integrated planning framework and provide mitigation strategies and recommendations for achieving sustainable urban development within the context of Urban Heat Islands. This helps to holistically solve UHI concerns, that can be developed by taking into account a variety of elements such as land use, building features, green areas, and surface attributes. This framework can be used to guide urban planners and policymakers in implementing measures to reduce UHI intensity and create more liveable and resilient urban environments by proposing mitigation strategies, such as Development Control Regulations (DCR), for existing and predicted hotspots. Thus, the ultimate objective of this study is to create a framework that incorporates UHI mitigation into urban planning methods, assuring long-term sustainability and increased quality of life for urban populations.

However, the study faces limitations due to the confined CMDA boundary, limited data availability, and the consideration of only a few controlled parameters for UHI mitigation. The influence of water bodies on UHIs is not within the study's scope, and the accuracy of software-stimulated models in delivering ground truths is a challenge. Despite these limitations, the research questions the influence of urban morphology on UHI generation and explores various mitigation strategies needed

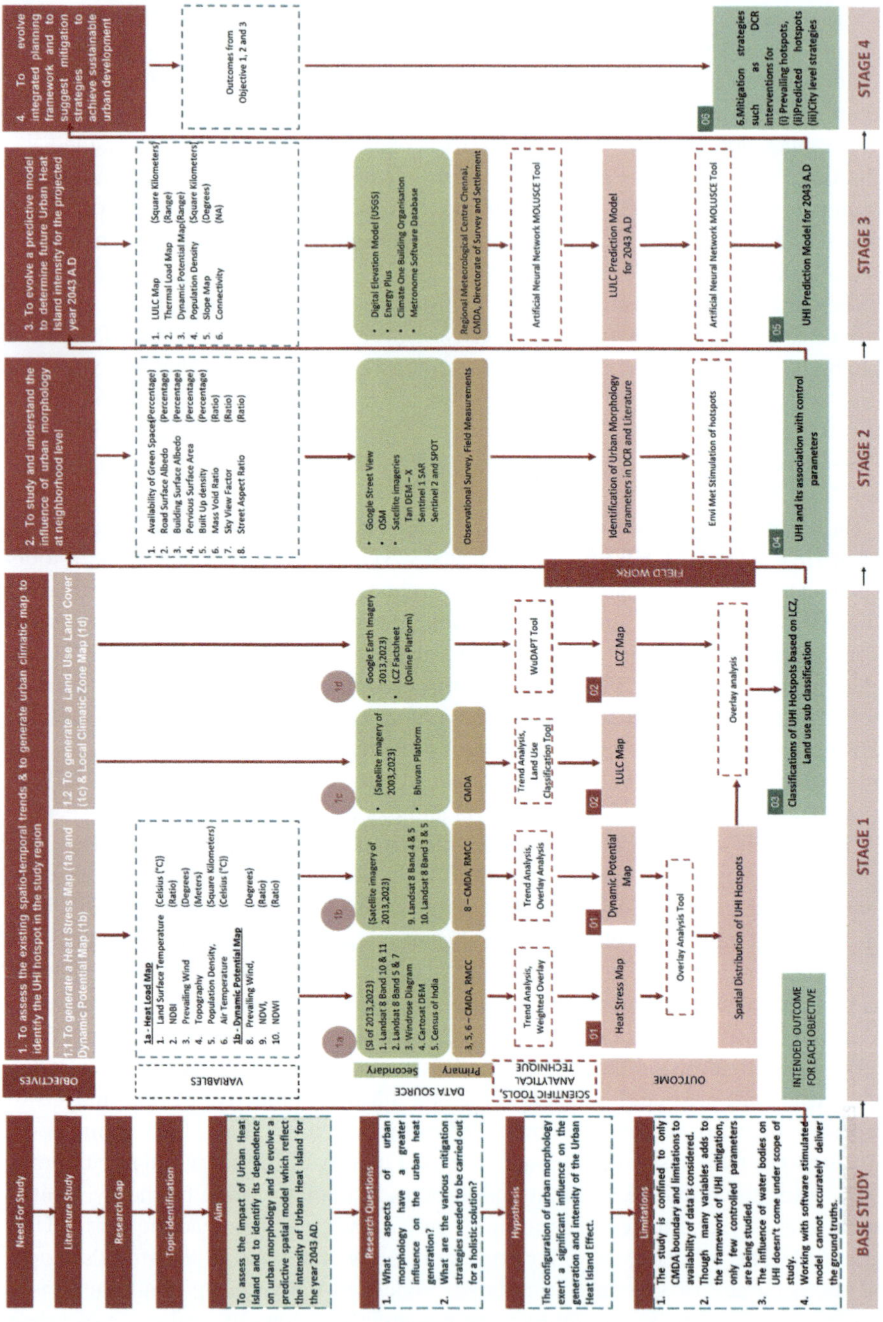

Fig. 5.1 Research framework to mitigate urban heat islands (*source* Authors 2024)

for a holistic solution. By utilizing scientific tools, analytical techniques, and data sources like satellite imagery, field measurements, and census data, the study aims to provide insights into UHI trends, predict future UHI intensity, and propose an integrated planning framework for sustainable urban development in the Chennai Metropolitan Area. To conduct research in Fig. 5.1, the data set will be collected during the field trip of 12 days from 3rd to 4th of February 2024, which was carried out to visit various departments and to conduct primary surveys. The primary data collection is carried out in Chennai Metropolitan Development Authority, Regional Meteorological Centre, Chennai, Directorate of Survey and Settlement. Along with these, primary observational surveys are conducted in the hotspots to understand the morphology of the study area for objective 2. The secondary data sources are satellite imageries of Landsat 8, DEM, LCZ Factsheet, and Sentinel. The future weather predictions can be referred to Energy Plus, Climate One Building Organisation, and Metronome Software Database.

5.3 Profile of Study Area

The case area for the study was identified Chennai, the capital of Tamil Nadu, is a thriving metropolis with a rich historical legacy and enduring significance in India's economic, political, and cultural landscapes. Formerly known as Madras, the city has undergone significant transformations post-independence, reflecting broader trends of urban development (Fig. 5.2).

The Chennai City Corporation, established in 1798, originally covered an area of 68 square kilometers and had a population of 5.40 lakh in 1901. By 2014, the population had grown to 70.81 lakh, and the city's area expanded to 426 square kilometers. The decadal growth rate has fluctuated between a minimum of 2.62% and a maximum of 64.97%. Chennai's climate is classified as a tropical wet and dry climate (Aw) under the Köppen climate classification system, indicating distinct wet and dry seasons. The city experiences hot, wet summers and dry winters. The rainy season typically spans from July to December, with the heaviest rainfall occurring between mid-October and mid-December. The dry season lasts from January to May, with the hottest temperatures recorded between late May and early June. Chennai's annual average temperatures range from 25.4 °C in January to 33.0 °C in May, while sea surface temperatures fluctuate between 26.5 °C in January and 30 °C in May. The city receives an average annual rainfall of approximately 1,235 mm.

Prevalent wind patterns play a vital role in shaping Chennai's climate. During the majority of the year, south-westerly winds prevail, transitioning to north-easterly winds during the remaining months. These wind patterns contribute to the region's overall weather patterns and help moderate seasonal temperature changes. Despite experiencing hot and humid conditions throughout the year, Chennai's coastal location prevents extreme variations in seasonal temperature. As a result, the city enjoys relatively stable climatic conditions, allowing residents to enjoy outdoor activities even during the warmest summer months.

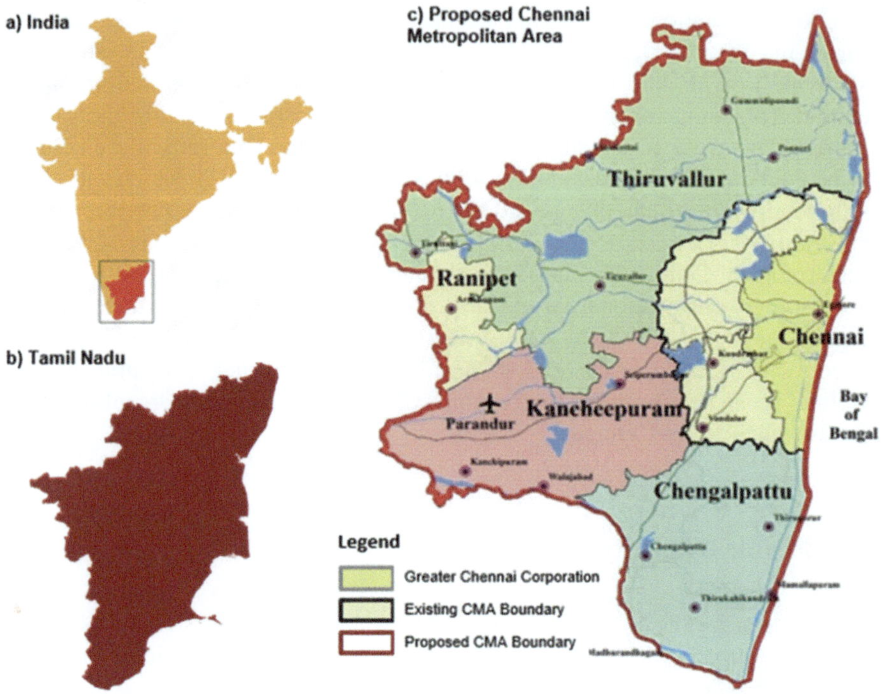

Fig. 5.2 Profile of Chennai (*source* CMDA)

The urbanization of the Chennai Metropolitan Area (CMA) is projected to undergo significant changes over the coming decades, leading to the transformation of land use patterns and built-up areas (Mathivanan and Madhavan 2019; Lata et al. 2021). A study conducted by Anna University in collaboration with the Indian Institute of Technology-Madras forecasts a substantial increase in the built-up area within the CMA. The study predicts that the built-up area, which covered 656 sq.km in 2021, will expand to 688 sq.km by 2031, 718 sq.km in 2041, and 750 sq.km by 2051. By the end of the century, the total built-up area is expected to reach 864 sq.km, constituting 72% of the total area of the CMA. This urban expansion will result in a decline in agricultural lands, with the extent of agricultural land projected to decrease from 346 sq.km to 289 sq.km by 2051. Additionally, the growth of urban areas and infrastructure demands will lead to the reduction of water bodies within the CMA, with the availability of water bodies expected to decrease from 153 to 145 sq.km by 2051. The study highlights the potential challenges that unchecked urban expansion may bring, including cycles of droughts and floods, seawater intrusion, groundwater depletion, and increased urban heat islands.

5.4 Materials and Methods

5.4.1 Urban Climatic Analysis

To determine how urbanization has affected the local climate, a variety of climatic parameters inside an urban environment are assessed and evaluated as part of the urban climatic study (Fig. 5.3).

The concept of the Urban Climatic Map (UC Map) emerged in the 1970s, pioneered by the Environmental Meteorological Working Group from the German Meteorological Society. This innovative approach aimed to bridge the gap between the scientific understanding of urban climatology and the practical application of this knowledge in urban planning and design. The initial studies conducted in German cities like Stuttgart and Kassel demonstrated the potential of integrating climatic maps with urban planning strategies, particularly in addressing air pollution and correlating climatic factors with the built environment. The UC Map serves as a comprehensive tool for visualizing and analyzing the thermal and wind environments within an urban setting, providing crucial information to guide sustainable urban development (Ren and Ng 2013; Kotharkar and Bagade 2020). At the core of

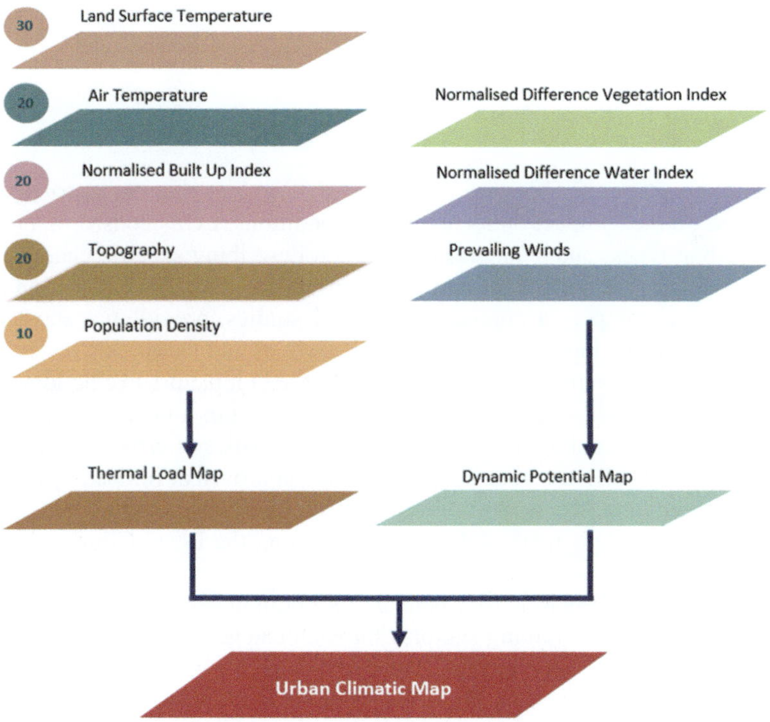

Fig. 5.3 Urban climatic map—methodology (*source* Authors, 2024)

the UC Map methodology is the concept of "climatopes"—spatial units on the map that represent areas with similar urban climatic conditions and features (Ng and Ren 2015).

A thermal load map can be created to visually represent and analyze the thermal characteristics of a specific area, to obtain the map, a weighted overlay analysis was performed to produce a thermal load map by including the layers that were discussed in the methodology flowchart such as Land Surface Temperature, Normalized Difference Built-up Index, Population Density, Topography and Air temperature with specifies weightages in the flowchart after normalization of each layer (Fig. 5.4).

A dynamic potential map is a type of map that reflects the changing and evolving nature of natural and human phenomena over time and involves the visualization of spatial data to represent the potential or suitability of different areas for specific purposes. The UCM provides a comprehensive picture of the spatial distribution of thermal comfort within the city. Areas with high thermal load and low dynamic potential would be most susceptible to heat stress. Conversely, areas with low thermal load and high dynamic potential would be relatively comfortable. CMA is experiencing significant urban heat island effects due to factors like urbanization and reduced greenery, leading to extreme heat zones that pose health risks to the local population. Utilizing tools like the UC Map can help identify and analyze these heat-prone areas within the CMA, supporting the development of targeted strategies for climate change adaptation and urban heat island mitigation.

5.4.2 Local Climate Zone

Local Climate Zones (LCZs) are a classification system that categorizes surface types based on their impact on local climate conditions. LCZs consist of 17 zones, with 10 urban types, and 7 natural types- each describing specific characteristics of surface cover, structure, and land use (Stewart and Oke 2012; Chen and Peng 2016). These zones play a crucial role in UHI studies, providing a standardized framework for analyzing urban temperature variations and facilitating the exchange of temperature observations. LCZs are defined based on factors like building types, vegetation cover, and ground surface properties, offering a detailed classification approach for understanding the micro-scale characteristics of urban landscapes.

The LCZ WUDAPT methodology refers to the approach developed by the World Urban Database and Access Portal Tools (WUDAPT) for creating Local Climate Zone (LCZ) maps. This methodology involves using the Local Climate (Figs. 5.5 and 5.6).

A Local Climate Zone Map for Chennai Metropolitan Area is generated using the WuDapt tool by giving training samples for each category and obtaining the final classification. From the generated LCZ map, it can be interpreted that Local Climate Zone 1, and Local Climate Zone 7 are negligible in the study area compared to other LCZ types area. The share of area for each classification under the region of interest is discussed in the following Table 5.1.

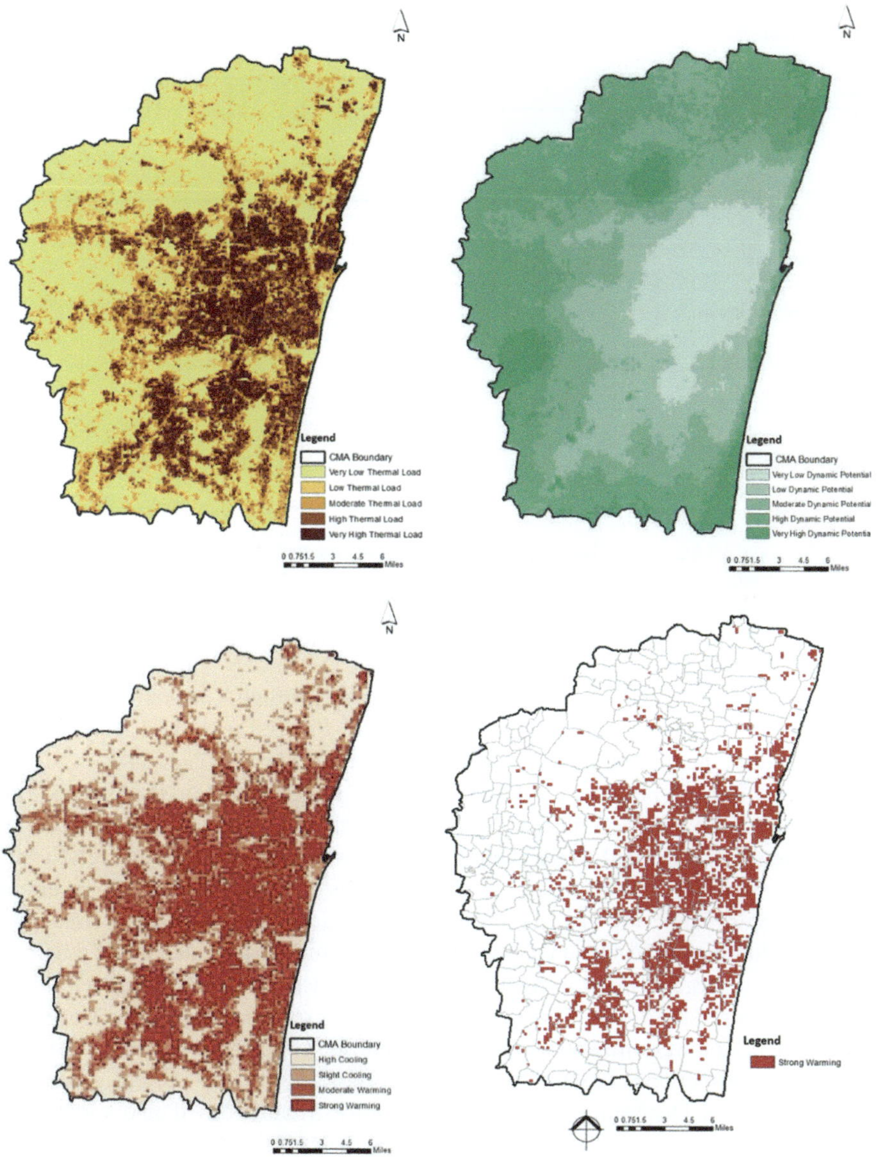

Fig. 5.4 Heat stress map of Chennai, 2023 (top left), dynamic potential map of Chennai, 2023 (top right), urban climatic map of Chennai, 2023 (bottom left), strong warming zones of Chennai, 2023 (bottom right) (*source* Authors, 2024)

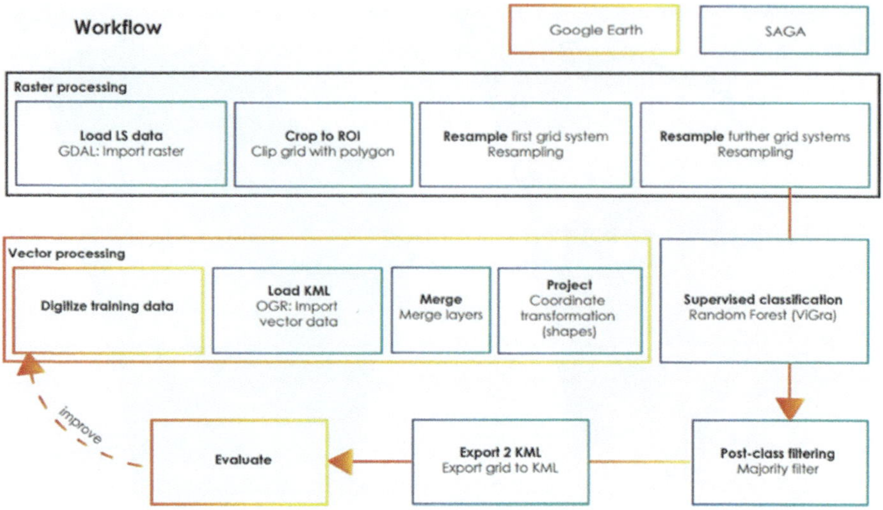

Fig. 5.5 Local climate zone classification—methodology (*Source* Michal Lehnert, 2021)

5.4.2.1 Exploration of Critical LCZs

To understand the relationship between the Local Climate Zones and Urban Heat islands, each layer of Local Climate Zones was overlayed with the LST of 2023 to understand the prevailing temperatures in each zone (Figs. 5.7 and 5.8).

The relationship between UHI and LCZ was studied based on the Land Surface Temperature and based on this LCZs which are subjected to high temperatures have been classified as critical LCZs, which will be used to understand the relationship between urban morphology and local climate zones. Understanding the local climate zones enables planners to identify areas that are particularly prone to heat stress, such as zones with high urban heat island intensity. This information can guide the implementation of targeted heat mitigation strategies, such as increasing green spaces, improving ventilation, and incorporating cool roofs or reflective surfaces to reduce surface temperatures and enhance thermal comfort. Cities may increase their resistance to the effects of climate change by incorporating LCZ data into urban planning procedures. Designing communities and infrastructure based on distinct zones' thermal properties can help reduce heat-related dangers, increase energy efficiency, and create more sustainable and liveable urban settings.

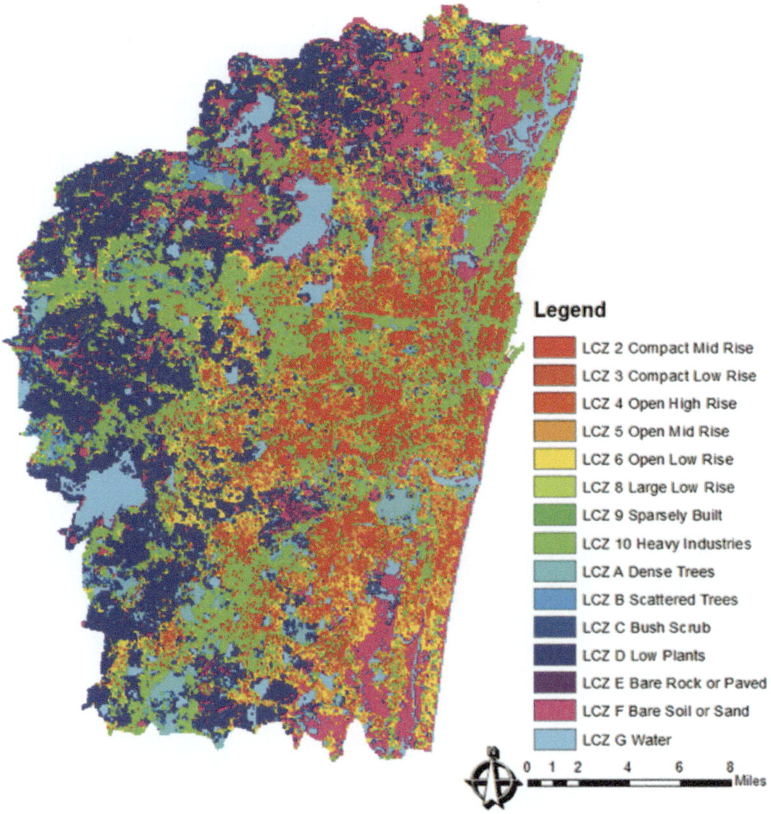

Fig. 5.6 Local climate zone classification of Chennai, 2023 (*source* Authors, 2024)

5.4.3 Urban Morphology and a Comparative Analysis of LCZs in Urban Core and Peri-Urban Regions

The relationship between urban morphology and the Urban Heat Island (UHI) effect is a key focus in urban climate research. Urban morphology, which measures the spatial structure, physical form, and internal relationships within a city, significantly influences the intensity and distribution of the UHI effect. Elements such as building density, vegetation coverage, building materials, height, compactness, and land cover types all play crucial roles in affecting UHI patterns. The geometry of urban structures, including buildings and street canyons, can limit wind flow and trap heat radiation, leading to higher temperatures in urban areas compared to rural surroundings. Additionally, the conversion of natural land cover into impervious surfaces, reduced latent heat flux, and greater heat absorption further intensify the UHI effect. A deeper understanding of how urban form contributes to UHI is vital for urban

Table 5.1 LCZ's area distribution

LCZ classification	LCZ name	Area (Sq. Km)
LCZ 2	Compact mid rise	92.92
LCZ 3	Compact low rise	43.82
LCZ 4	LCZ 4 open high rise	0.07
LCZ 5	Open mid rise	90.41
LCZ 6	Open low rise	110.7
LCZ 7	Lightweight low rise	0
LCZ 8	LCZ 8 large low rise	0.48
LCZ 9	LCZ9 sparsely built	8.65
LCZ 10	LCZ 10 heavy industry	266.99
LCZ A	Dense trees	37.27
LCZ B	LCZB scattered trees	27.8
LCZ C	LCZC bush	1.15
LCZ D	LCZD low plants	244.13
LCZ E	LCZE bare rock or paved	2.29
LCZ F	LCZF bare soil or sand	178.36
LCZ G	LCZ G water	84.98
	Total area	**1189.95**

Source Authors, 2024

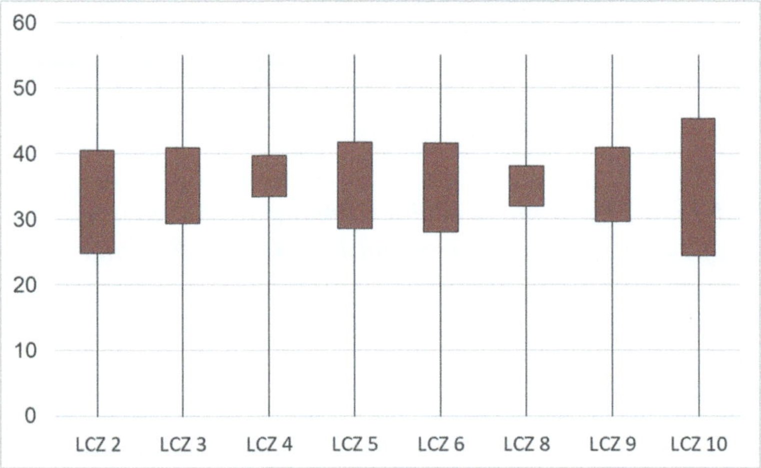

Fig. 5.7 Temperature range of built LCZs (*source* Authors, 2024)

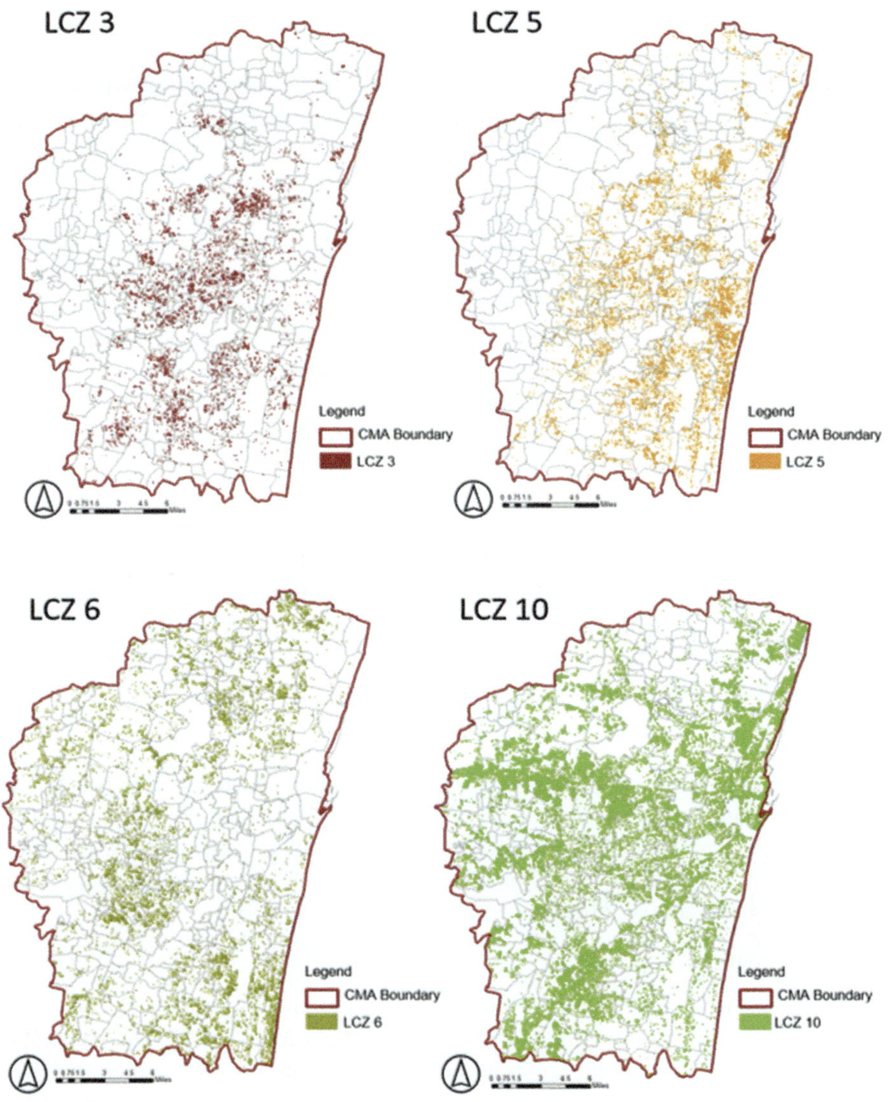

Fig. 5.8 Critical LCZs in Chennai (*source* Authors, 2024)

planning, climate mitigation, and developing sustainable, resilient cities in response to the challenges posed by climate change.

5.4.3.1 Simulation Model for Selected LCZs

Nine locations were selected with 6 neighborhoods in the core regions with different land uses and 3 neighborhoods in the peri-urban region. ENVI- met Simulations were made taking into all the possible input data. The simulations used epw file for obtaining climate-related data. Typically, May is one of the hottest months in the city. To obtain the microclimate ENVI- Met software was employed, with urban microclimates being produced in the spaces interface, climate data being included as an epw file in the ENVI- guide, and the final Simulation occurring in the ENVI-core programming. The results of these Simulations—Potential Air Temperature (12 am, 12 pm), Sky View Factor (SVF) and Mean Radiant Temperature (MRT) can be examined via the Leonardo interface (Fig. 5.9 and Table 5.2).

From the undertaken comparative study of Potential Air Temperature, Sky View Factor and Mean Radiant Temperature in nine different Local Climate Zones, the following inferences have been made.

1. Material Albedo

Surfaces with higher albedo values reflect more sunlight and absorb less heat, leading to lower surface temperatures and reduced heat retention in within buildings. The higher the Solar Reflective Index of a material, the higher is the thermal comfort within the buildings. The change in surface albedo of cool and pavements can alter the temperature in the region.

2. Pervious materials

Pervious materials like pervious concrete, porous asphalt, and permeable pavers, can increase the amount of vegetation, and reduce the number of heat-absorbing surfaces. This lowers surface temperatures, reduces the amount of heat absorbed by buildings and pavement, and improves air quality. Additionally, previous materials can help recharge groundwater supplies and reduce stormwater runoff.

3. Green Spaces

Green spaces are more efficient at mitigating heat compared to built-up areas, and the reduction in high-value areas is smaller in scenarios where green spaces are preserved. Green spaces and urban heat islands have an inverse correlation. Thus, areas with more trees like Anna Nagar, and Besant Nagar are cooler compared to VOC Nagar and T Nagar.

4. Sky View Factor

High SVF values typically correspond to areas with more open spaces, less building density, and greater availability of vegetation. These areas tend to have better ventilation and more efficient cooling through natural processes like convection and evapotranspiration. Therefore, there's an inverse relationship between SVF and UHI is absorbed.

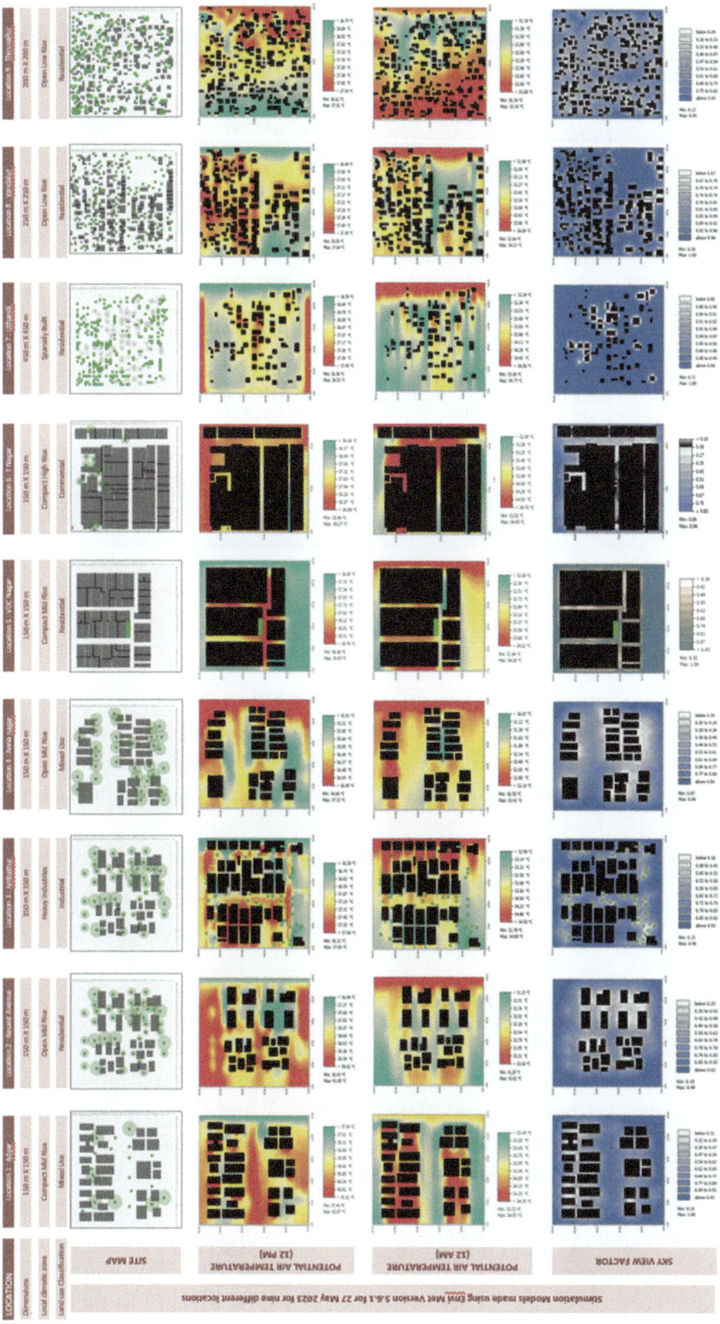

Fig. 5.9 ENVI-met simulation of selected LCZs *Source* Generated by Authors in ENVI–Met 5.6.1

Table 5.2 Potential air temperature of selected LCZ zones

No	Location name	LCZ classification	Land use classification	Potential air temperature (12 am)		Potential air temperature (12 pm)	
				Min	Max	Min	Max
1	Adyar	Compact mid rise	Mixed use	33.22	34.55	36.9	39.03
2	Besant Nagar	Open mid rise	Residential	31.2	33.82	36.59	38.53
3	Ambattur	Heavy industries	Industrial	32.78	34.88	37.44	43.07
4	Anna Nagar	Open mid rise	Mixed use	30.58	33.41	34.85	37.32
5	VOC Nagar	Compact mid rise	Residential	31.94	34.28	35.99	39.27
6	T Nagar	Compact high rise	Commercial	33.06	34.73	36.43	40
7	Uthandi	Sparsely built	Residential	32.82	34.95	36.31	37.95
8	Vandalur	Open low rise	Residential	32.64	34.23	36.85	37.64
9	Thiruvallur	Open low rise	Residential	30.39	33.3	36.62	37.82

(*Source* Authors, 2024)

5. Aspect Ratio

Aspect ratio of buildings directly affects the obstruction of the sky from a given point within an urban area. High aspect ratio buildings tend to have a larger footprint relative to their height. As a result, they can create more shadow areas and obstruct views of the sky from ground level, leading to lower SVF values in their immediate vicinity.

These factors will be researched to arrive at recommendations and strategies to mitigate Urban Heat Islands.

5.4.4 Future Prediction of Heat Stress Zones

The simulations of Land Use and Land Cover (LULC) change and predictions were performed using the Artificial Neural Network (ANN) and Cellular Automata (CA) models. The ANN was employed to calculate the transition probabilities for LULC and Urban Climatic Map changes, utilizing multiple output neurons to simulate various LULC transformations within the ANN-CA framework. Cellular Automata were then used to model LULC and Urban Climatic changes by applying the transition probabilities generated through the ANN learning process. The entire analytical procedure was executed using QGIS, with its MOLUSCE module facilitating the ANN-CA modeling process (Fig. 5.10).

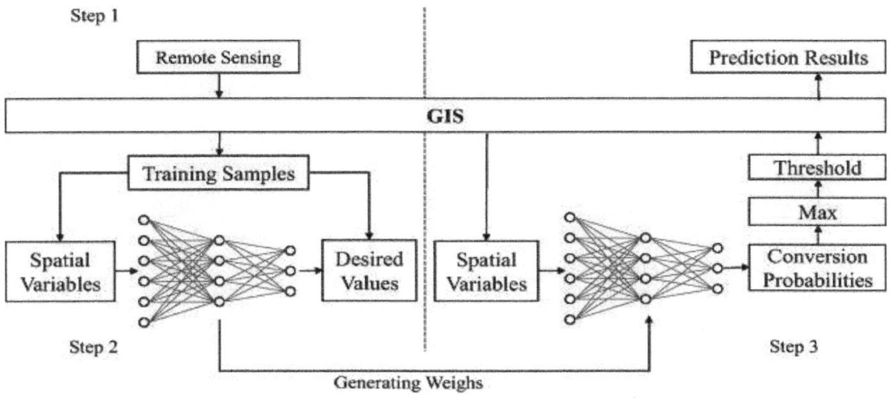

Fig. 5.10 Artificial neural network based cellular automaton (ANN-CA) model structure (*source* (Quanmei Gong, 2023))

5.4.4.1 Prediction of Land Use Land Cover for 2043 A.D.

To develop the prediction model for Land Use and Land Cover (LULC), input LULC layers were generated using the maximum likelihood classification method, a supervised classification approach. This method assumes that the statistics for each class in each spectral band are normally distributed and calculates the probability that a given pixel belongs to a specific class based on these statistics. Each pixel is then assigned to the class with the highest probability, taking into account factors such as the mean measurement vector, divergence, standard deviation, and probability for each class. By leveraging the mean and covariance matrix of each class, the maximum likelihood classifier determines the likelihood of a pixel belonging to a particular class and assigns it accordingly. This approach is particularly effective for categorizing pixels into various land cover classes based on their spectral properties, making it an essential tool for image classification and land use mapping in remote sensing applications. Using this classification method, Land Use and Land Cover maps for the Chennai Metropolitan Area were created for the years 2003, 2013, and 2023 (Fig. 5.11).

The ANN-CA (Artificial Neural Network—Cellular Automata) analysis was utilized to predict land use/land cover (LULC) changes and urban climate dynamics in the study area. By integrating Artificial Neural Networks with Cellular Automata, this approach aimed to enhance the accuracy and reliability of forecasting LULC changes. The model leveraged input data from 2003 and 2023 LULC maps along with explanatory input layers to simulate future scenarios for the year 2043 AD, capturing the spatial distribution of land use changes over time.

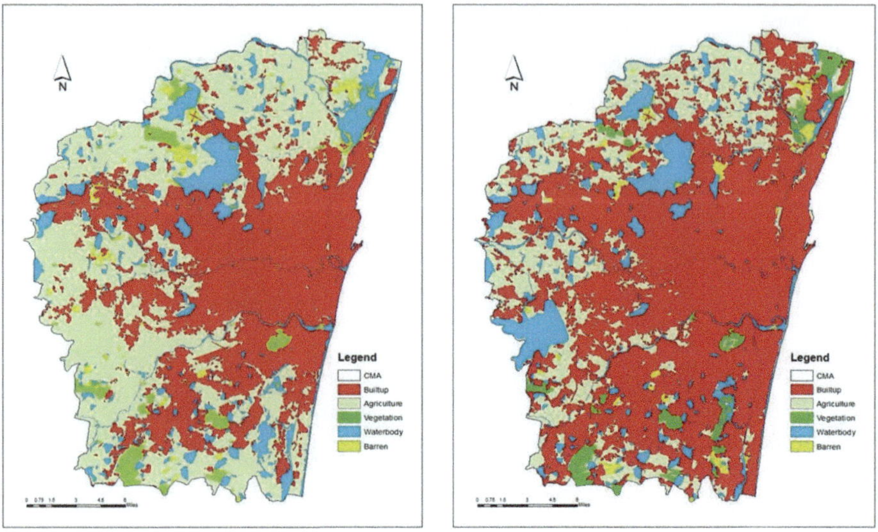

Fig. 5.11 Land use land cover of CMA—2003 (left), 2023 (right) *Source* Generated by Authors 2024

5.4.4.2 Prediction of Heat Stress Zones for 2043 A.D.

Subsequently, the same methodology was applied to develop a predictive model of Urban Climatic Map for the projected year of 2043 A.D. The Urban Climatic Map for 2003, 2013 and 2023 were generated by incorporating the methodology in Fig. 5.5. The overall analytic procedure is carried out by QGIS and its MOLUSE module were utilized for the ANN-CA modeling. The model leveraged input data from 2003 and 2023 Urban Climatic maps along with explanatory input layers of UC Map of 2013 to simulate future scenarios for the year 2043 AD, capturing the spatial distribution of climate zones in the study region. This prediction model helps us to identify the extreme heat stress zones for the future and to formulate effective mitigation measures and also to phase the mitigation strategies based on their intensity of heat exposure (Fig. 5.12).

The prediction model for LULC and heat stress maps can give useful information and aid decision-making in various domains. In this context, these models may assist predict future urban growth trends and guide informed decisions about land use management, infrastructure development, and green space preservation. This knowledge is critical for reducing the UHI effect and developing more sustainable and resilient communities.

Furthermore, understanding LULC changes and heat stress distribution might help build appropriate adaptation measures to climate change consequences like extreme heat occurrences. By identifying sensitive locations, public health officials may launch targeted actions to protect vulnerable people and prevent heat-related

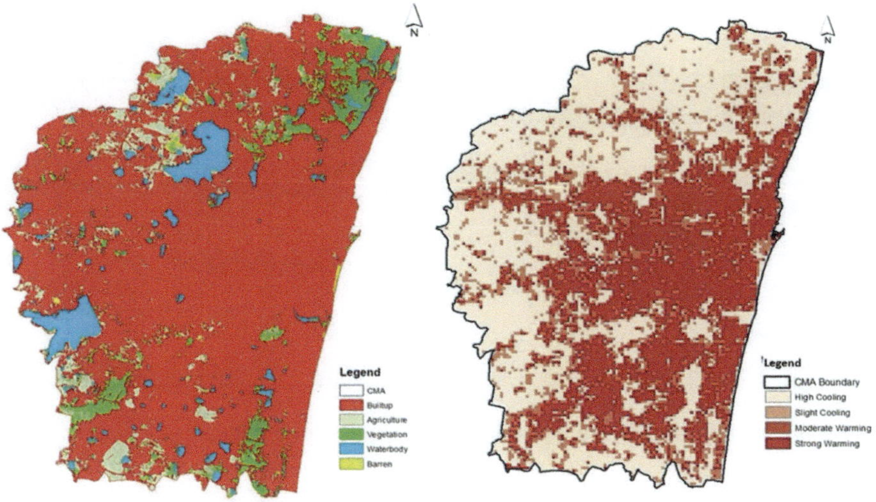

Fig. 5.12 Predicted LULC map of CMA, 2043 A.D—Left predicted heat stress zones of CMA, 2043 A.D—right (*source* Authors, 2024)

diseases. Furthermore, predictive models can help with Environmental management by emphasizing heat-stress-prone locations and guiding conservation efforts to protect natural ecosystems and biodiversity in the face of urbanization and climate change. This comprehensive approach to Environmental management can aid in the delicate balance between urban growth and ecological preservation.

5.5 Results and Discussion

5.5.1 Mitigation Strategy to Curb UHI

The phenomenon known as the "Urban Heat Island" (UHI) effect occurs when temperatures in urban areas are greater than in nearby rural areas. The high percentage of impermeable surfaces, the absence of flora, waste heat from human activity, and urban geometry are some of the reasons contributing to this temperature differential. Many strategies have been suggested and put into action to mitigate the UHI effect and will fit in one of the following broad categories (Fig. 5.13).

To reduce urban heat islands, this mitigation plan uses an area-based, zone-based, and point-based approach. The two main categories of mitigation are cool material-based solutions and nature-based solutions. These two interventions, coupled with changes to urban morphological factors like construction material, density, building height, and pavement material, would change the land use and land cover of the CMA region. Through all of these interventions, the building's total thermal comfort

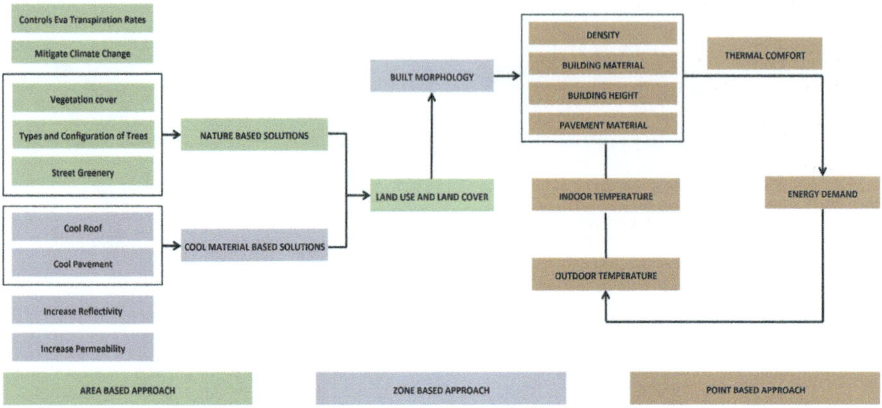

Fig. 5.13 An overview of UHI mitigation strategies (*source* Authors, 2024)

will be increased, lowering inside temperatures and, in turn, the building's energy consumption, which will also indirectly lower outside temperatures.

5.5.2 Neighborhood Level Mitigation Scenarios and Results

However, numerous studies focus on the relationship between urban form and UHI. There have been relatively few studies in India that examine the impact of mitigation techniques on urban temperatures (Rajashree Kotharkar 2020). As a result, several generalized scenarios were developed, and Environmental Simulations were produced in crucial LCZs to better understand the influence on the study region. This will help us prioritize our mitigation methods depending on their effectiveness (Table 5.3 and Fig. 5.14).

5.5.3 Characteristics of Critical LCZ Wards

To demonstrate and comprehend the effectiveness of mitigation techniques in lowering temperature in the study region. Two essential villages have been selected for the comprehensive execution of strategies. These villages were chosen based on the most critical LCZs within the CMA region: LCZ 10 (Heavy Industry); LCZ 5 (Open Midrise). Mannur village was chosen for LCZ 10 (Heavy Industries), as the region of interest fits entirely within the LCZ 10 categorization and industrial use in nature. For LCZ 5 (Open Mid Rise), Vivekanadapuram village in Mylapore—Triplicane taluk was studied. The region of interest in this zone is classified as LCZ 5 with residential land use (Figs. 5.15 and 5.16).

Table 5.3 Mitigation scenarios to curb UHI effect

No	Scenario	LCZ 10 Mannur		LCZ 5 Mylapore	
		Surface albedo	Vegetation	Surface albedo	Vegetation
1	Base case	Roof Albedo 0.5 Wall Albedo 0.4	As per site features	Roof albedo 0.3 Wall albedo 0.4	As per site features
2	Building level (cool roof)	Albedo of roof is increased from 0.5 to 0.9	Same as Base case	Albedo of roof is increased from 0.3 to 0.9	Same as base case
3	Street level (cool pavement)	Street material is changed from asphalt to light concrete Albedo of asphalt—0.1 Albedo of light concrete—0.5	No changes	Street material is changed from asphalt to light concrete Albedo of asphalt—0.1 Albedo of light concrete—0.5	No changes
4	Street level (cool pavement + trees)	Street material is changed from asphalt to light concrete Albedo of asphalt—0.1 Albedo of light concrete—0.5	Trees of various crowns based on the width of the street	Street material is changed from asphalt to light concrete Albedo of asphalt—0.1 Albedo of light concrete—0.5	Trees of various crowns based on the width of the street

(*source* Authors, 2024)

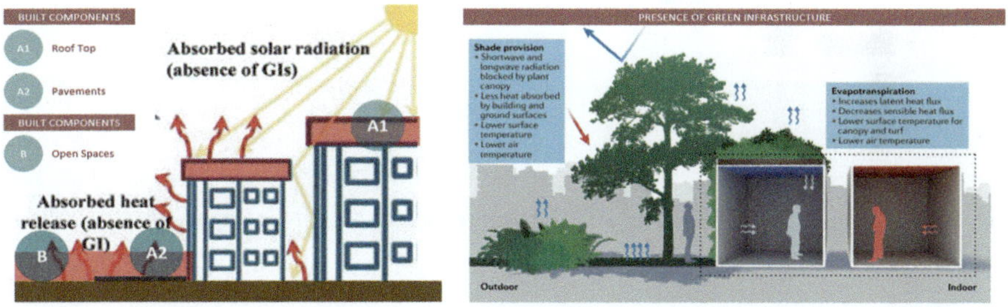

Fig. 5.14 UHI mitigation strategies (*source* Authors (left), *(Nyuk Hien Wong, 2021)* (right))

For the mentioned locations, ENVI-met Simulations were made taking into all the possible input data. The Simulations used epw file for obtaining climate-related data. Typically, May is one of the hottest months in the city. Based on the sun path diagram, it is observed that 5am recorded the lowest temperature whereas 3 pm recorded the highest temperature in the month of May. To obtain the microclimate ENVI- Met software was employed, with urban microclimates being produced in the spaces interface, climate data being included as an epw file in the ENVI- guide, and the final Simulation occurring in the ENVI—core programming. The results of

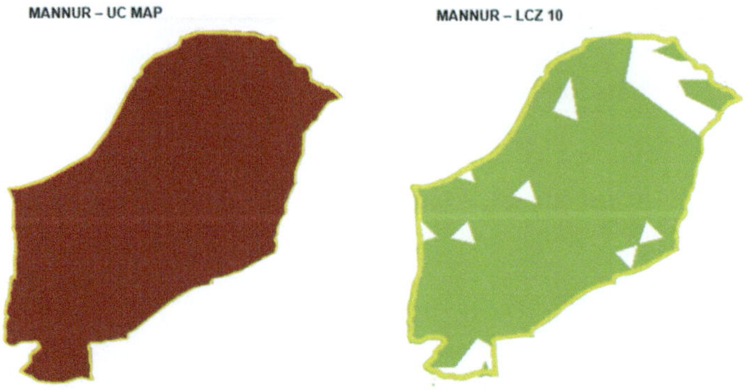

Fig. 5.15 Mannur UC map (left), Mannur LCZ 10 Map (left) (*source* Authors, 2024)

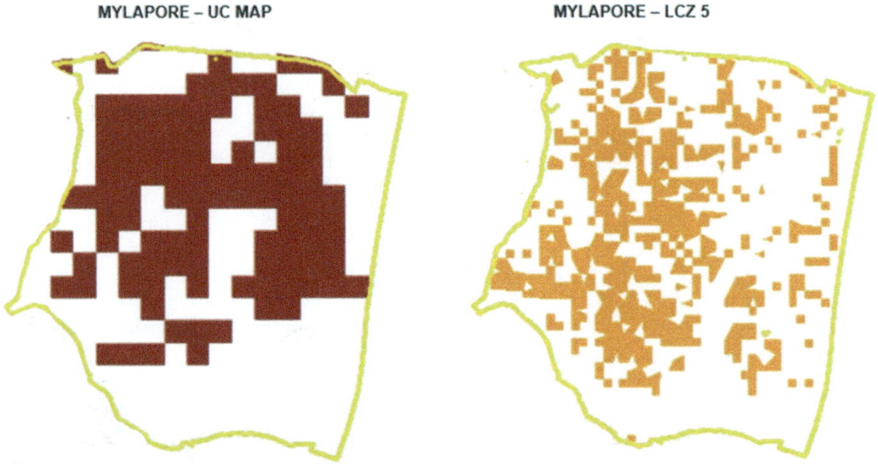

Fig. 5.16 Mylapore Taluk UC Map (left), Mylapore Taluk LCZ 5 Map (right) (*source* Authors, 2024)

these Simulations—Potential Air Temperature, Sky View Factor and Mean Radiant Temperature can be examined via the Leonardo interface (Tables 5.4 and 5.5).

The above ENVI-met model iterations of the base case and the other three mitigation scenarios of cool roof, cool streets and cool streets with plantation of trees show that both Scenario 2 and 3 of implementing cool roofs and cool streets respectively are showing positive results in reducing the potential air temperature of the neighbourhood. Whereas Scenario 4 of incorporating both cool streets and tree plantation exhibits the highest positive relationship to reduced potential air temperature of the study area. It can be noticed that LCZ 10 (Mannur) have a lesser cooling effect compared to that of LCZ 5 (Vivekanadhapuram) due to various factors such as land use, open space ratio etc., it is seen that in both the cases of LCZ, a combination

Table 5.4 Characteristics of Mannur and Vivekandhapuram

Characteristics of LCZ 10		Characteristics of LCZ 5	
Low-rise and mid-rise industrial structures		Open arrangement of mis-rise buildings	
Land cover mostly paved or hard-packed		Abundance of pervious land surface	
Metal, steel and concrete construction materials		Concrete, steel, stone and glass materials	
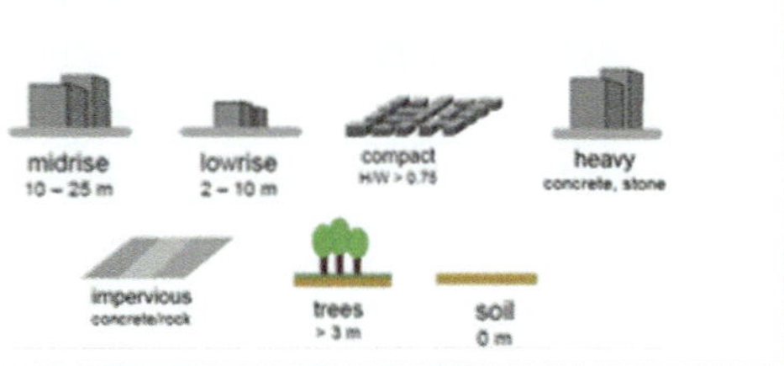		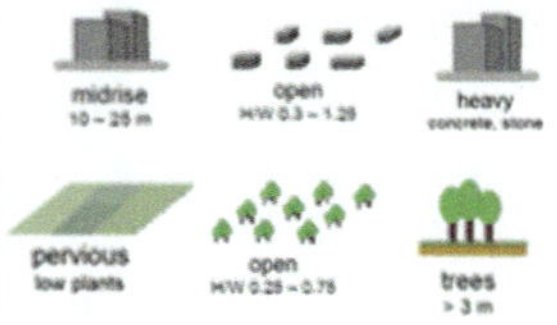	
Critical study ward for LCZ 10		Critical study ward for LCZ 5	
Village number	87	Village number	112
Village name	Mannur	Village name	Vivekanadapuram area
Taluk	Ambattur	Taluk	Mylapore–Tripilicane
District	Thiruvallur	District	Chennai
Municipality	Ambattur	Municipality	Chennai
Local body	Chennai corporation	Local Body	Chennai corporation
Critical study area for LCZ 10		Critical study area for LCZ 5	

of cool street and plantation of trees scenario performs better compared to the other scenario (Figs. 5.17 and 5.18).

This makes us understand that cool roofs, cool streets and tree plantations are effective mitigation strategies and should be incorporated at the earliest. Cool roofs and cool streets can be implemented in the short term due to their easy installation, immediate results and cost-effectiveness with very less maintenance. Whereas tree plantation can be used as a midterm or long-term process, due to the rate of growth of trees, maintenance, availability of plantation area etc. Apart from mitigating the heat, these mitigation recommendations also possess various advantages which are discussed below in the table.

Table 5.5 Characteristics of Mannur and Vivekanadhapuram and PAT

Location	Scenario	Mitigation strategies	Potential air temperature (5 am)		Potential air temperature (3 pm)	
			Min	Max	Min	Max
LCZ 10 Mannur	Scenario 1	Base case	28.57	31.67	39.6	42.02
	Scenario 2	Cool roof	28.37	31.55	39.49	41.90
			−0.2	−0.12	−0.11	−0.12
	Scenario 3	Cool streets	28.19	31.55	39.43	41.88
			−0.38	−0.12	−0.17	−0.14
	Scenario 4	Cool streets and trees	28.04	31.43	38.92	41.55
			−0.53	−0.24	−0.68	−0.65
LCZ 5 Vivekanadhapuram	Scenario 1	Base case	27.11	30.16	36.55	38.95
	Scenario 2	Cool roof	27.08	29.9	35.88	37.84
			−0.03	−0.26	−0.67	−1.11
	Scenario 3	Cool street	26.18	29	34.9	36.98
			−0.93	−1.16	−1.65	−1.97
	Scenario 4	Cool streets and trees	26.05	28.69	34.57	36.82
			−1.06	−1.47	−1.98	−2.13

5.6 Recommendations

India being a data-sparse country, needs to first check on data availability and accuracy for prevailing data, then find what are the policy barriers, as well as analyze the scope of implementation based on the resource constraints. To make the policy effective on the ground it's important to involve communities, identify vulnerable communities and have a functional operational and management committee to check on the implementation effectiveness. The formation of a committee under the Ministry of Environment Forest and Climate Change and Chennai Metropolitan Development Authority to study urban heat islands, which will analyze and address the impact of urbanization on local climates. The committee's role is to conduct in-depth research, gather data, and analyze the factors contributing to the urban heat island effect in a specific city or region and to come up with an Urban Heat Action Plan (Fig. 5.19).

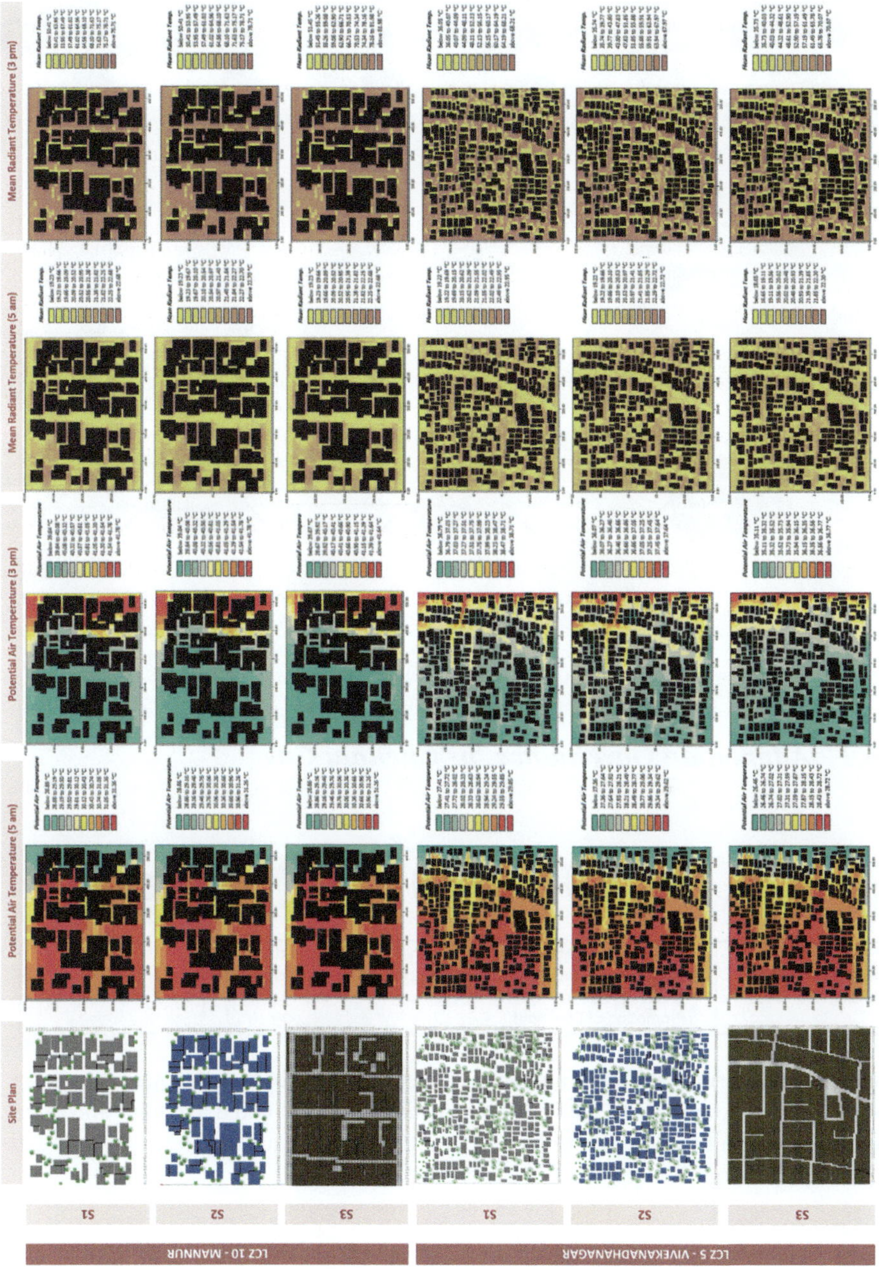

Fig. 5.17 Base case and mitigation scenario strategies for critical LCZ's (*Source* Authors, 2024)

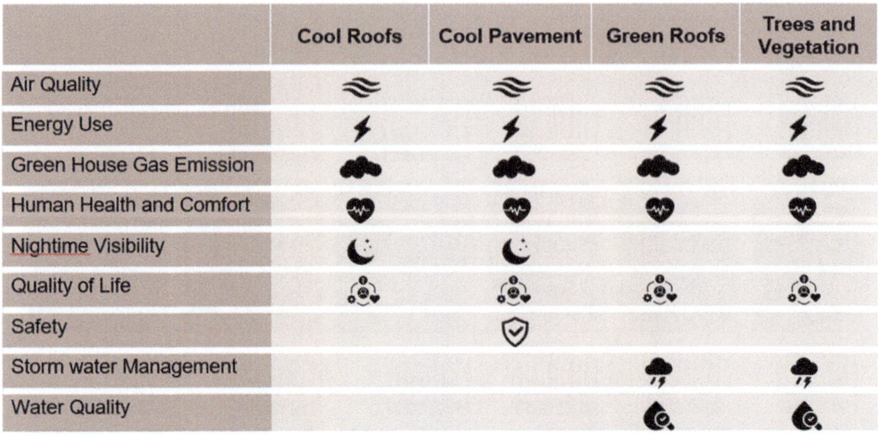

Fig. 5.18 Benefits of mitigation strategies

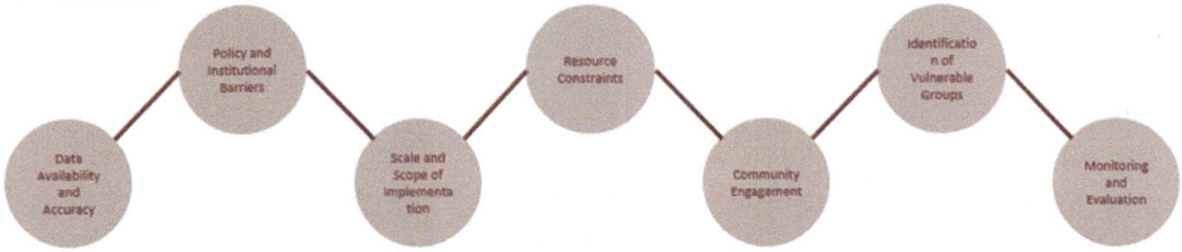

Fig. 5.19 Recommendation framework for urban heat action plan (*source* Authors, 2024)

5.6.1 Adaptation Governance for Chennai City

Since these government bodies are already been a part of the city adaptation plan for the Heat Wave Action Plan, these can be considered as the nodal agencies and other line agencies can be identified to incorporate the new initiatives (Chen 2014; Shakti Sustainable Energy Foundation 2012) (Fig. 5.20).

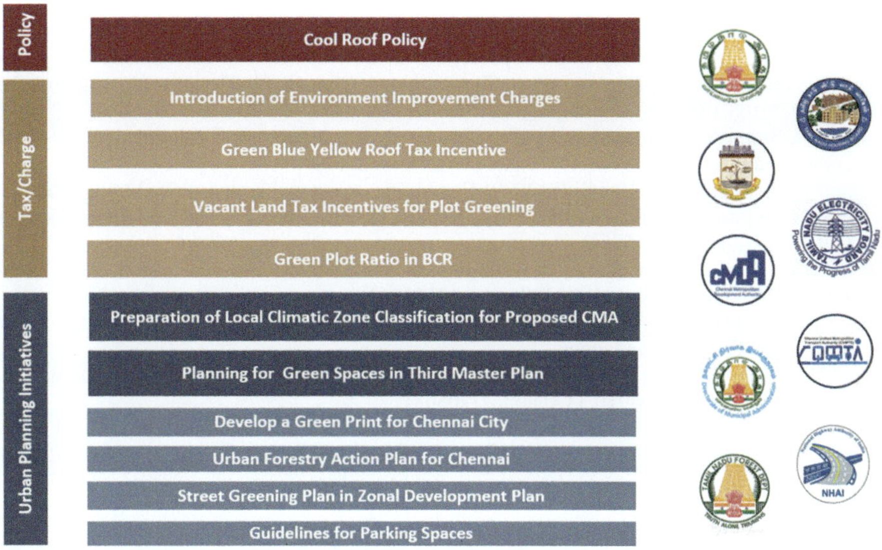

Fig. 5.20 Adaptation governance for Chennai city (*source* Authors, 2024)

5.7 Conclusions

The chapter on Urban Heat Island (UHI) mitigation developed a comprehensive climate-sensitive planning framework to address UHI intensity in urban environments. This framework focuses on assessing and mitigating UHI impacts through multi-scale planning strategies, incorporating various mitigation measures such as reforestation, green and cool roofs, cool pavements, and the creation of green spaces. The study highlights the importance of dynamic modeling in urban planning, demonstrating the effectiveness of implementing these mitigation strategies across urban areas. It emphasizes the need to modify surface characteristics within the urban fabric to reduce UHI effects and calls for policy reforms in urban development, construction practices, and land use to adapt to the challenges posed by climate change effectively.

Declarations

Conflict of Interest: The Authors declare no competing interests.

References

Aria M, Cuccurullo C (2017) Bibliometrix: an R-tool for comprehensive science mapping analysis. J Informet 11(4):959–975. https://doi.org/10.1016/j.joi.2017.08.007

Arnfield J (2003) Urban heat island mitigation and urban planning: the case of Mexicali, B.C., Mexico. Atmos Environ 37(4):470–474

Chen C (2014) The citespace manual. Drexel University, College of Computing and Informatics
Chen YC, Peng TP-T (2016) A simple approach for the development of urban climatic maps based on the urban characteristics in Tainan, Taiwan. Urban Clim 18:32–43. https://doi.org/10.1016/j.uclim.2016.01.003
Cleugh H, Smith MS, Battaglia M (2011) Climate change: science and solutions for Australia. CSIRO Publishing
Demuzere M, Souchère S, Hamdi R (2020) Combining expert and crowdsourced training data to map urban form and functions for the continental US. Nat Commun 11(1):4734. https://doi.org/10.1038/s41467-020-18565-2
Di Sabatino S, Buccolieri R (2020) Multiple-scale nature of urban heat island. Atmosphere 11(11):1143. https://doi.org/10.3390/atmos11111143
Shakti Sustainable Energy Foundation (2012) Cool roof for cool Delhi: design manual. Shakti Sustainable Energy Foundation
Imran HM, Islam MI (2021) Causes, modeling, and mitigation of urban heat island. Atmosphere 12(5):511. https://doi.org/10.3390/atmos12050511
Jain R, Singh T (2022) Environmental impact and mitigation benefits of urban heat island effect: a systematic review. Archit Eng des Manag 18(4):325–343. https://doi.org/10.1080/17452007.2021.1934986
Kaloustian N, Bou-Zeid E (2016) Local climatic zoning and urban heat island in Beirut. Energy Build 114:110–119. https://doi.org/10.1016/j.enbuild.2015.06.057
Kotharkar R, Bagade A (2020) A systematic approach for urban heat island mitigation strategies in critical local climate zones of an Indian city. Sustain Cities Soc 53:101889. https://doi.org/10.1016/j.scs.2019.101889
Lata K, Saha SK, Ramamurthy A, Chundeli FA (2021) Smart global megacity: chennai sustainable development framework. In: Vinod Kumar T (eds) Smart global megacities. Advances in 21st century human settlements. Springer, Singapore. https://doi.org/10.1007/978-981-16-2023-2_4
Lehnert M, Szymanowski M, Sládková S (2021) Mapping local climate zones and their applications in European urban environments: a systematic literature review and future development trends. ISPRS Int J Geo Inf 10(8):530. https://doi.org/10.3390/ijgi10080530
Mathivanan M, Madhavan K (2019) Monitoring spatio-temporal dynamics of urban and peri-urban land transitions using ensemble of remote sensing spectral indices—a case study of Chennai metropolitan area, India. Environ Monit Assess 191(8):491. https://doi.org/10.1007/s10661-019-7591-2
Ng E, Ren C (eds) (2015) The urban climatic map: a methodology for sustainable urban planning. Taylor & Francis
Reba M, Seto KC (2020) A systematic review and assessment of algorithms to detect, characterize, and monitor urban land change. Remote Sens Environ 242:111746. https://doi.org/10.1016/j.rse.2020.111746
Ren C, Ng E (2013) The application of urban climatic mapping to the urban planning of high-density cities: the case of Kaohsiung, Taiwan. Cities 31:1–16. https://doi.org/10.1016/j.cities.2013.01.003
Saputra MH, Setiawan H (2019) Prediction of land use and land cover changes for north sumatra, Indonesia, using an artificial-neural-network-based cellular automaton. Sustainability 11(11):3024. https://doi.org/10.3390/su11113024
Sidiqui P, Bhat PA (2022) Urban heat island vulnerability mapping using advanced GIS data and tools. Curr Sci 122(5):761–770
Stewart ID, Oke TR (2012) Local climatic zones for urban temperature studies. Bull Am Meteor Soc 93(12):1879–1900. https://doi.org/10.1175/BAMS-D-11-00019.1
Wong NH, Chen L (2021) Greenery as a mitigation and adaptation strategy to urban heat. Nat Rev Earth Environ 2(8):593–605. https://doi.org/10.1038/s43017-021-00150-x

Chapter 6
Integrated Simulation Modeling and Geospatial Analysis for Assessing Flood and Urban Heat Vulnerability

Krithik B., Apoorva Chandra, Mohammed Firoz C., and Kuladeep Kumar Sadevi

Abstract Urban centers in the Global South face substantial challenges as a result of climate change, necessitating strategic approaches for enhancing climate resilience. This chapter focuses on Thrissur, a rapidly urbanizing region in Kerala, India, presenting an integrated analysis framework for assessing flood and urban heat vulnerabilities and evaluating future climate change scenarios. This study used HEC-RAS for unsteady flow analysis to assess flood risk under various climate change scenarios, considering current and future land use and rainfall patterns. By simulating different flood occurrences, this hydraulic modeling technique aids in identifying high-risk locations and possible impact zones. Concurrently, this chapter integrates land-use data, urban characteristics, and multitemporal satellite images to investigate urban heat island (UHI) phenomena using advanced geospatial analysis tools. In conclusion, based on the results of flood risk modeling, urban heat island analysis, and climate change projections, this study provides a comprehensive understanding of the vulnerabilities of Thrissur and culminates in strategic and informed decision-making.

Keywords Climate resilience · Geospatial analysis · Flood risk management · Urban heat island · HEC-RAS simulation modeling

6.1 Introduction

Urban centers are facing unprecedented issues due to climate change, with cities in the Global South being especially vulnerable (Revi et al. n.d.). Flooding and the consequences of urban heat islands (UHIs) stand out among the numerous risks associated

Krithik B. (✉) · A. Chandra · M. Firoz C. · K. K. Sadevi
Department of Architecture and Planning, National Institute of Technology Calicut, Kozhikode, India
e-mail: krithikb2000@gmail.com

L. K. Dashora et al. (eds.), *Urban Sustainability*, Disaster Risk Reduction,
https://doi.org/10.1007/978-981-96-4890-0_6

with climate change as serious threats to urban resilience and sustainability, particularly in fast-growing areas (Rosenzweig et al. 2018; Pelling et al. 2018). Despite being often studied independently, each of these phenomena requires an understanding of urban climate resilience architecture to be completed (Carter et al. 2015). The study was carried out in Kerala's Thrissur district, which has a distinct topographical and urban feature. As of the 2011 census, the district has 3,121,200 residents, 9.35% of the state's total population. Much of this change in land use has come from the urban boom, with Kerala experiencing urban expansion of 158.2% and agricultural area loss of 15.52% between 1990–2020, primarily affecting traditional agricultural and natural lands (Vijay and Varija 2024). The city is extremely vulnerable to flooding during monsoon seasons due to its location within the Thrissur Kole Wetland system and its growing urbanization (Karunakaran Prasanna et al. 2020). The urbanization of the city has intensified the urban heat island effect; in the summer, temperatures in the Thrissur urban core can be up to 3–4 °C higher than in the surrounding rural areas (State Action Plan on Climate Change and Human Health).

The study aims to assess Thrissur's flood and urban heat vulnerability and to conduct comprehensive risk assessments across different scenarios, considering impact severity.

The landscape of Thrissur District is dominated by a combination of residential, commercial, and industrial lands spread throughout agricultural lands and natural features. The urban heat island effect has been intensified by urbanization due to increasing built-up densities and decreasing green spaces. Flood risk, vulnerability and urban heat island impacts have never been assessed in the context of Thrissur city, its unique topographic settings, rising rates of urbanization and climatic sensitivity make it an ideal location for multiple scenarios integrated assessment of urban climate risks.

6.2 Literature Review

6.2.1 Flood Risk Assessment

Recent advancements in flood risk assessment techniques have highlighted the necessity of integrated strategies that go beyond conventional hydrological models. The idea of multi-dimensional flood vulnerability assessment was invented by (Jha et al. n.d.), combining economic, social, and physical aspects. Based on this foundation, Kunzewicz et al. (2014) proposed a dynamic framework for flood risk that incorporates climatic trends along with changes in urban form. Advanced hydraulic modeling methodologies have transformed the urban flood simulation domain, especially one produced by HEC-RAS from the Hydrologic Engineering Centre (Teng et al. 2017a). These models can replicate complex flood dynamics under various climate change scenarios, which provides essential information for the adaptive evolution of urban environments. The methods of flood risk assessment have also been refined via more recent research by Vojković et al. (2016) and Papathoma-Köhle et al. (2019), who integrated additional elements into flood risk assessments, namely, urban morphology, GSI, and socioeconomic vulnerabilities.

6.2.2 Urban Heat Island Assessment

The UHI phenomenon is increasingly a matter of interest within the field of urban climatology. Emerging studies examined the extensive implications of UHI on urban energetic processes, ecology, and public health. (Oke et al. 2017).

A seminal review of two decades of UHI research laid the groundwork for understanding the complex interplay between urban form, materials, and local climate.

Urban heat islands (UHIs) exhibit both temporal and spatial variability. Li et al. (2020) and Heaviside et al. (2017a) have both noted that increased population density, rapid rates of urbanization, and urban morphology are all factors that influence UHI intensity. In their research, Santamouris (2015) and Akbari and Kolokotsa (2016a) have corroborated heat retention via surface material (e.g., highways) and urbanization as having a higher heat retention effect. Findings regarding urban heat implications stress the importance of UHI mitigation interventions in urban development.

The increasing issue arises from land surface change, waste heat from energy use, and urban design that captures heat (Arnfield 2003).

Within an urban context, UHI intensity can vary temporally both during day and night as well as spatially from area to area, relative to urban morphology, population density, and weather (Heaviside et al. 2017b).

The interaction of surface albedo and urban heat islands (UHIs) is complex; particularly, when considering the road networks in urban climatology which deserve scrutiny. Push factors such as impermeable surfaces, lack of vegetation, highway-level urbanization, and heat from transportation can suggest that the highest temperatures will occur along state and national highways. Road characteristics, through their thermal properties and attributes, influence local temperature climates. The ultimate ground surface, both in bulk and in composition is needed to understand the thermal effect, as surface characteristics can create a large impermeable surface that modifies the urban thermal environment (Santamouris 2014). The heat capacity and thermal conductivity of asphalt—concrete is used less frequently for roads—allows asphalt to absorb, store, and subsequently release substantial heat throughout the day and night (Akbari and Kolokotsa 2016b).

In this context, surface albedo or solar reflectance presents an important factor for heat absorption and UHI. Dark asphalt will exhibit a low albedo and thus absorb more heat than light-albedo roads, such as concrete (Li et al. 2013) intensifying the UHI phenomenon of noticeably increased temperatures upon and around roads.

6.3 Materials and Methods

6.3.1 Study Area

Thrissur is one of the 14 districts within Kerala, India's southernmost state, situated in the central part of the state. It is bounded by the Western Ghats on the east and the Arabian Sea on the west. Thrissur is known as the cultural capital of Kerala and a major hub for trading, business, and education (Figs. 6.1 and 6.2).

The study has been conducted under three levels: District level, Municipal corporation, and Urban Agglomerated region. The major focus area is the urban agglomerated area which is shown in Fig. 6.3.

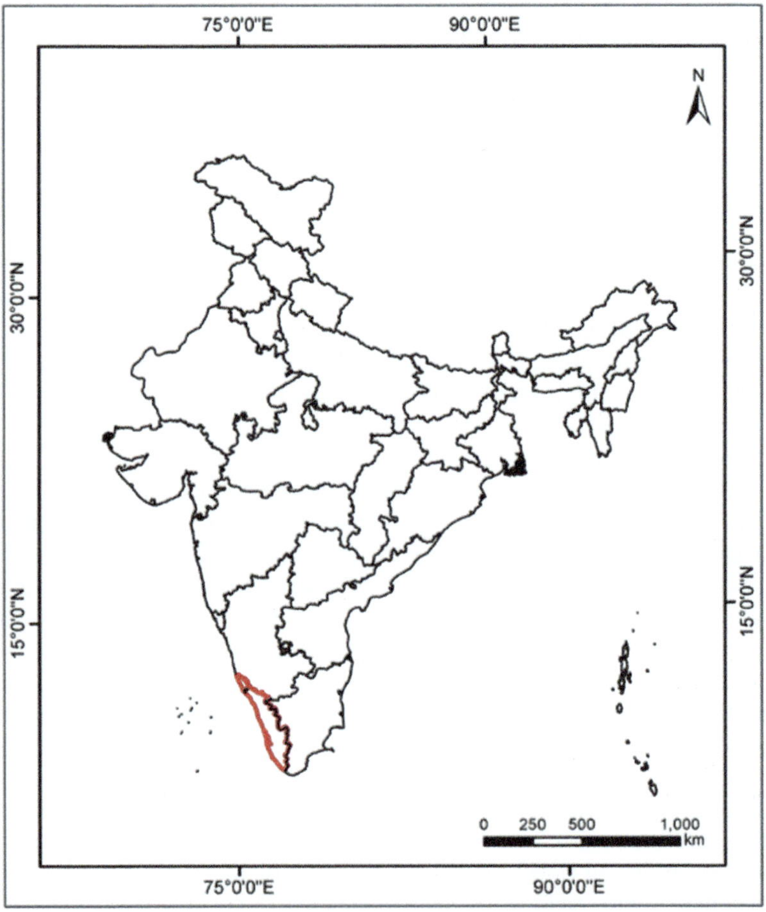

Fig. 6.1 Location map of Kerala (*Source* Kerala geoportal)

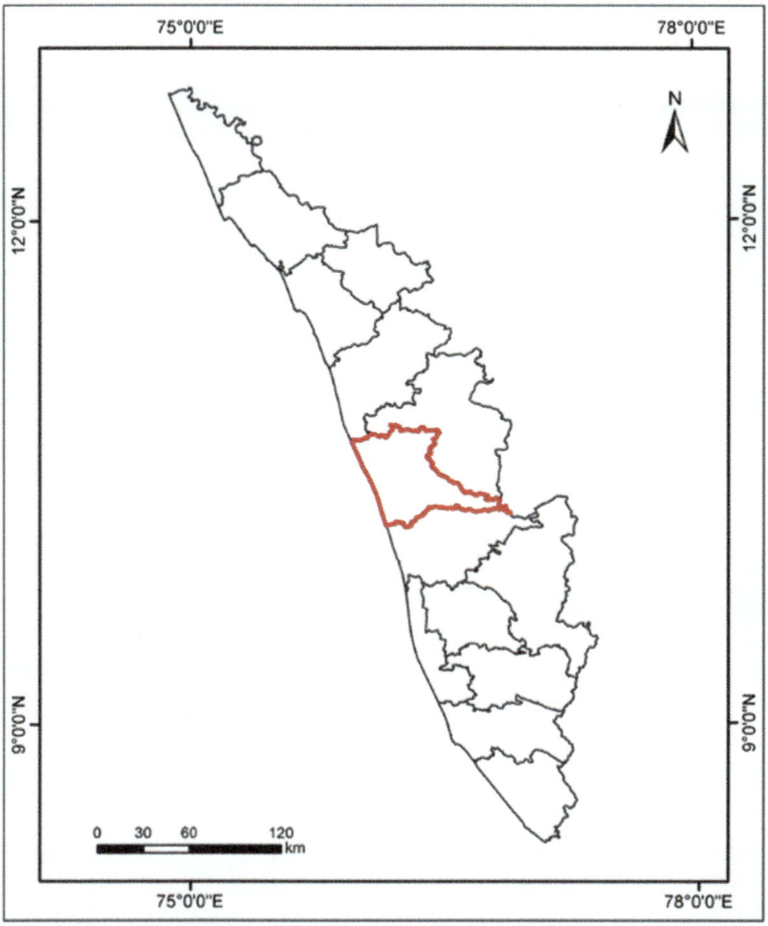

Fig. 6.2 Location map of Thrissur district (*Source* Kerala geoportal)

District level: At the district level the population is 3,110,327, according to the 2011 census, with a population density of 1,026 people per square kilometer.

The landscape of the district extends from coastal lowlands in the west to the midlands in the center, and this variation contributes extensively to climate patterns, drains, and land use (District Survey Report of Minor Minerals 2016).

The Kole wetland system is an important feature of Thrissur's landscape, and it is designated as a Ramsar site with an area of about 13,632 hectares (shown in Fig. 6.4). The Kole wetlands play an important role in the drainage basin hydrology of the region, provide flood regulation services, and support a high level of biodiversity, but urban outstretch and changes to land use patterns are increasingly applying pressure on wetlands. In total, the area of water quality wetlands has decreased by ten percent between 1967 and 2012 (Karunakaran Prasanna et al. 2020).

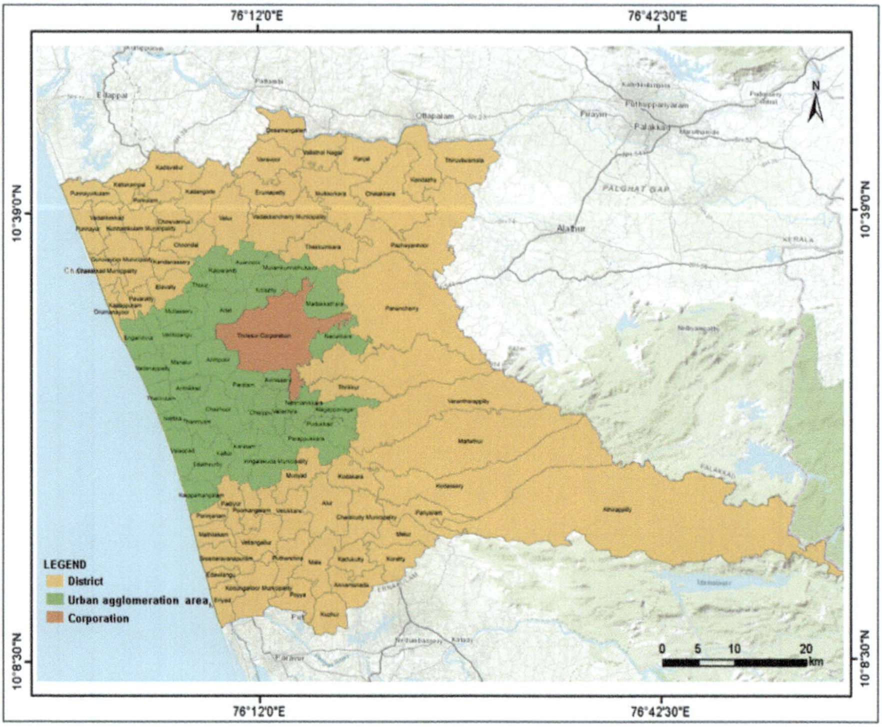

Fig. 6.3 Thrissur urban agglomeration area and corporation area (*Source* Census 2011)

Thrissur experiences a tropical monsoon as an average yearly rainfall of 2,310 mm to 3,955 mm is received in the area. It is noted that approximately 70% of the rainfall received occurs during the southwest monsoon season from June until September (District Survey Report of Minor Minerals 2016). This considerable amount of annual rainfall, with the additional factor of the area's hilly terrain being often urbanized, renders Thrissur vulnerable to flooding.

Municipal Corporation: The Thrissur Municipal Corporation had a population of 3,15,957 according to the census of 2011. The Thrissur Municipal Corporation is recorded as currently the second-largest city corporation in Kerala.

Urban Agglomerated Region: Thrissur is recognized as the 3rd largest urban agglomerated region, in the state of Kerala, India, following Kochi and Kozhikode. The total population of the Thrissur UA region is marked as 1,861,269 people, males tabulated as 879,643 and females tabulated as 981,626.

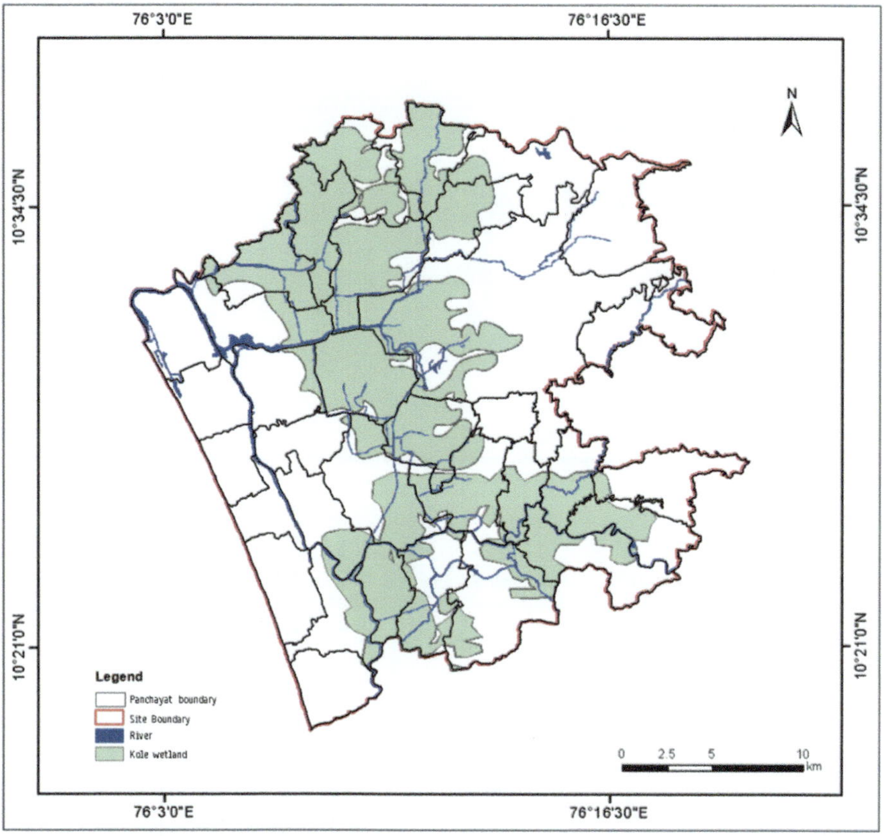

Fig. 6.4 Kole wetland in Thrissur (*Source* Kerala IDRB)

6.3.2 Methodology

This research will follow a comprehensive, multi-stage methodology in assessing the climate resilience of Thrissur district in recognizing flood risk and urban heat vulnerability. The overall methodological framework is as shown in the Fig. 6.5.

The methodology is divided into two key components, each for an identified aspect of urban climate vulnerability:

1. Flood Simulation Modeling
2. Urban Heat Vulnerability Analysis.

The first Phase describes flood risk assessment following unsteady flow analysis in the HEC-RAS system. The analysis is described through data input, model development, and simulation scenarios are considered to capture both current and future flood risk under the various climate change scenarios and land uses. The result was

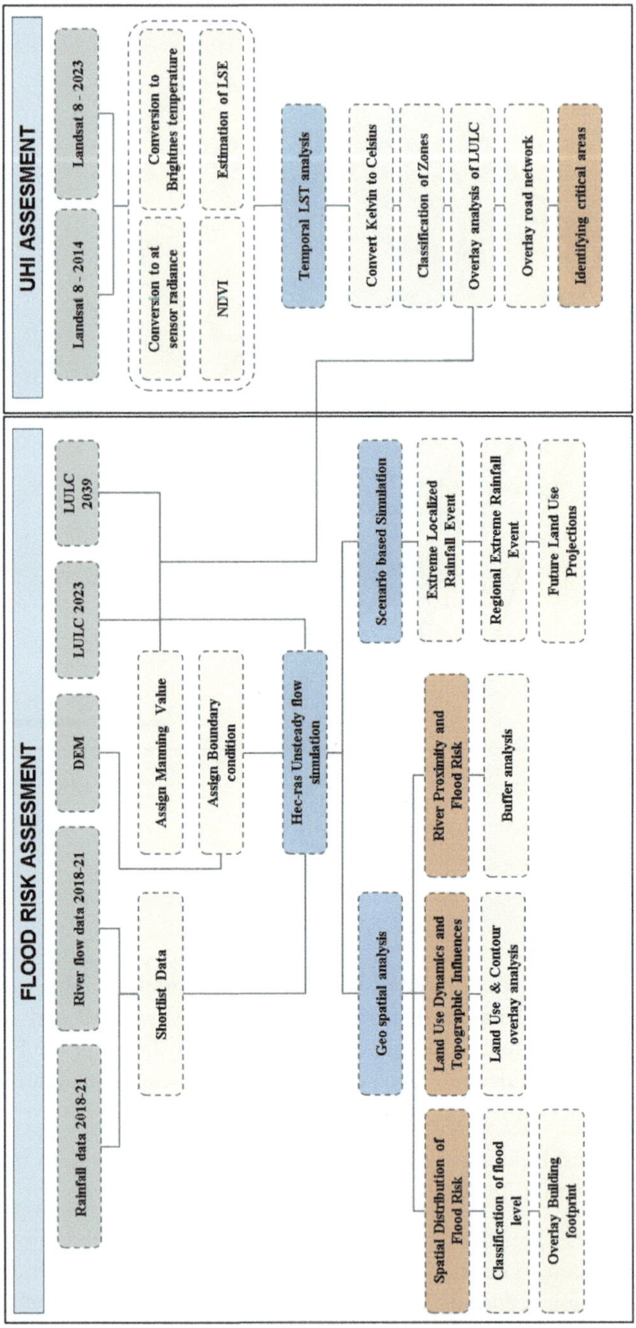

Fig. 6.5 Methodology

further refined through analysis at the panchayat level in urban agglomerated areas and at the ward level in the corporation area.

While in Phase Two, the Urban Heat Island (UHI) analysis, is outlined as a method for the assessment of heat vulnerability for both the Thrissur district and across the urban landscape, while following a comprehensive, multi-stage process of assessment of heat vulnerability. The study results are structured in three sections. The first part of the study visualizes the temporal analysis of the entire delineated region. The second part examines the impact of land use factors on urban heat islands and the third part identifies key areas within the urban environment that require interventions.

The study integrates advanced remote sensing techniques with geospatial analysis to provide a nuanced understanding of the UHI phenomenon and its temporal dynamics.

6.3.3 Data Collection and Processing—Flood Assessment

The initial phase of the study involved the acquisition and preparation of crucial datasets. Data on river flow and historical precipitation were gathered. The position of the river gauge station is illustrated in Fig. 6.6. Figure 6.7 displays the accessible river flow information and emphasizes the shared monthly data across three river datasets. Additionally, a high-resolution Land Use Land Cover (LULC) dataset for 2023 and a Digital Elevation Model (DEM) of the research area were obtained. To aid in evaluating flood impact on urban structures, building footprint information was also procured.

6.3.3.1 Hydraulic Model Setup

A hydraulic model was developed using the Hydrologic Engineering Center's River Analysis System (HEC-RAS). This model is frequently employed in flood simulation research (Teng et al. 2017b). The model encompassed river geometry, cross-sections, and boundary conditions. Based on the river channel's physical attributes, Manning's roughness coefficients were designated. Additionally, impervious values were assigned to various land cover types utilizing the 2023 LULC data, as shown in Table 6.1.

6.3.3.2 Flood Simulation Process

The flood simulation process involved several key steps:

1. Terrain Mapping: A Digital Elevation Model (DEM) for 2019 was produced by combining two tiles, providing an accurate depiction of the study area.

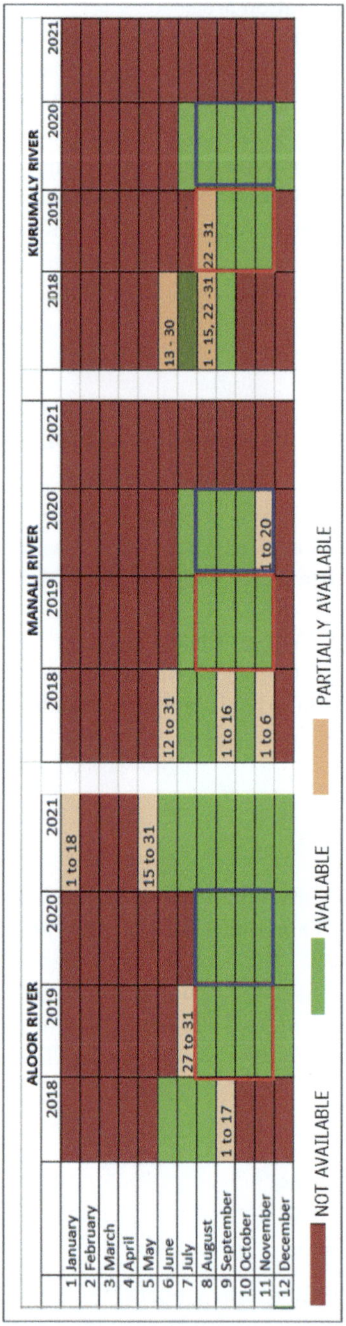

Fig. 6.6 River flow data. (*Source* Kerala IDRB)

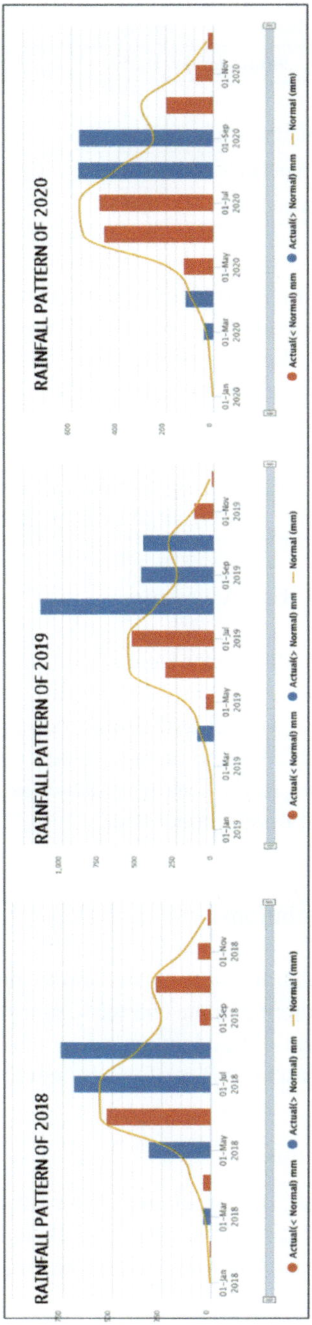

Fig. 6.7 Rainfall pattern (*Source* India meteorological department)

Table 6.1 Manning's and percent impervious values for different land cover types

Land cover type	Manning's value	Percent impervious value
Built-up	0.03	100
Agriculture land	0.05	40
Dense vegetation	0.07	50
Waterbody	0.035	10
Barren land	0.013	70
Wetland	0.086	20

2. Boundary Condition Assignment: To define the study region and set hydrological parameters at its boundaries, boundary conditions were meticulously established. Flow hydrographs were applied for upstreams, and for streams downstream, typical depth conditions were established (Costabile and Macchione 2015).
3. Unsteady Flow Analysis: From August 2 to November 2, 2019, an unsteady flow analysis was carried out with a computational interval of one hour. Because of the extraordinarily heavy rainfall during this period, a more accurate depiction of extreme weather conditions was possible (Vojtek and Vojteková 2016).

6.3.3.3 Spatial Analysis and Risk Assessment

To assess the risk to different built environment operational areas, mapped simulations of the flooding were correlated with footprint building data as points. Floodwater levels from the flood simulations were classified into flood risk areas and buildings' potential for flooding was evaluated based on their location within a given flood risk area. This analysis was extended to the Panchayat level as well as the Corporation level, aggregating and assessing flood impact data.

6.3.3.4 Scenario-Based Simulations

A series of scenario-based simulations were carried out to improve the understanding of possible flood hazards in different climate change scenarios. Simulations provide a framework for determining potential flood vulnerability in future and provide important information regarding the susceptibility of the system to stimulus. The scenarios addressed were the following:

(i) **Future Land Use Projections**

To assess the potential flood risk, the simulation was performed using the land use land cover of 2039. By incorporating these projected changes, the simulation offers insights into the evolving flood risk landscape, allowing for proactive urban planning and flood management strategies.

(ii) **Extreme Localized Rainfall Event**

The simulation analyzes the impacts of an extreme total rainfall event in Thrissur by employing the maximum recorded daily rainfall of 230 mm/day sustained for seven days. The simulation was further integrated with the river flow conditions at maximum flow conditions to illustrate worst-case scenario that can occur.

(iii) **Regional Extreme Rainfall Event**

To analyze the extreme scenario, a simulation was carried out with the entire Kerala region's maximum daily precipitation ever recorded (398 mm/day), over a week under maximum river flow. This scenario illustrates a regional-scale and intense Rainfall scenario impacting the region. It gives us evidence of what would occur as a result of a regional-scale extreme weather event in the study area, particularly in light of the interactions between different hydrological systems (Kreibich et al. 2015).

6.3.4 Data Collection and Processing—Land Surface Map

Geographic information systems and remote sensing were used to create the Land Surface Temperature map for the study region. The procedure is depicted in Fig. 6.5.

6.3.4.1 Data Acquisition

The LST map for this study was created using Landsat 8 satellite images. Images taken between March and May that had no more than 10% cloud cover were selected to ensure 30 × 30 m resolution, seasonal uniformity, and the best possible visibility. The Landsat images were downloaded from the USGS Earth Explorer website.

6.3.4.2 Data Processing

a. Convert Digital Numbers to At-Sensor Spectral Radiance: The first step in processing satellite images is to convert digital numbers to sensor spectral radiance. This conversion uses different Landsat 8 band equations, implemented using band math in ArcGIS 10.8. The radiometric scale is standardized by this process for use with different sensors and platforms (Chander et al. 2009).
b. Convert Spectral Radiance to Temperature in Kelvin: Pre-launch calibration constants are then used to convert spectral radiance to temperature in Kelvin for thermal band data.
c. Convert Kelvin to Celsius: These temperatures are then converted from Kelvin to Celsius by deducting 273.15. This conversion makes the temperature data more relatable and usable for most applications.

6.3.4.3 Index Calculation

a. **Calculate NDVI (Normalized Difference Vegetation Index):** This metric is used to assess the biomass, stress tolerance, and health of the vegetation. Higher scores on the index, which goes from -1 to $+1$, indicates healthier vegetation. This approach facilitates understanding of the vegetation's spatial distribution in the study area.
b. **Calculate NDBI (Normalized Difference Built-up Index):** This index is used to identify developed and urban areas. Like the NDVI, its values range from -1 to $+1$, where higher scores correspond to more urbanized areas (Zha et al. 2003).
c. **Land Surface Emissivity (LSE) Estimation:** LSE is a significant surface parameter obtained from emitted radiance estimated. It is determined using Normalized Difference Vegetation Index (NDVI) values and vegetation proportion (PV). This process addresses challenges related to radiometric calibration and cloud detection (Sobrino et al. 2004).

Land Surface Temperature (LST) Calculation: The calculation of Land Surface Temperature (LST) that incorporates brightness temperature and land surface emissivity. This computation takes into account factors such as albedo, vegetation coverage, and soil moisture content. LST plays a vital role in influencing the distribution of energy between soil and vegetation, and it impacts surface air temperature (Sobrino et al. 2004; Bala et al. 2023) (Fig. 6.8).

6.4 Result and Discussion

6.4.1 Flood Assessment

The flood vulnerability assessment of Thrissur district employed a comprehensive methodology to analyze and interpret flood risk across the region. The methodology of the study was distinguished by three main elements.

Firstly, the simulated flood levels were divided into three classes for improved interpretation: low (<1 m), medium (1–3 m), and high (>3 m) as illustrated in Fig. 6.9. This categorization system made it easier to visualize and comprehend the degree of flooding in various district areas.

Secondly, the flood inundation map was combined with the point data that indicated the locations of the buildings. Through this integration, the spatial distribution of structures and simulated flood levels can be directly correlated, offering vital information about the possible effects on infrastructure.

Lastly, the percentage of buildings affected at each flood risk level was calculated. This offered a clear measure of the scale of impact and formed an essential basis for subsequent vulnerability assessments.

Flood risk analysis to identify the affected building at two different levels: Urban Agglomerated area and Corporation area (Fig. 6.10).

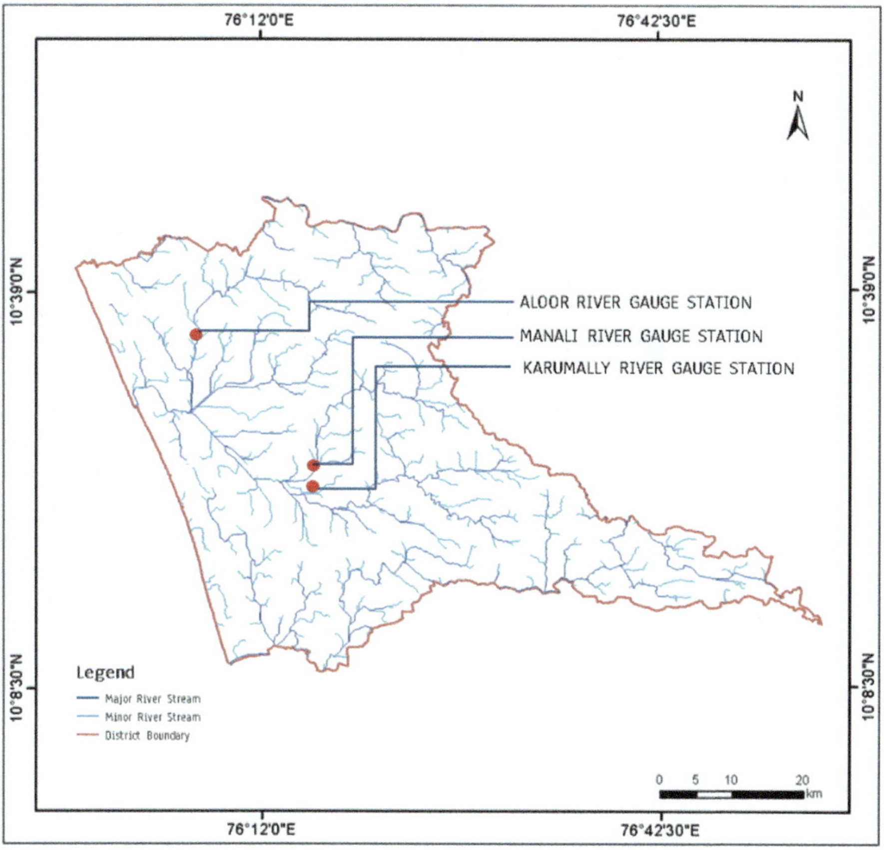

Fig. 6.8 River Guage station location (*Source* Kerala IDRB)

6.4.1.1 Spatial Distribution of Flood Risk—Urban Agglomerated Area

From the spatial analysis of flood risk, it was observed a heterogeneous distribution of flood risk throughout the Thrissur district, with the corporation area emerging as the severely affected area. We see the evidence of a greater risk here in the urban core with 25,511 reported buildings affected by flood risk.

Out of the 25,511 reported buildings in the corporation area, five panchayats (Madakamthara, Edangaiyur, Anthikkad, Natikka, and Edathiruthi) reported that greater than or equal to 20% of their buildings were affected by floods. The observed distribution of flood risk according to location leads to the consideration of flood risk management measures within and around urban areas.

Additionally, the affected area of the corporation varied with the northern part of the corporation emerging as particularly impacted, primarily due to converging factors such as topographic conditions, wetlands, and two major rivers. Based on this evidence, we can observe the intersection between the landscape features connecting

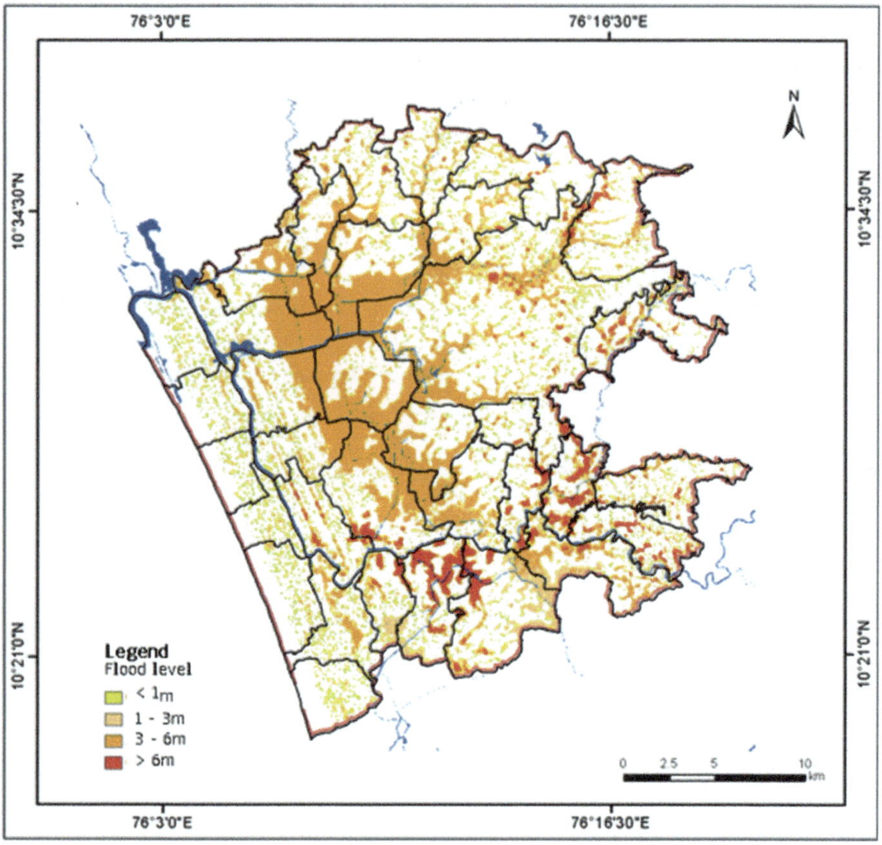

Fig. 6.9 Flood level map

hydrological networks to topographic systems echoing the role of topography and hydrological networks in shaping flood vulnerability landscapes.

Land Use Dynamics and Topographic Influences

The panchayat-specific analysis revealed intricate relationships between land use, topography, and flood vulnerability, highlighting the complex interplay of factors contributing to flood risk in Thrissur district. The Fig. 6.11 shows the identified critical location for detailed analysis.

Figures 6.12 and 6.13 show the comparison of flood vulnerability and land use in Nanmankkara and Alangappanagar panchayats, the combination of steep slopes and proximity to the Manali River significantly increased flood risk for vegetation,

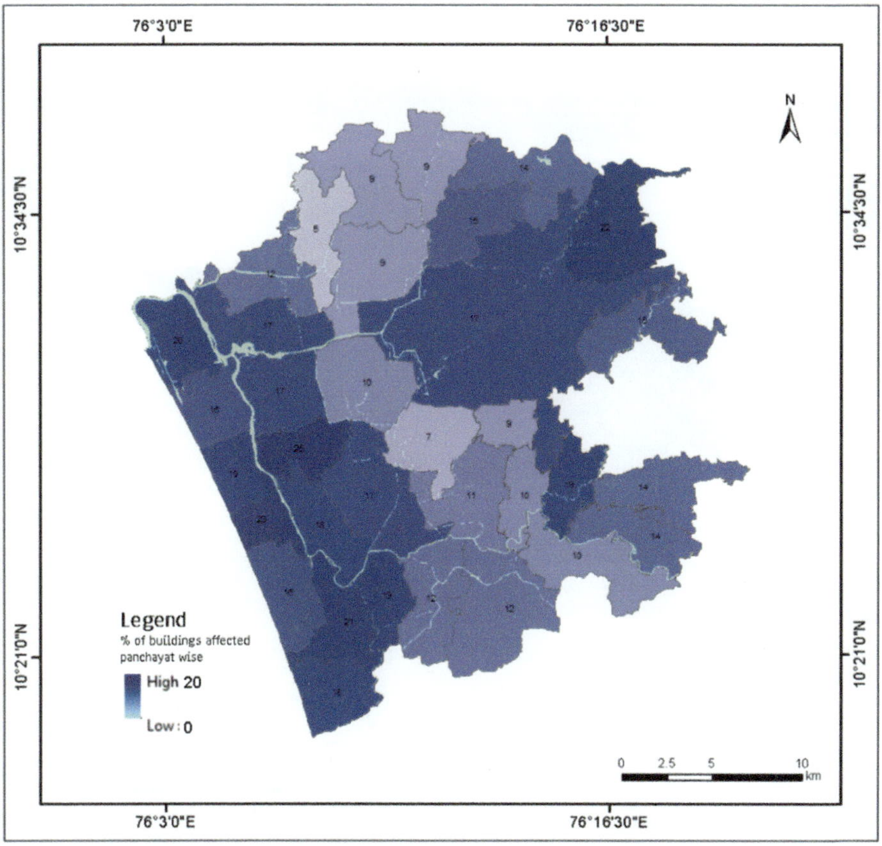

Fig. 6.10 Panchayat wise flood affected building%

wetland, and built-up areas. The steep slope topography accelerates river flow, exacerbating flood risks in these areas, particularly for buildings located near the river banks. A similar case was also found in Madakkathara panchayat.

Low-lying areas, particularly those where the rivers converge, become more susceptible. This can be well explained from Fig. 6.14. Arimpoor panchayat which primarily contains areas with dense vegetation and agricultural lands are at high risk because of its low elevation in the Kole wetland and immediate proximity to the Puzhakkal and Thanikkudam rivers.

The analysis revealed critical locations across various panchayats, each with unique risk profiles:

1. Kaippamangalam: The built-up areas and vegetated lots in this area are being exposed to risks due to the region's steep slopes and location downstream of the river.

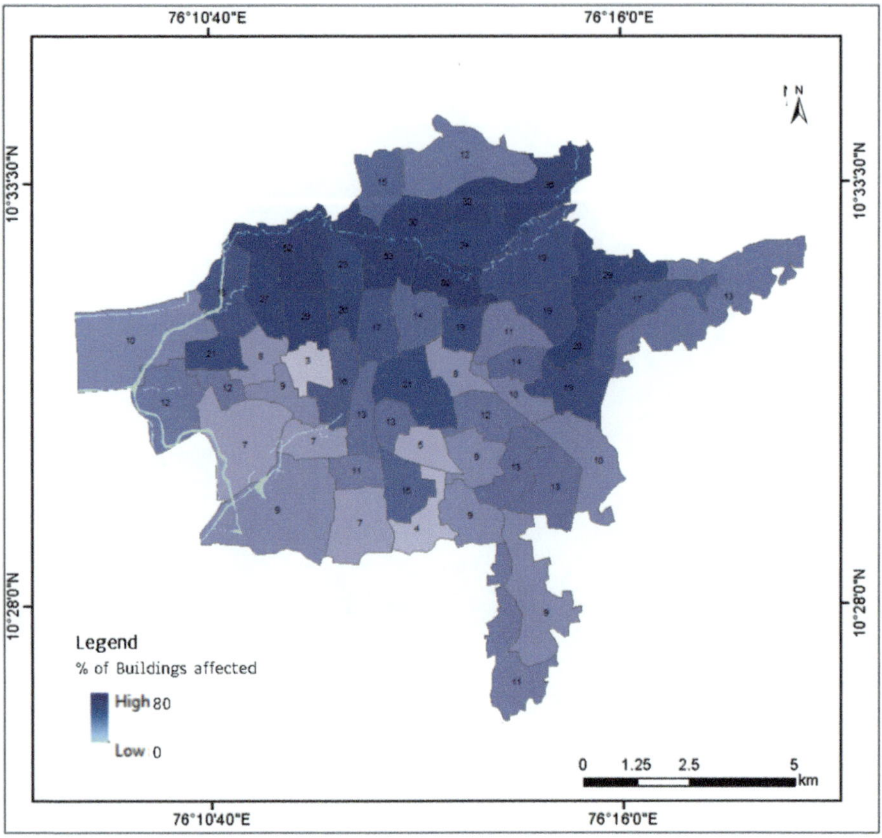

Fig. 6.11 Flood affected building % in corporation

2. Natikka, Thaniyam, and Valappad: The vulnerability of these coastal Panchayats is amplified by lower elevations, proximity to the river in a region, and increased surface runoff from the roads in coastal low elevation areas.
3. Engandiyur: This area is primarily affected by vegetated lots with lesser risks, but vulnerability increased with the downstream inhibiting location of the Puzhakkal and Thanikkudam rivers.

6.4.1.2 Spatial Distribution of Flood Risk—Corporation Area

Figure 6.15 shows the flood level zones, percentage of buildings affected ward wise along with the ward number of the corporation area, where several wards showed distinct vulnerability patterns:

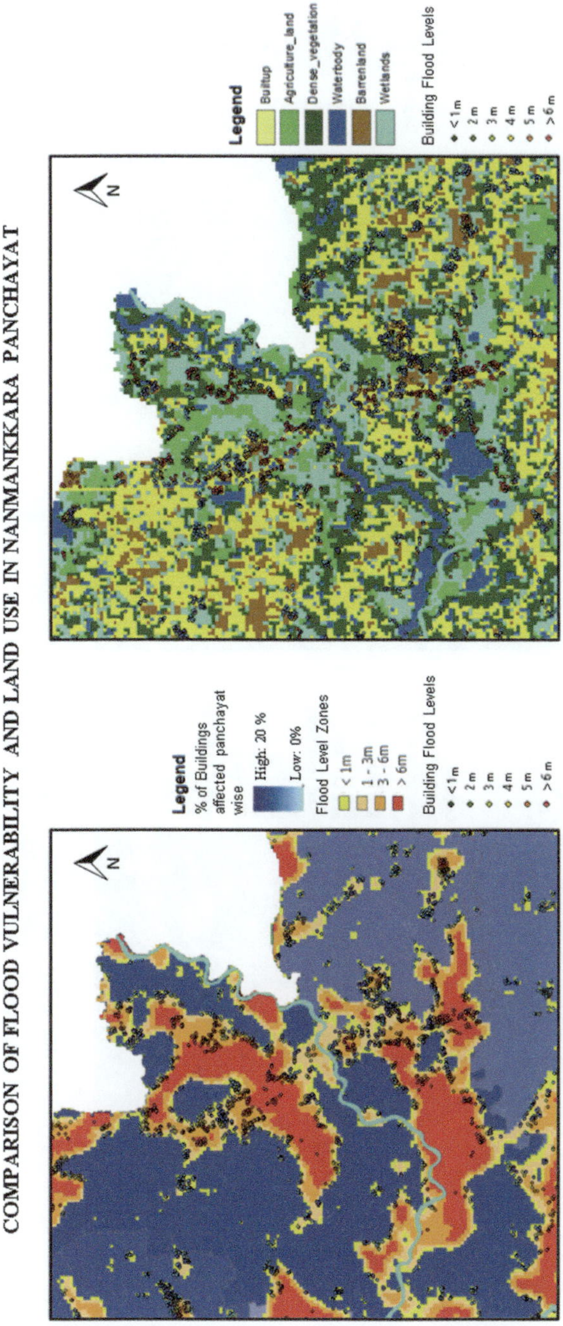

Fig. 6.12 Comparison of flood vulnerability and land use in Nanmankkara panchayat

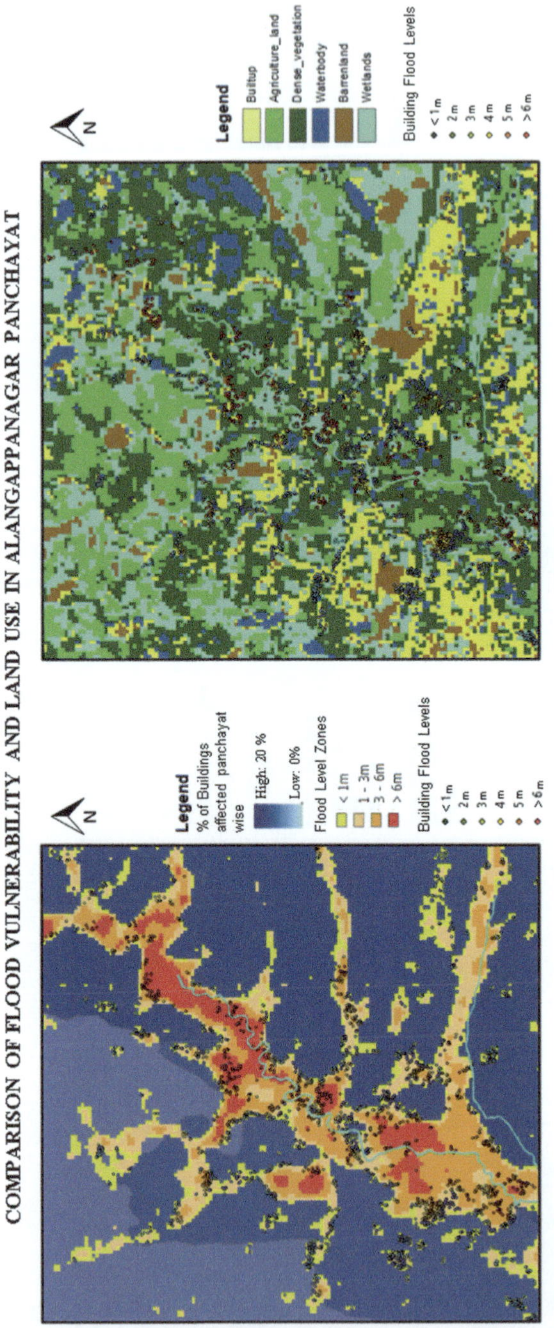

Fig. 6.13 Comparison of flood vulnerability and land use in Alangappanagar panchayat

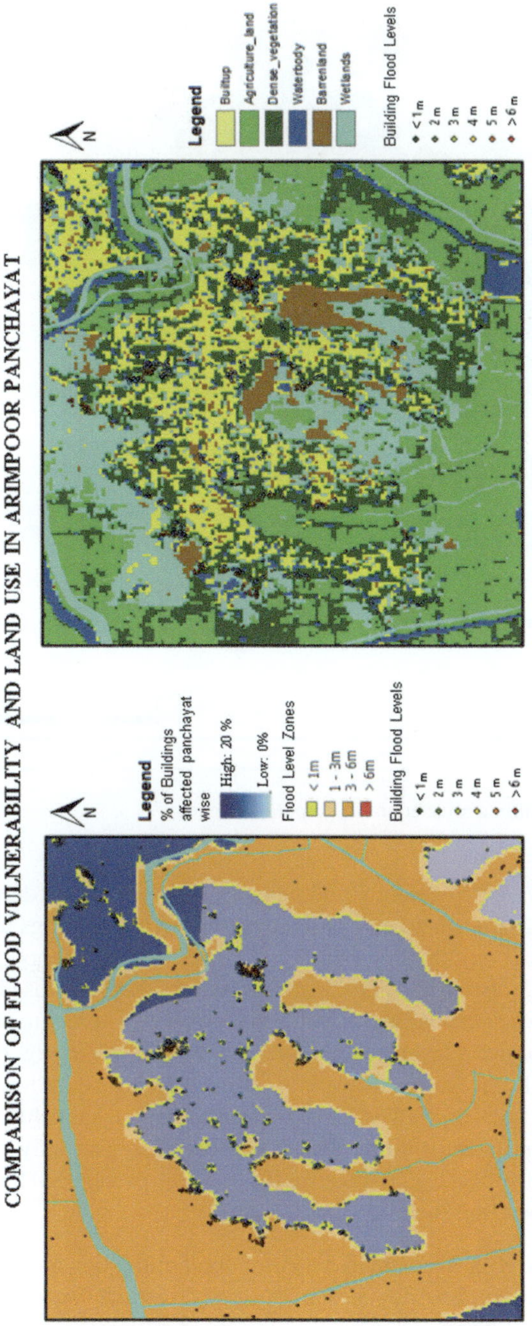

Fig. 6.14 Comparison of flood vulnerability and land use in Arimpoor panchayat

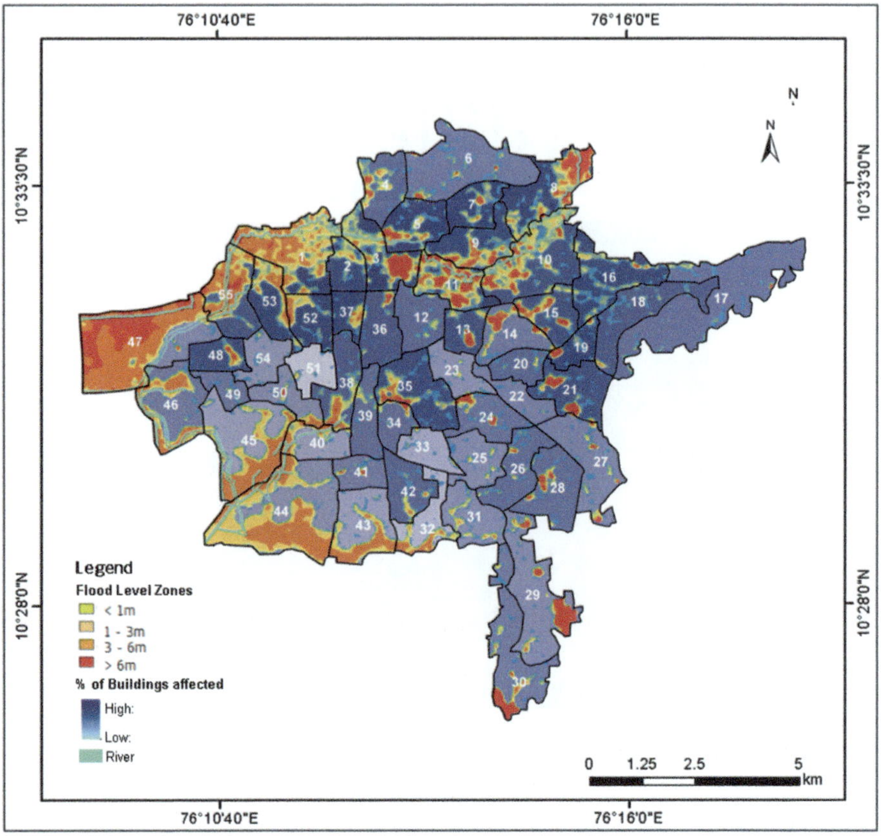

Fig. 6.15 Ward-wise building affected in the corporation level

1. Wards 31, 43, and 44: Vegetation and built-up areas are at risk due to the lower elevation of the Kole wetland area and the proximity of buildings to agricultural land. The comparison of the flood level and satellite image is shown in Fig. 6.16
2. Wards 35 and 23: Built-up areas and vegetation face increased risk due to steep slopes between roads, highlighting the impact of urban topography. The comparison of the flood level and satellite image is shown in Fig. 6.17
3. Ward 47: Agricultural land is especially vulnerable due to its low elevation in the Kole wetland area, as well as its proximity to the Puzhakkal River.
4. Wards 8, 10, 11, and 3: Built-up areas were noted to be significantly at risk because of steep topography that accelerates river flow and the presence of buildings along riverbanks and near the Thanikkudam River (Fig. 6.18).

The function of the Kole Wetland as a natural flood mitigation measure was noted to be a significant finding, and its function of flood overflow control highlights the importance of preserving and enhancing natural flood buffers.

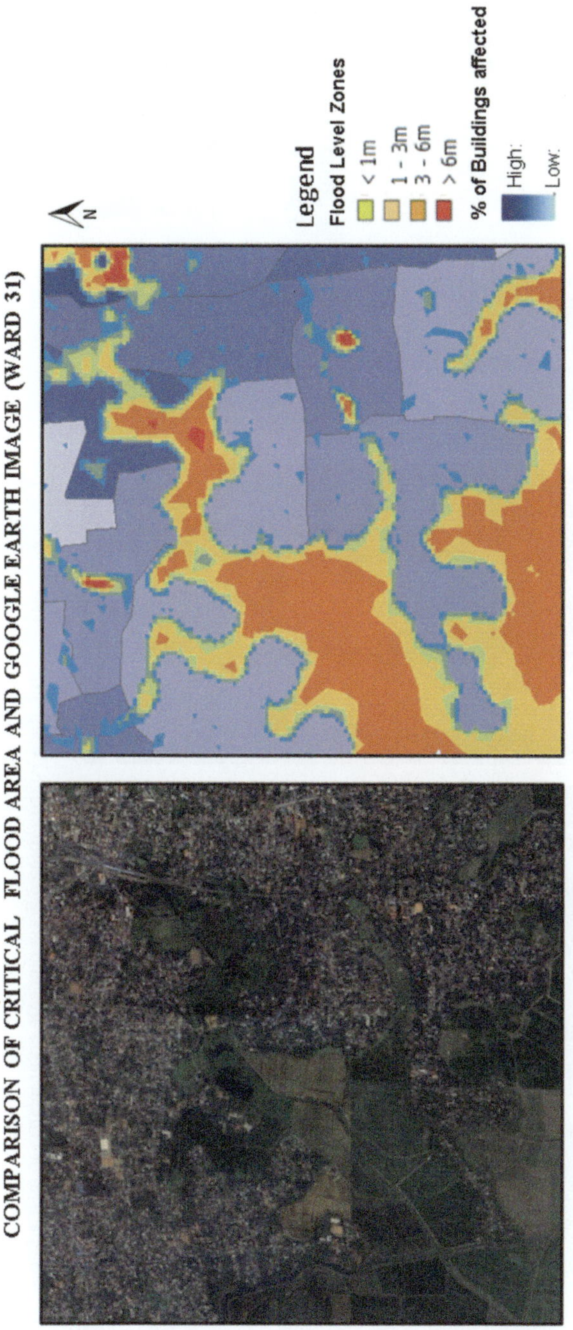

Fig. 6.16 Comparison of critical flood area and google earth image (Ward 31)

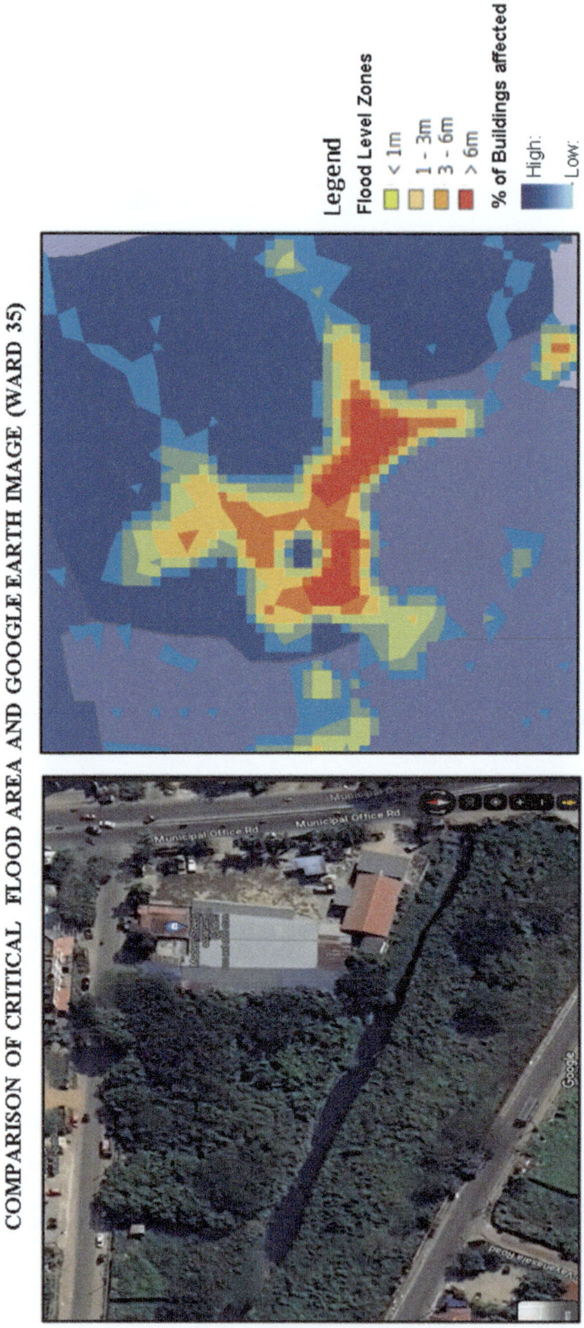

Fig. 6.17 Comparison of critical flood area and google earth image (Ward 35)

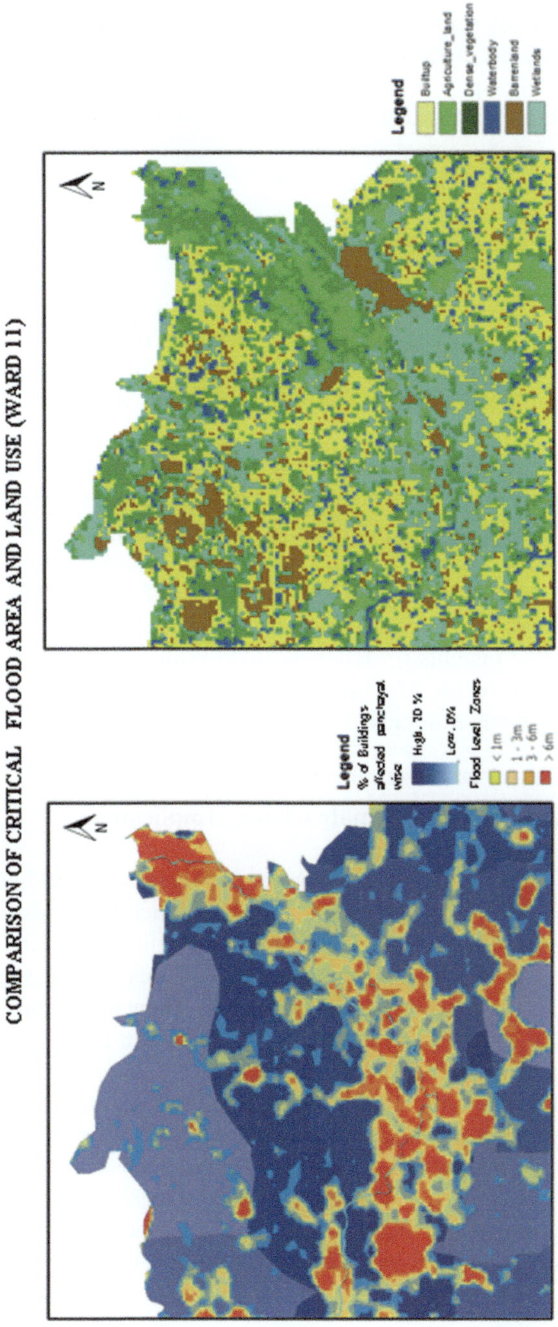

Fig. 6.18 Comparison of critical flood area and land use (Ward 11)

This detailed analysis of land use and topographic influences on flood risk aligns with the growing body of literature emphasizing the need for context-specific flood risk assessments. Recent studies have highlighted the importance of considering local topography, land use patterns, and hydrological characteristics in developing effective flood management strategies. Moreover, the observed interplay between built-up areas, vegetation, and water bodies underscores the need for integrated land use planning in flood management strategies.

6.4.1.3 River Proximity and Flood Risk

The buffer analysis of river proximity as shown in the Fig. 6.19 revealed a strong correlation between proximity to water bodies and flood risk. A total of 89,331 buildings were identified as affected, with a significant proportion (63%) located within 500 m of rivers. This distribution pattern, where 29% of affected buildings fall within the 200–500 m range from rivers, suggests that flood risk extends considerably beyond immediate floodplains. These results align with a study by Nones (2017) that highlights the necessity of taking extended flood impact zones into account when assessing risk and designing metropolitan areas (Fig. 6.20).

The concentration of affected buildings at various distances from rivers (10% within 50 m, 9% within 50–100 m, 15% within 100–200 m, and 29% within 200–500 m) provides a nuanced understanding of flood risk distribution. This granular analysis offers valuable insights for tailoring flood mitigation strategies to specific risk zones, as advocated by (de Moel et al. 2015) in their comprehensive review of flood risk assessment methodologies.

6.4.1.4 Future Scenarios and Climate Change Implications

The scenario-based simulations, including projections based on 2039 Land Use/Land Cover (LULC) and extreme rainfall events, provided crucial insights into potential future flood risks as shown in Fig. 6.21. The observation that increased rainfall intensity affects water levels more significantly than floodplain extent suggests a need for adaptive strategies focusing on water level management and improved drainage systems. This finding aligns with recent research on urban flood resilience in the context of climate change (Pregnolato et al. 2017) (Fig. 6.22).

The simulations incorporating extreme rainfall scenarios as shown in Fig. 6.23 (up to 398 mm/day) revealed vulnerabilities in the current urban fabric that may be exacerbated by climate change (Directorate Central Water Commission 2018). This aligns with global climate change projections that predict an increase in extreme weather events and underscores the urgency of incorporating climate change considerations into urban flood management strategies (Wilby and Keenan 2012).

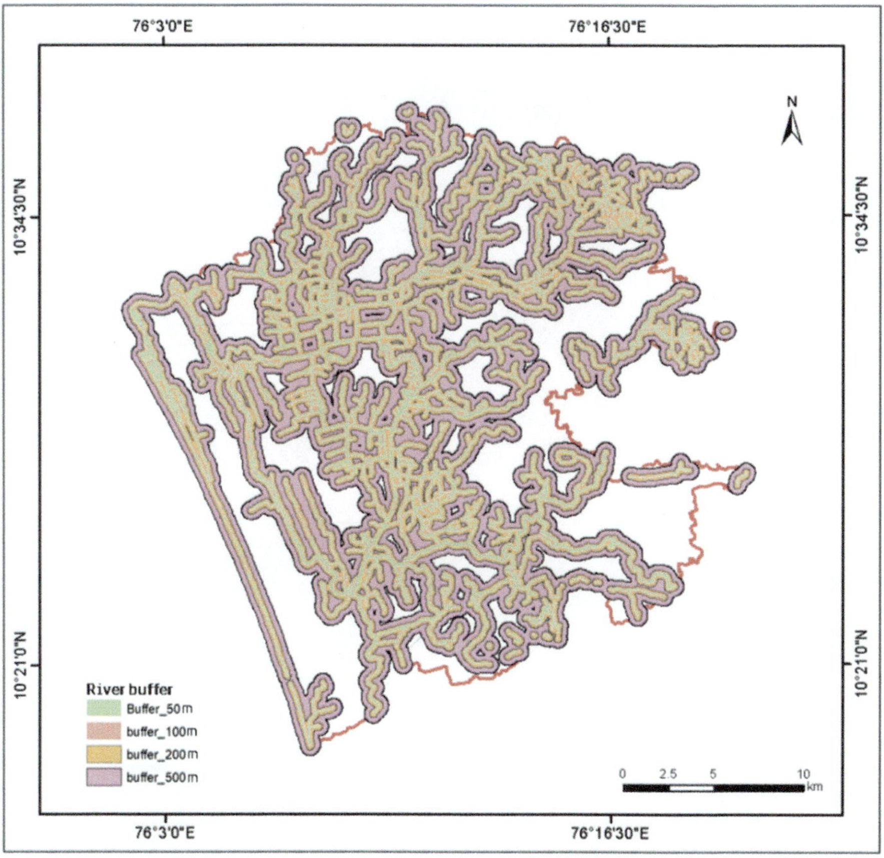

Fig. 6.19 Buffer analysis

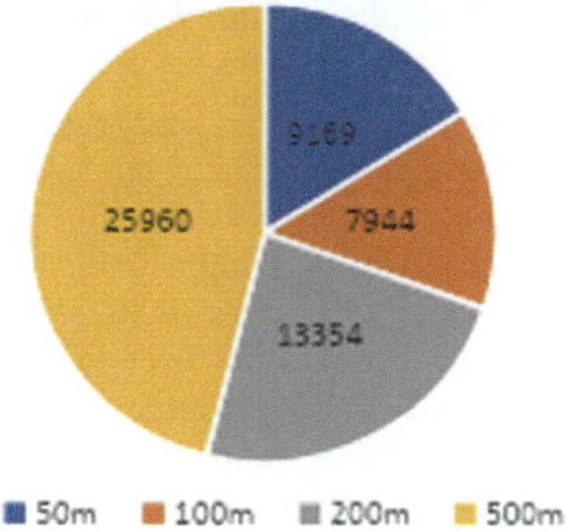

Fig. 6.20 Number of buildings in buffer zones

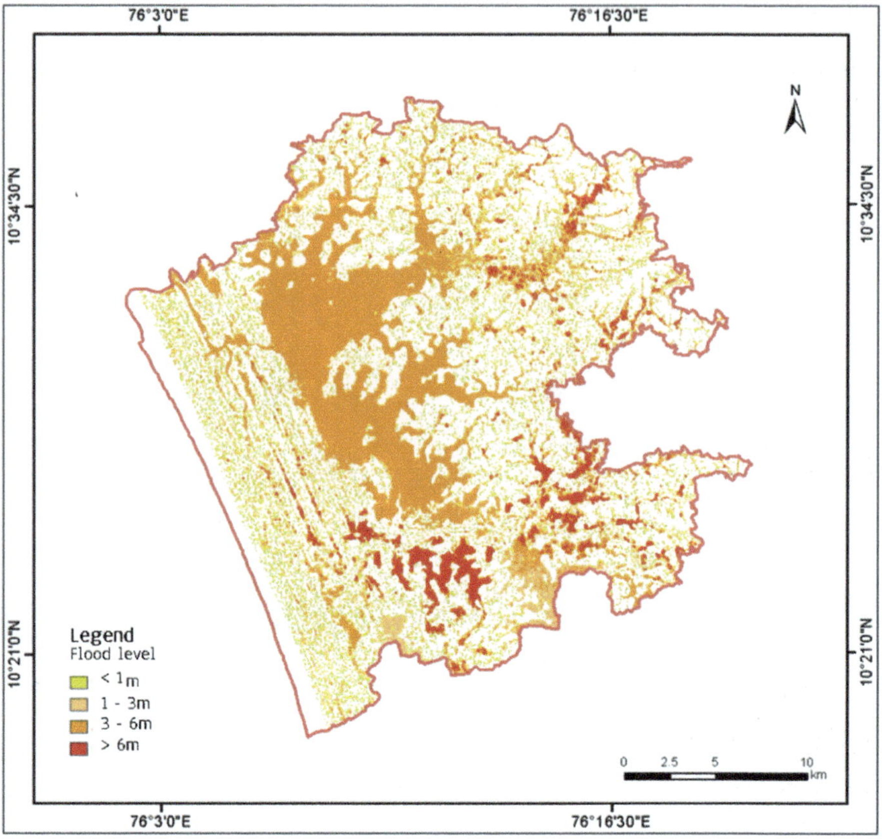

Fig. 6.21 Flood level based on 2039 LULC

6.4.2 *Urban Heat Island*

The initial phase of the study focuses on comparing the urban heat map of 2014 and 2023 which was taken as the base year for analysis. The comparison of the urban heat island map shows a significant change of shift in thermal heat mainly within the urban area. Starting with the lowest recorded temperature in 2014, the increase from 14 to 22 degrees in 2023, it shows a shift to a warming trend as presented in Figs. 6.24 and 6.25. Even the traditionally low-temperature zones of the district started to experience warmer temperature within a span of 9 years.

Moreover, the comparison with the highest temperature further shows the intensity of the heat island. In 2014, the peak temperature recorded was 41 °C whereas in 2023 the temperature raised to 46 °C. The increase in the upper limit of the temperature shows the heightened thermal profile of the urban area, influenced by the factors like urbanization, climate change and land-use change.

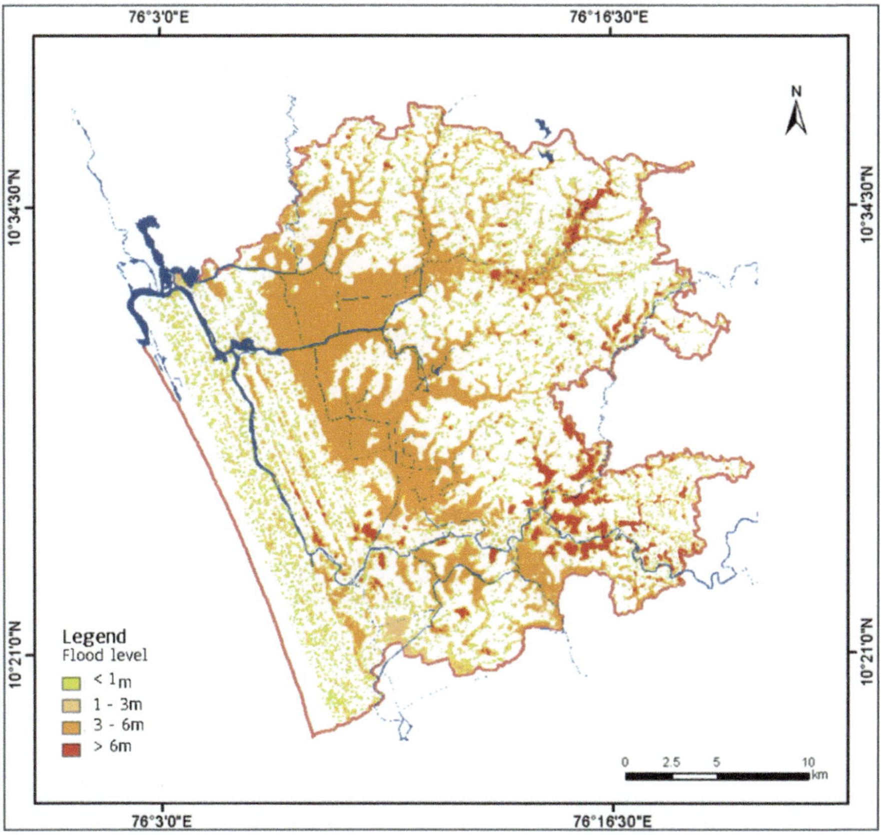

Fig. 6.22 Flood level based on localised rainfall (230 mm/day) for one week

Figures 6.24 and 6.25 show a decrease in the low-temperature area by 92% in 2023 and an increase in high-temperature area by 98.14%.This substantial reduction in the low-temperature area is associated with a variety of factors, including overall warming trends, changes in atmospheric conditions, alterations in local land use.

The expansion of high-temperature areas can significantly affect local ecosystems and human activities. It also influences agricultural practices, as crop yields and growing seasons may shift in response to changing thermal conditions (Lobell 2014).

Additionally, changes in temperature zones have socio-economic consequences, influencing aspects like energy consumption and infrastructure planning. Shifts in temperature patterns can significantly impact energy demand, particularly for heating and cooling in urban areas (Sailor 2011).

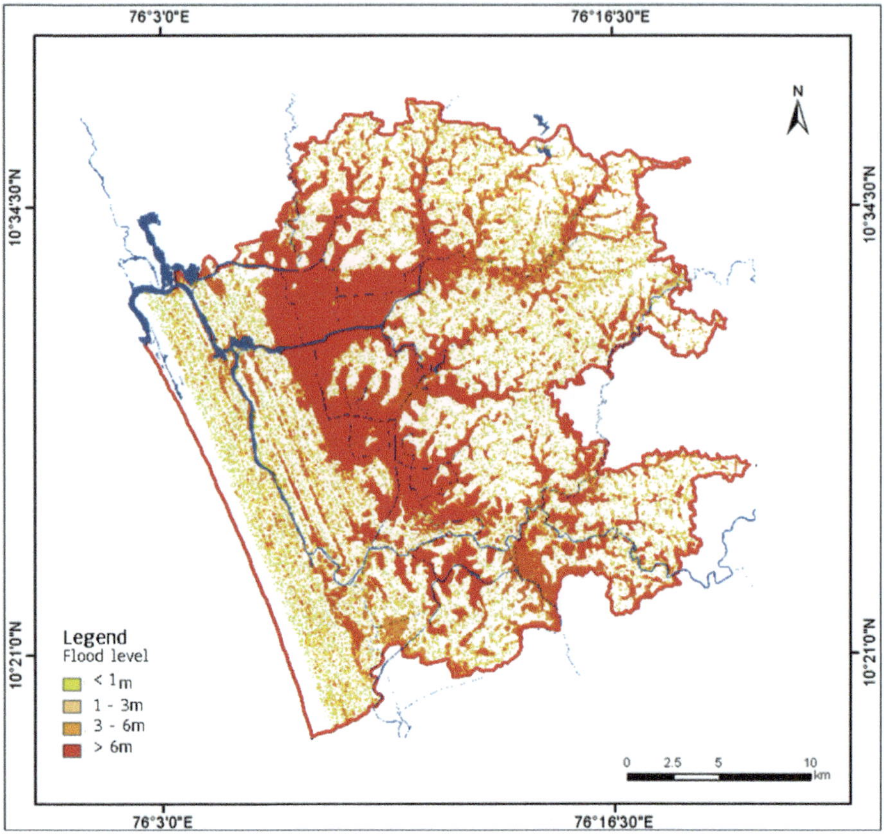

Fig. 6.23 Flood level based on regionalized rainfall (398 mm/day) for one week

6.4.2.1 Urban Heat Island and Land Use

The thermal landscape of Thrissur shows varied temperature profile levels and the city has been divided into three zones—low, medium, and high-temperature area.

The analysis showed defined temperature zones in Thrissur and demonstrated considerable differences over the study area. This section shall look into the temperature classification first followed by the observed trends to depict the trend of temperature levels and its evolution within the city with time.

Temperature Classifications

In order to examine the thermal patterns properly in Thrissur, the urban area was classified into three separate temperature zones namely:

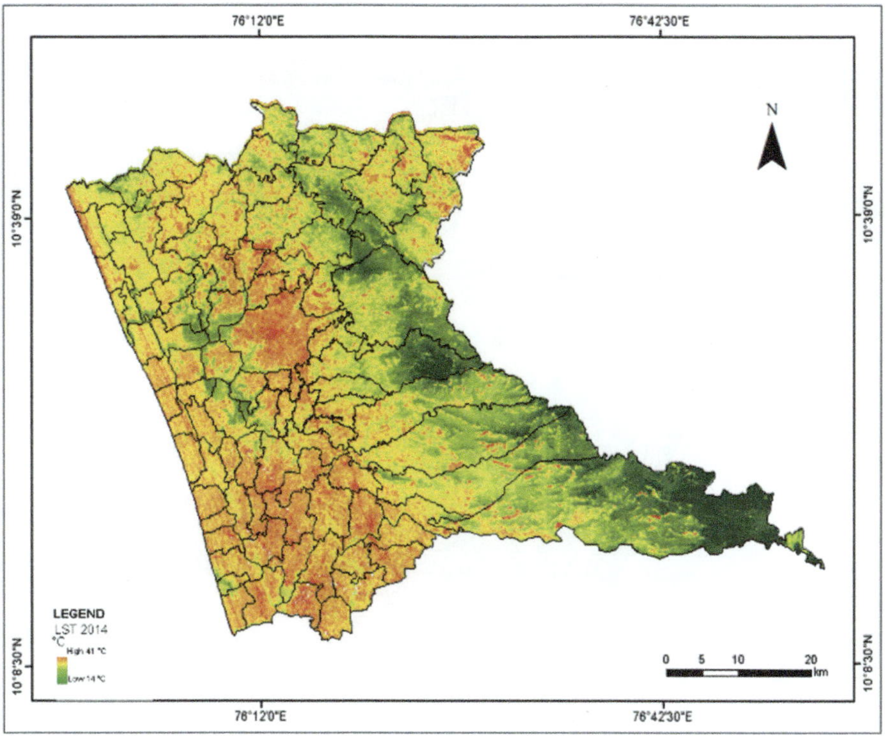

Fig. 6.24 District level land surface temperature—2023

- **Low-temperature region (21–25 °C)**: The range represents the cooler part of a corporation. These places often possess characteristics to alleviate heat build-up, like green spaces, water bodies, or less urbanized areas.
- **Medium-Temperature Zone (25–28 °C)**: The medium temperature range shows a mild increase in warming. These areas are likely exhibiting a slow increase in temperature trends throughout the study. It is an area that often functions as a buffer zone between the cooler and warmer areas of the city, areas where mixed land-use dominates or in a process of urbanization.
- **High-Temperature Zone (28–36 °C)**: This range signifies the warmer temperatures within the corporation area. These zones are typically associated with dense urban development, high levels of anthropogenic heat release, and a predominance of heat-absorbing surfaces. As highlighted by Heaviside et al. (2017b) such high-temperature zones in cities are often the focus of UHI mitigation efforts due to their potential impacts on human health and urban livability.

The temperature classifications revealed significant changes in the temperature profile of Thrissur over the nine years from 2014 to 2023 in the urban agglomerated area as shown the Figs. 6.26 and 6.27.

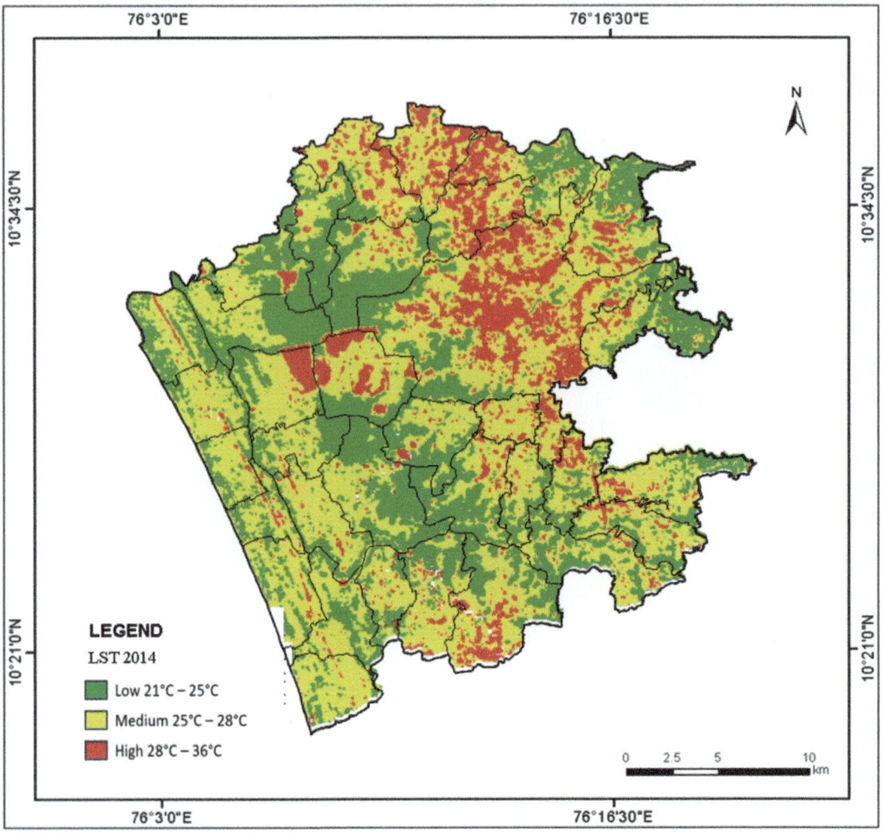

Fig. 6.25 Land surface temperature map—2014 (Urban agglomeration area)

1. Change in Highest Temperatures: The analysis indicates a 12.19% increase in the area experiencing the highest temperatures. This substantial expansion of high-temperature zones suggests that regions within Thrissur, particularly those already characterized by warmer conditions, have undergone a measurable warming trend.
2. Change in Lowest Temperatures: The most dramatic shift was observed in the lowest temperature range, with an impressive 57.14% increase. This substantial rise in the baseline temperatures of traditionally cooler areas is particularly noteworthy. It suggests that even areas that previously served as thermal refuges within the urban environment are experiencing significant warming.

A comparison of land use in Thrissur in 2014 across different temperature zones is shown in the Fig. 6.28

- **Low-Temperature Zone (21–25 °C, 559.92 km²)**: The largest land area, covering 559.92 km², falls within the low-temperature zone. This suggests that a substantial

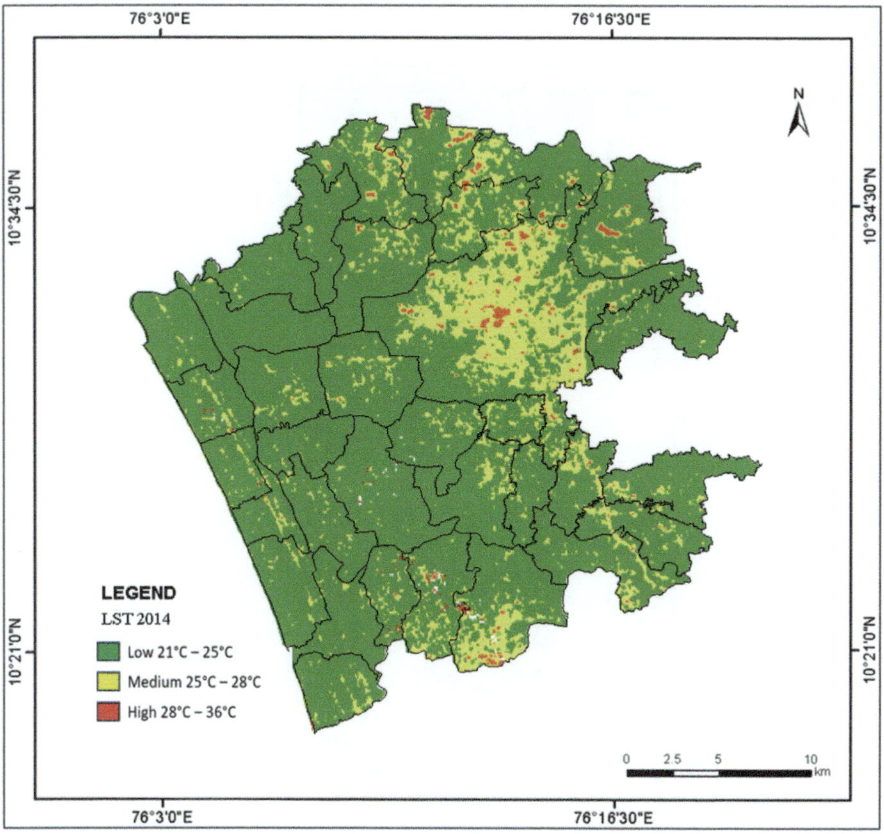

Fig. 6.26 Land surface temperature map—2023 (Urban agglomeration area)

portion of Thrissur experiences relatively lower temperatures ranging between 21 and 25 °C. Areas characterized by low temperatures include green spaces, water bodies, or regions with higher vegetation cover. These areas were likely contributing to the overall cooling effect within the city.

- **Medium-Temperature Zone (25–28 °C, 113.05 km^2)**: A comparatively smaller extent, spanning 113.05 km^2, falls within the medium-temperature zone. This indicates areas with temperatures ranging between 25 and 28 °C.

The medium-temperature zone include a mix of residential, commercial, and industrial areas. The moderate temperatures suggest a balance between urban development and natural elements.

- **High-Temperature Zone (28–36 °C, minimal area):** The high-temperature zone is confined to a minimal area, suggesting that only a small portion of Thrissur experiences higher temperatures ranging between 28 and 36 °C.

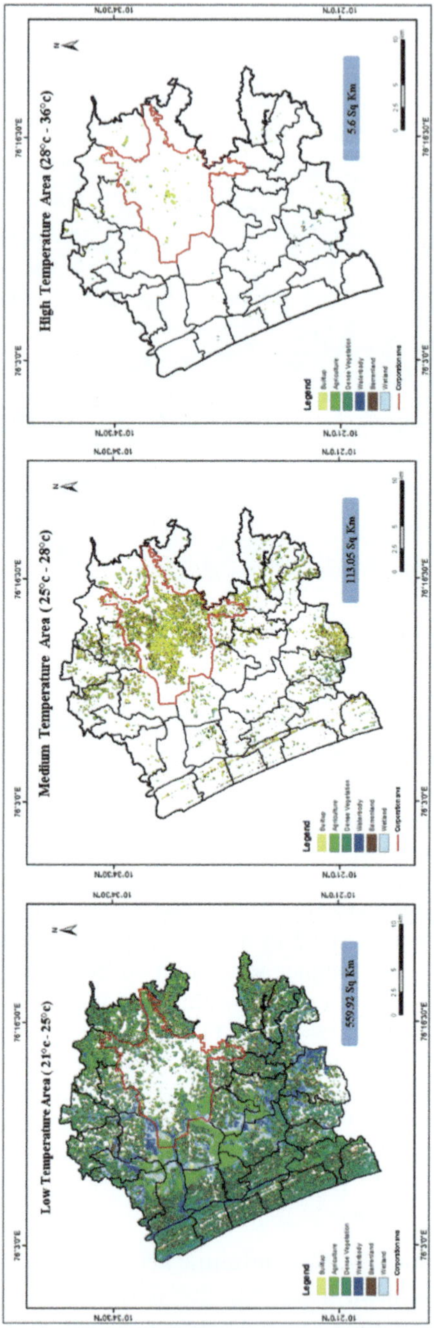

Fig. 6.27 Comparison of land use under different temperature classification—2014

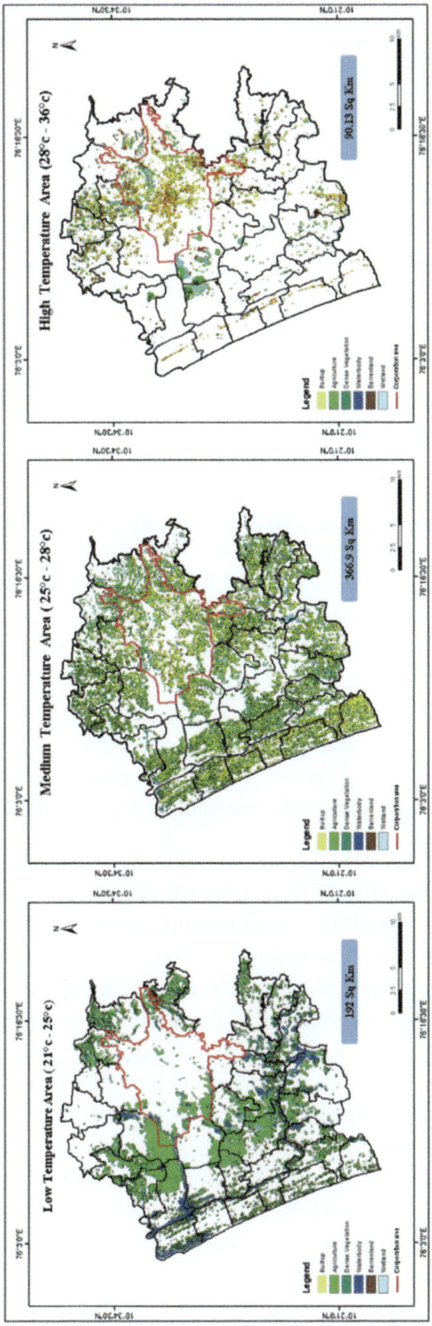

Fig. 6.28 Comparison of land use under different temperature classifications—2023

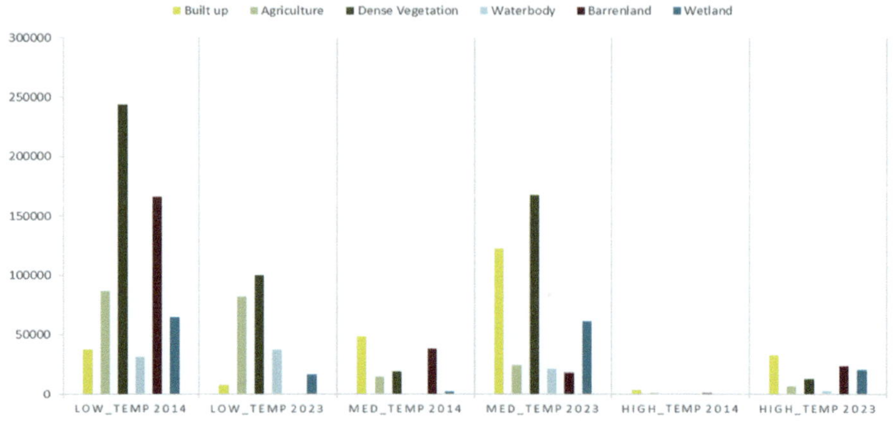

Fig. 6.29 Variation in the land use as in the three different temperature zones

Areas with high temperatures included urban heat islands, densely built-up urban areas, or locations with limited green cover. These regions are more susceptible to heat-related challenges.

Comparison of land use in Thrissur in 2023 across different temperature zones is shown in the Fig. 6.29

In 2023, 192 km^2 is under the low temperature, medium temperature covers an area of 366.9 km^2 and high temperature covers an area of 90.13 km^2.

While comparing 2014 and 2023 there is a notable shift in temperature zone distributions from 2014 to 2023, with a decrease in the low-temperature zone and an increase in both the medium and high-temperature zones.

The graphs presented in the Fig. 6.30 shows the variation in the land use as in the three different temperature zones.

- **Low-Temperature Zone (21–25 °C)**: The 4% decrease in built-up areas in the low-temperature shows a potential trend toward preserving natural landscapes or limiting urban sprawl in low-temperature regions.
- **Medium-Temperature Zone (25–28 °C)**: The 2.5% increase in built-up areas and an 8.4% increase in dense vegetation in the medium-temperature zone signify a complex interplay between urban development and environmental conservation. Balancing these factors is crucial for sustainable growth and maintaining biodiversity in regions with moderate temperatures.
- **High-Temperature Zone (28–36 °C)**: The absence of major built-up areas in the high-temperature zone in 2014 (only 5.6 km^2) indicates that these areas were not extensively developed at that time. However, the significant increase to 90.13 km^2 in 2023 is due to the expansion of urbanization and infrastructure in areas characterized by higher temperatures.

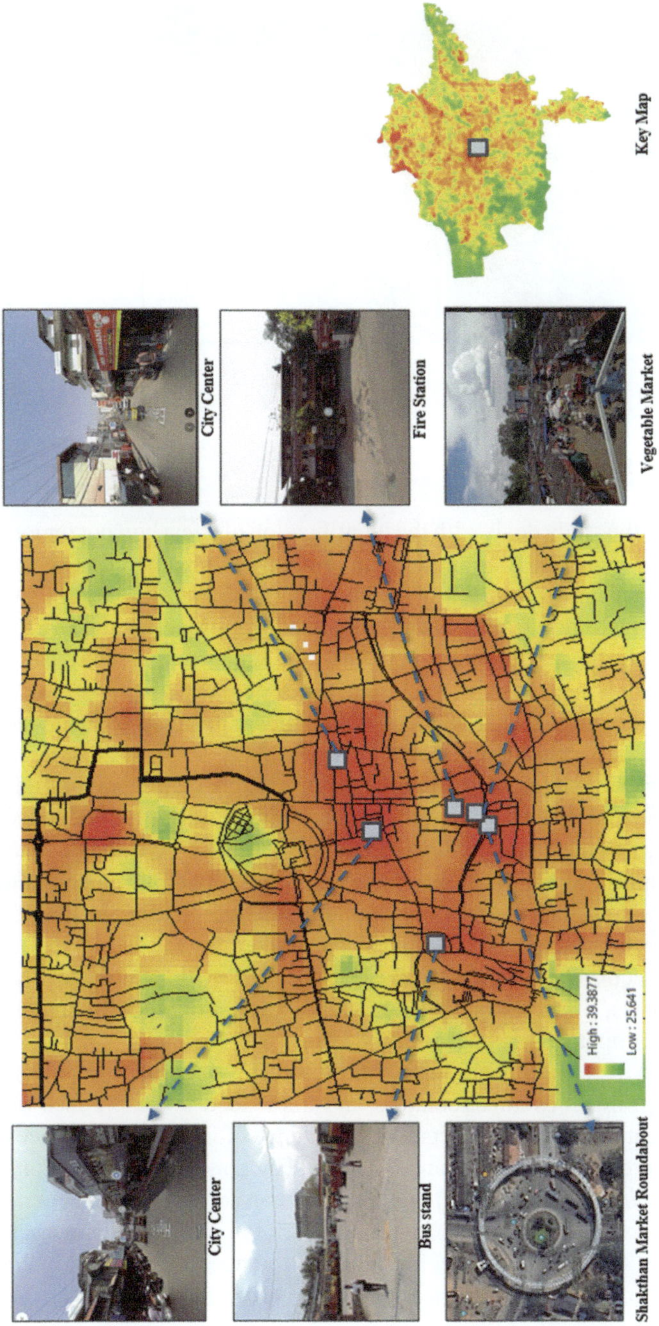

Fig. 6.30 Key hotspot areas

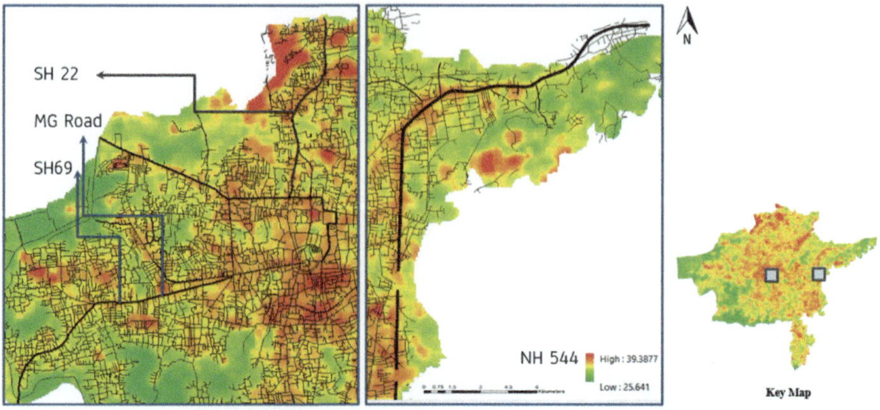

Fig. 6.31 Major roads under the high temperature zone

Identification of Hotspot Regions

The third part of the study deals with the identification of a few hotspot regions, which required interventions based on the LST map which was prepared. This analysis is crucial for pinpointing areas that require immediate intervention as presented in the Fig. 6.31 to mitigate urban heat island effects and improve overall urban thermal comfort. Through the careful analysis of the LST data, several key locations were identified as major hotspot regions, including the city center, bus stand, market roundabouts, fire station, and market area. These hotspots share common characteristics that contribute to their elevated temperatures.

The city center emerged as a significant hotspot, characterized by high-density development and intense human activity. This urban core typically experiences higher temperatures due to the concentration of buildings, reduced vegetation, and anthropogenic heat emissions Similarly, the main bus terminal and its surroundings were identified as another hotspot, with elevated temperatures attributed to the constant flow of vehicles, extensive paved surfaces, and limited green spaces.

Major intersections, particularly those associated with market areas, showed higher LST values, often suffering from traffic congestion, lack of shade, and the heat-absorbing properties of road surfaces.

The common factors contributing to these hotspots include the congestion of buildings, which impedes air circulation and traps heat; a notable lack of vegetation, reducing natural cooling effects; the prevalence of impervious surfaces such as asphalt and concrete, which absorb and retain heat; increased traffic emissions, adding to the anthropogenic heat load; and heightened human activity, which generates additional heat. These findings underscore the complex interplay between urban design, human activities, and local climate, highlighting the need for targeted interventions in these areas to mitigate heat stress and improve urban livability.

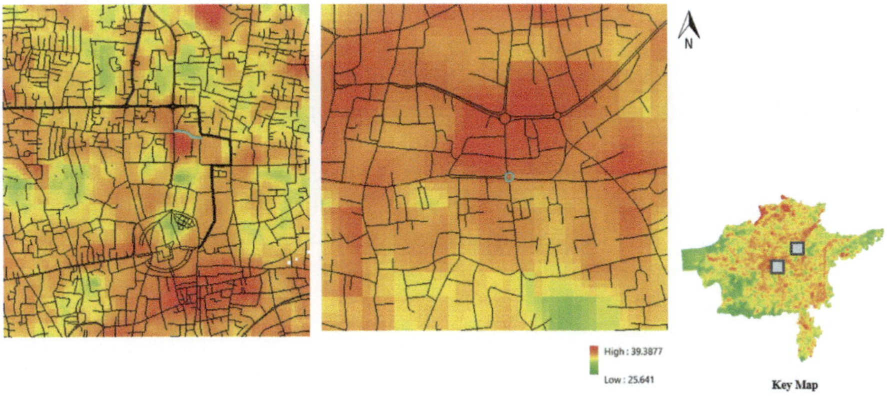

Fig. 6.32 Municipal corporation road and roundabout

6.4.2.2 Urban Heat Island Along Major Roads and Highways

There is a high surface temperature along the major roads: NH544 and along State highway 69 and 22 as shown in Fig. 6.32.

The temperature is even higher at the roundabouts: Mathrubhumi roundabout, MG Road, Town hall Road, Stadium Road, Municipal office Road, Market Road, College Road.

The factors contributing to the urban heat island (UHI) effect around roundabouts are akin to those influencing elevated temperatures along state and national highways. Common elements include issues like congestion, insufficient vegetation, and the prevalence of impervious road surfaces. The heat generated by traffic, combined with the limited permeability of road materials, leads to increased heat absorption and retention, mirroring the conditions observed in roundabouts. Similarly, the lack of greenery exacerbates the warming effect, and the built environment's conFig. 6.uration contributes to heat-trapping, creating an environment reminiscent of UHI conditions found near roundabouts.

6.5 Conclusion

This study provides a comprehensive analysis of existing and future climate-related risks in Thrissur, with a particular emphasis on flood vulnerability and the urban heat island (UHI) effect. The findings reveal large regional variations in flood risk, with the corporate zone being the most vulnerable location. The study shows the complex interaction between land-use practices, geographic features, and flood susceptibility by identifying the major hotspots across several panchayats. Proximity to water bodies appeared as the major predictor of flood risk, emphasizing the significance of careful design in riverside constructions. Wetlands were found to have a moderating effect, boosting aquifer recharge during extreme flood scenarios. Additionally, through the temporal analysis, it was found that the land surface temperature between 2014 and 2023 revealed a warming trend due to the rapid urbanization and land-use changes in the recent years. Key UHI hotspots were identified in the dense developed area and areas with extensive impervious surfaces. This integrated assessment provides a robust foundation for climate-adaptive urban planning in Thrissur and serves as a valuable template for similar assessments in other rapidly urbanizing regions facing comparable climate-related challenges.

6.6 Recommendations

The findings underscore the importance of integrated urban development that can address both flood risk and urban heat challenges, as well as the creation of location-specific measures. The identification of high-risk zones and UHI hotspots improves community engagement by facilitating focused awareness campaigns and localized adaptation strategies, ensuring that communities in susceptible areas are more informed and prepared for climate-related disasters. This study offers a thorough evaluation; however, additional research is required regarding the long-term effects of climate change, the efficacy of adaptation strategies, the socio-economic aspects of climate vulnerability, and the capacity of innovative technologies to improve urban climate resilience. This integrated evaluation establishes a solid basis for climate-adaptive urban design in Thrissur and acts as a useful model for analogous assessments in other swiftly urbanizing areas encountering similar climate-related issues.

References

Akbari H, Kolokotsa D (2016a) Three decades of urban heat islands and mitigation technologies research. Energy Build 133:834–842. https://doi.org/10.1016/j.enbuild.2016.09.067

Akbari H, Kolokotsa D (2016b) Three decades of urban heat islands and mitigation technologies research. Energy Build 133:834–842. https://doi.org/10.1016/j.enbuild.2016.09.067

Arnfield AJ (2003) Two decades of urban climate research: a review of turbulence, exchanges of energy and water, and the urban heat island. Int J Climatol 23(1):1–26. https://doi.org/10.1002/joc.859

Bala RS, Sathyanarayanan S, Jeyaparakash ST (2023) Understanding the relationship between urban heat island and urban vegetation by reviewing three decades of satellite images for the Chennai Metropolitan Area. IOP Conf Seri Earth Environ Sci 1210(1). https://doi.org/10.1088/1755-1315/1210/1/012029

Carter JG, Cavan G, Connelly A, Guy S, Handley J, Kazmierczak A (2015) Climate change and the city: building capacity for urban adaptation. Prog Plan 95:1–66. https://doi.org/10.1016/j.progress.2013.08.001

Central Water Commission (2018) Study report: Kerala floods (August 2018). Hydrology (S) Directorate, Central Water Commission. Retrieved from Kerala State Disaster Management Authority website. https://sdma.kerala.gov.in/wp-content/uploads/2020/08/CWC-Report-on-Kerala-Floods.pdf

Centre for Environmental & Occupational Health, Climate Change & Health (CEOH&CCH), National Centre for Disease Control. State action plan on climate change and human health: Kerala (SAPCCHH Version 2.1). NCDC, Ministry of Health & Family Welfare, Government of India. Retrieved from NCDC website. https://www.ncdc.mohfw.gov.in/wp-content/uploads/2024/05/8-Kerala-SAPCCHH-Version-2-1.pdf

Chander G, Markham BL, Helder DL (2009) Summary of current radiometric calibration coefficients for Landsat MSS, TM, ETM+, and EO-1 ALI sensors. Remote Sens Environ 113(5):893–903. https://doi.org/10.1016/j.rse.2009.01.007

Costabile P, Macchione F (2015) Enhancing river model set-up for 2-D dynamic flood modelling. Environ Model Softw 67:89–107. https://doi.org/10.1016/j.envsoft.2015.01.009

de Moel H, Jongman B, Kreibich H, Merz B, Penning-Rowsell E, Ward PJ (2015) Flood risk assessments at different spatial scales. Mitig Adapt Strat Glob Change 20(6):865–890. https://doi.org/10.1007/s11027-015-9654-z

Department of Mining and Geology, Government of Kerala (2016) District survey report of minor minerals (except river sand): Thrissur District. Department of Mining and Geology. Retrieved from Kerala DMG website. https://dmg.kerala.gov.in/wp-content/uploads/2020/01/dsr_thi.pdf

Heaviside C, Macintyre H, Vardoulakis S (2017a) The urban heat island: implications for health in a changing environment. Curr Environ Health Rep 4(3):296–305. Springer. https://doi.org/10.1007/s40572-017-0150-3

Heaviside C, Macintyre H, Vardoulakis S (2017b) The urban heat island: implications for health in a changing environment. Current Environ Health Rep 4(3):296–305. https://doi.org/10.1007/s40572-017-0150-3

Jha AK, Bloch R, Lamond J (n.d.) Cities and flooding a guide to integrated urban flood risk management for the 21st century

Karunakaran Prasanna C, Kizhakkekara Vannadil P, Pattath Ayyappan V (2020) Wetland conservation and sustainable development in Kerala, India, 1–12. https://doi.org/10.1007/978-3-319-71065-5_115-1

Kreibich H, Bubeck P, Van Vliet M, De Moel H (2015) A review of damage-reducing measures to manage fluvial flood risks in a changing climate. Mitig Adapt Strat Glob Change 20(6):967–989. https://doi.org/10.1007/s11027-014-9629-5

Kundzewicz ZW, Kanae S, Seneviratne SI, Handmer J, Nicholls N, Peduzzi P, Mechler R, Bouwer LM, Arnell N, Mach K, Muir-Wood R, Brakenridge GR, Kron W, Benito G, Honda Y, Takahashi K, Sherstyukov B (2014) Le risque d'inondation et les perspectives de changement climatique mondial et régional. Hydrol Sci J 59(1):1–28. https://doi.org/10.1080/02626667.2013.857411

Li H, Harvey JT, Holland TJ, Kayhanian M (2013) Erratum: the use of reflective and permeable pavements as a potential practice for heat island mitigation and stormwater management. Environ Res Lett 8(4):015023. https://doi.org/10.1088/1748-9326/8/4/049501

Li Y, Schubert S, Kropp JP, Rybski D (2020) On the influence of density and morphology on the urban heat island intensity. Nat Commun 11(1). https://doi.org/10.1038/s41467-020-16461-9

Lobell DB (2014) Climate change adaptation in crop production: beware of illusions. Glob Food Sec 3(2):72–76. https://doi.org/10.1016/j.gfs.2014.05.002

Nones M (2017) Flood hazard maps in the European context. Water Int 42(3):324–332. https://doi.org/10.1080/02508060.2016.1269282

Oke TR, Mills G, Christen A, Voogt JA (2017) Urban climates. Cambridge University Press. https://doi.org/10.1017/9781139016476

Papathoma-Köhle M, Cristofari G, Wenk M, Fuchs S (2019) The importance of indicator weights for vulnerability indices and implications for decision making in disaster management. Int J Disaster Risk Reduct 36. https://doi.org/10.1016/j.ijdrr.2019.101103

Pelling M, Leck H, Pasquini L, Ajibade I, Osuteye E, Parnell S, Lwasa S, Johnson C, Fraser A, Barcena A, Boubacar S (2018) Africa's urban adaptation transition under a 1.5° climate. In Curr Opin Environ Sustain 31:10–15. Elsevier B.V. https://doi.org/10.1016/j.cosust.2017.11.005

Pregnolato M, Ford A, Wilkinson SM, Dawson RJ (2017) The impact of flooding on road transport: a depth-disruption function. Transp Res Part D: Transp Environ 55:67–81. https://doi.org/10.1016/j.trd.2017.06.020

Revi A, Satterthwaite DE, Aragón-Durand F, Corfee-Morlot USA J, Kiunsi RB, da Silva J, Dodman D, Maskrey A, Pahwa Gajjar S, Balbus J, Cardona OD, Sverdlik A, Barros V, Dokken D, Mach K, Bilir T, Chatterjee M, Ebi K, Estrada Y, MacCracken S et al (n.d.). Mark pelling (UK). William Solecki

Rosenzweig C, Solecki W, Romero-Lankao P, Mehrotra S, Dhakal S, Bowman T, Ibrahim SA (2018) Climate change and cities: second assessment report of the urban climate change research network: summary for city leaders. In: Climate change and cities: second assessment report of the urban climate change research network, pp xvii–xlii. Cambridge University Press. https://doi.org/10.1017/9781316563878.007

Sailor DJ (2011) A review of methods for estimating anthropogenic heat and moisture emissions in the urban environment. Int J Climatol 31(2):189–199. https://doi.org/10.1002/joc.2106

Santamouris M (2014) On the energy impact of urban heat island and global warming on buildings. Energy Build 82:100–113. https://doi.org/10.1016/j.enbuild.2014.07.022

Santamouris M (2015) Analyzing the heat island magnitude and characteristics in one hundred Asian and Australian cities and regions. Sci Total Environ 512–513:582–598. https://doi.org/10.1016/j.scitotenv.2015.01.060

Sobrino JA, Jiménez-Muñoz JC, Paolini L (2004) Land surface temperature retrieval from LANDSAT TM 5. Remote Sens Environ 90(4):434–440. https://doi.org/10.1016/j.rse.2004.02.003

Teng J, Jakeman AJ, Vaze J, Croke BFW, Dutta D, Kim S (2017a) Flood inundation modelling: a review of methods, recent advances and uncertainty analysis. In: Environmental modelling and software, vol 90, pp 201–216. Elsevier Ltd. https://doi.org/10.1016/j.envsoft.2017.01.006

Teng J, Jakeman AJ, Vaze J, Croke BFW, Dutta D, Kim S (2017b) Flood inundation modelling: A review of methods, recent advances and uncertainty analysis. In: Environmental modelling and software, vol 90, pp 201–216. Elsevier Ltd. https://doi.org/10.1016/j.envsoft.2017.01.006

Vijay A, Varija K (2024) Spatio-temporal classification of land use and land cover and its changes in Kerala using remote sensing and machine learning approach. Environ Monit Assess 196(5). https://doi.org/10.1007/s10661-024-12633-y

Vojinovic Z, Hammond M, Golub D, Hirunsalee S, Weesakul S, Meesuk V, Medina N, Sanchez A, Kumara S, Abbott M (2016) Holistic approach to flood risk assessment in areas with cultural heritage: a practical application in Ayutthaya, Thailand. Nat Hazards 81(1):589–616. https://doi.org/10.1007/s11069-015-2098-7

Vojtek M, Vojteková J (2016) Flood hazard and flood risk assessment at the local spatial scale: a case study. Geomat Nat Haz Risk 7(6):1973–1992. https://doi.org/10.1080/19475705.2016.1166874

Wilby RL, Keenan R (2012) Adapting to flood risk under climate change. Prog Phys Geogr 36(3):348–378. https://doi.org/10.1177/0309133312438908

Zha Y, Gao J, Ni S (2003) Use of normalized difference built-up index in automatically mapping urban areas from TM imagery. Int J Remote Sens 24(3):583–594. https://doi.org/10.1080/01431160304987

Chapter 7
Assessing the Environmental Vulnerability of Coastal City- Kochi, Kerala State, India

Adinarayanane Ramamurthy, Agnes Mariam Xavier, and Faiz Ahmed Chundeli

Abstract Coastal cities are at the forefront of environmental vulnerability due to their geographical location. They face many hazards, including severe flooding, storm surges, and rising river flows. These events can devastate coastal regions, causing significant infrastructural damage and disrupting livelihoods. This chapter focuses on developing a comprehensive approach for evaluating the environmental vulnerability of Kochi, a coastal city in Kerala, India. Kochi, like many coastal cities, is threatened by environmental hazards like flooding, storm surges, and socio-environmental challenges arising from its geographical location. These threats are exacerbated by the presence of sensitive locations and environmental factors. The study aims to analyze the city's distinct features and environmental factors, including existing pollution levels, shoreline change using the Digital Shoreline Analysis System (DSAS) tool, and the physical susceptibility of the area to environmental hazards. This analysis underscores the need for a comprehensive approach to managing the city's growth and expansion. The study acknowledges that coastal city vulnerability increases due to demographic and economic expansion into sensitive areas. By integrating methodologies for coastal vulnerability assessment, adaptation strategies, and socio-environmental considerations, the study aims to enhance Kochi's preparedness and resilience to environmental challenges. This comprehensive approach will inform the development of effective strategies for managing future environmental risks and ensuring the city's sustainable development.

Keywords Cities · Vulnerability · Environment · Floods · Storm surge

---

A. Ramamurthy (✉) · A. M. Xavier · F. A. Chundeli
School of Planning and Architecture Vijayawada, Vijayawada, Andhra Pradesh, India
e-mail: adinarayanane@spav.ac.in

A. M. Xavier
e-mail: agnesmariamxavier@gmail.com

F. A. Chundeli
e-mail: faizahmed@spav.ac.in

L. K. Dashora et al. (eds.), *Urban Sustainability*, Disaster Risk Reduction,
https://doi.org/10.1007/978-981-96-4890-0_7

7.1 Introduction

Assessment of environmental vulnerability is important to assess and quantify damage and the extent of potential destruction under current scenarios and future conditions of an area. This helps to prepare mitigative measures at the local or city level in tune with the changing global conditions. Therefore, to evaluate environmental vulnerability, a comprehensive methodology is important to understand and manage the risks associated with climate change and other hazards. Coastal areas being one of the several affected and highly vulnerable areas, several studies have focused on developing methodologies to assess coastal vulnerability each with different parameters as seen through the studies by (Rahman 2023; Barros et al. 2022; Tanim and Goharian 2023; Basheer Ahammed et al. 2016). These methodologies typically take into account various factors such as geomorphology, coastal slope, shoreline erosion rates, sea-level rise, wave height, tide range, vegetation width, and socio-economic aspects to assess vulnerability (Murali et al. 2013) Environmental vulnerability assessments especially in coastal regions are complex and a comprehensive analysis requiring diverse data sources and analytical techniques offers a holistic understanding of the risks involved (Alias and Wadhwa 2023).

7.2 Urban Dynamics and Environmental Vulnerability

Urban dynamics in coastal cities like Kochi, India, are influenced by various factors such as urbanization trends, urban initiatives, diverse ecosystem services, ambient air quality, and health issues like metabolic syndrome as seen through Krishnapriya (2020), Anujan et al. (2022a), Rahman (2023) and Bindu et al. (2021). These studies shed light on the complexities of urban growth and their environmental impacts. The rapid urbanization and development in coastal regions, where a significant portion of the global population resides, highlight the importance of understanding and addressing environmental vulnerabilities in these cities (Torresan et al. 2012; Ružić et al. 2019).

In the Indian context of coastal cities, several upcoming studies are focusing on urban agglomerations and their implications for future urbanization trends. Understanding the urban space in cities like Kochi is crucial, as it helps define urban macro-structures and assess the development of specific urban morphologies over time (Anujan et al. 2022b; Ramamurthy et al. 2024). Initiatives like the Atal Mission for Rejuvenation and Urban Transformation play a significant role in shaping urban dynamics and socioeconomic implications in cities like Kochi (Rahman 2023). In coastal urban settings like Kochi in India, the scenario is diverse and complex. The growth pattern in this context provides insights into the interactions between residents and their urban environment, highlighting the importance of green spaces and sustainable urban planning (Bindu et al. 2021). The economic impact of events such as COVID-19 on urban form and services in cities like Kochi underscores the vulnerability of urban economies to external shocks (Alias and Wadhwa 2023).

Evaluating how building explosions affect ambient air quality in coastal urban areas such as Kochi is crucial to comprehending the effects on the environment and public health. To comprehend the effects on the environment and public health, it is crucial to evaluate how building explosions affect the ambient air quality in urban coastal cities like Kochi. Understanding the effects on the environment and public health that arise from building explosions in urban coastal cities such as Kochi requires evaluating the effect of these explosions on ambient air quality. These studies show that urbanization and environmental vulnerability in coastal cities like Kochi, India, are multifaceted and require a comprehensive understanding of urban growth, spatial planning, economic resilience, and environmental sustainability to ensure sustainable development and well-being of urban residents. This chapter looks into a methodology to assess this vulnerability incorporating the spatial aspects specific to the unusual development in Kochi with respect to its environmental assets from an urban standpoint.

7.3 Development of Research Methodology

7.3.1 Profile of the Study Region

The study area consists of the entire Kochi Municipal Corporation and 10 Coastal Gram Panchayats. The coastal gram panchayats include Elamkunnapuzha, Njarackal, Nayarambalam, Edavanakkad, Pallipuram, Kuzhupilly, Mulavukkad, Chellanam, Kumbalangi and a part of Cheranallur. Hence, the population of the part of Cheranallur is calculated based on the number of households assuming that the household size of 4 (Fig. 7.1).

The city of Kochi is prone to several hazards. Natural, geological, biological, and manmade hazards are among the concerns. Arya and Lancelet (2021) has studied the impact of Urban Expansion and Rising surface Temperature in Kochi. These risks have the potential to inflict human harm, including death or injury, property damage, social and economic upheaval, and environmental degradation. In the event of rainfall, the region experiences monsoon rain, albeit the pattern of the rain has changed throughout time. The frequency of extremely heavy rainfall has increased and is occurring at 2.04 times annually during the past decade, whereas the frequency of heavy rainfall has decreased.

Because of the area's frequent rainfall, it has also seen some of the worst floods in the state throughout the years. Recently, it emerged that in the event of severe rainfall, flash flooding occurs. The floods of 1924 also known as the Great Flood of'99 occurred as a result of heavy rainfall more than 64% of the average causing damages in the floodplain of river Periyar encompassing the whole city. Similar floods of 1961 and 2013 caused water logging in several areas. The floods of 2018

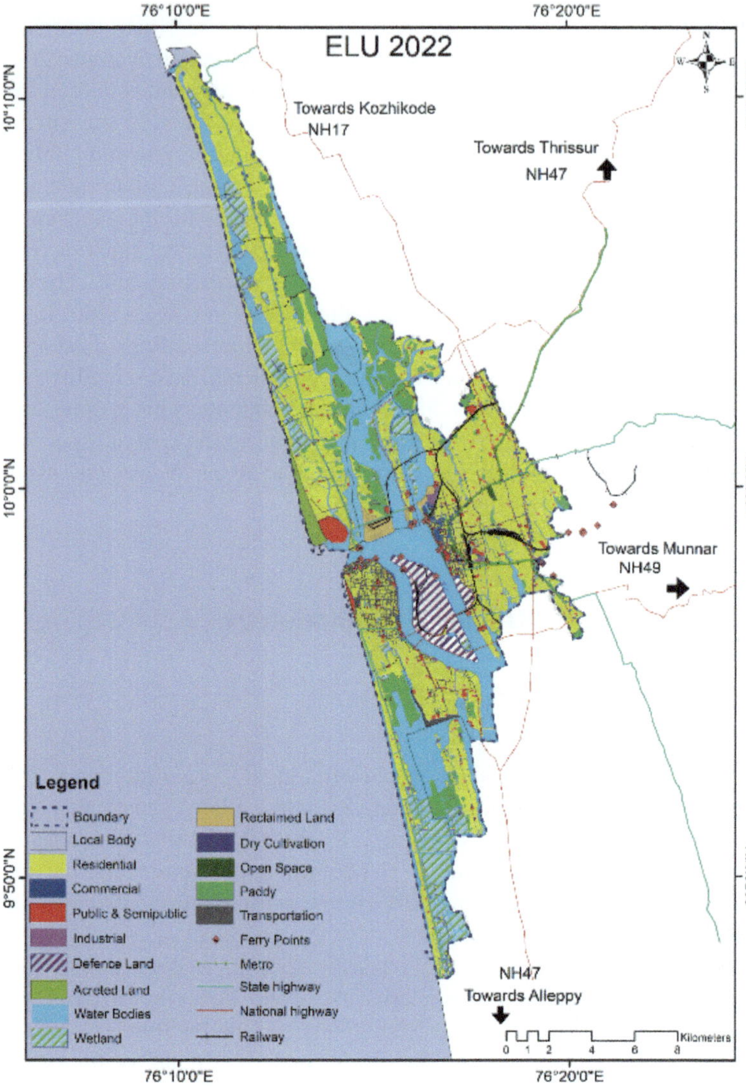

Fig. 7.1 Study region

throughout the state of Kerala caused the most damage surpassing all previously recorded floods causing loss of lives and property and severe financial losses. This event paved the way for more awareness about flood control through city planning interventions and more importance on the physical and environmental vulnerability of the city of Kochi.

The floods in the area are caused by a variety of factors. Between 1985 and 2015, the built-up area in the Ernakulam district increased by 212%, while water bodies decreased by more than 17%. The high increase in built-up area and the fall in water bodies are two elements leading to the flood's devastating consequences. These Kochi's landscapes were formerly wetlands, which emerged as a result of geomorphological processes. They were located near the point where a lake drains into the ocean, depositing clay and silt. Building roads in this region hampered the natural flow of water. The main canals of Kochi can be navigated by small and medium-sized boats. The secondary canals were formerly the city's natural drainage system for flooding, but they are now in severe degradation owing to silting and garbage dumping and no longer work as intended. Construction debris from infrastructure projects, in addition to polluting the canals and backwaters, adds to considerable waterlogging and flood-like situations. The flood-prone areas of the study region have been given above. The area is also susceptible to earthquakes and comes under zone 3 where the maximum expected intensity is 5.6 M. Although the occurrence of earthquakes has been not very frequent, there have been records of past earthquakes over the years (Tables 7.1 and 7.2).

Ernakulam district features a 46-km-long coastline and includes 12 coastal communities. Approximately 69% of the coastline in Cochin Taluk has experienced degradation, making the entire area vulnerable to coastal erosion. The accretion of Vypin Island, a key barrier island, has led to a seaward shift in the coastline. Between 2000 and 2014, Vypin Island's shoreline moved significantly eastward, resulting in land erosion. The tsunami of 2004 may have contributed to these changes. However, cumulative changes between 1970 and 2014 have led to accretion near the estuary mouth, with the bar mouth extending 1.75 km into the sea, significantly narrowing the estuary's opening. The shoreline at Elangunnapuzha in Vypin receded by approximately 954 m, with a 1.36 km recession at the southernmost tip. Vypin Island and Willington Island are the primary locations affected by marine erosion in the area.

Table 7.1 Flood-prone area of study region

Cherayi	Pallippuram	Cochin district
Munambam–Velangannippallikkusameepam	Kuzhuppilly	
Chathangadu–Aniyilkadappuram		
Veeliyathamparambu, Puthenkadappuram	Edavanakkadu village	
Aarattuvazhikadappuram	Nayarambalam	
Chappa Kadappuram	Njarakkal village	
Light House	Elangunnappuzha village	Ernakulam district
St John Padam	Puthu Vyppu village	
Soudi, Mundan Veli	Fort Cochin	
Manassery, Cheriyakadavu	Rameswaram village	
Kandakkadavu, Kannamali	Palluruthi village	
	Kumbalangi village	
	Chellanam	

Table 7.2 Earthquake-prone area of study region and history of earth quakes in the study region

Si. no	Location	Intensity	Magnitude	Date
1	Idamalayar	–	2.1	2006
2	Muvattupuzha	–	1.2	2005
3	Malayattoor_Ermakulam	III–IV	3.5	2001
4	Idamalayar	III	3.0	1986
5	Arakuzha, Muvattupuzha	–	4.2	1971
6	Arakuzha, Muvattupuzha	–	3.7	1972
8	Thiruvengulam, Kanayannur	–	4.5	1972
9	Cochin	VI	5.0	07/26/1953
10	Vypin	VII	5.7	11/11/1941
Major islands	*1970 (km^2)*	*1990 (km^2)*	*2000 (km^2)*	*2014 (km^2)*
Vypin Island	3.3276	5.2987	6.8036	6.9654
Willington Island	6.4664	6.4372	7.4745	7.9326

Source Ernakulam district disaster management plan 2015

Additionally, Chellanam, situated in the southern part of the study area, has experienced significant coastal erosion due to factors like rising sea levels, low-lying lands, and other natural influences.

7.3.2 Methods of Standardization and Quantification of Environmental Sustainability

To evaluate the environmental susceptibility of a geographical or administrative region, it is important to consider the affecting factors. Detailed study and analysis of the change in environmental assets are required. These are often specific to the urban area or region under consideration. Therefore, much thought has to be given to the factors considered and the importance that they play in the area. By utilizing methodologies such as Multi-Criteria Decision Analysis (MCDA) or the Analytic Hierarchy Process (AHP), urban planners can systematically evaluate and prioritize sustainability criteria, leading to more informed decision-making processes and sustainable urban development outcomes.

7.3.3 Mechanism to Establish Vulnerability Assessment

Ecologically sensitive areas within metropolitan regions are the foundation of sustainable urban development, playing a critical role in maintaining ecological balance and ensuring urban environmental and ecological security. Rapid urban expansion encroaches upon these sensitive areas, posing significant threats to the safety of urban ecosystems. By utilizing appropriate methodologies, we can define the scope of Eco-Environmental Sensitivity Analysis and set urban expansion boundaries to support rational and sustainable development. Coastal regions, where residents are heavily impacted, are particularly vulnerable. Pollution exerts pressure on these ecosystems both ecologically and economically. Coastal and marine ecosystems often include coral reefs, seagrass beds, and mangroves, which are especially sensitive to environmental changes. In addition to these ecosystems, areas vulnerable to pollution include coastal towns and regions where activities like fishing, tourism, and fish or salt pond management occur. To assess environmental changes or degradation over time, key factors affecting these regions were analyzed.

Assessing the water quality of the region gives interesting results. The groundwater was analyzed through the Water Quality Index of 9 selected wells around the study area that were analyzed for the years 2015, 2018, and 2022. Utilizing the WQI can assist decision-makers and stakeholders in assessing the appropriateness of water sources for different purposes such as drinking water supply, agricultural irrigation, and ecological health (Makki and Manii 2020; Artiningrum 2023; Simonyan 2021). The parameters taken into consideration were. Fluoride content, Iron Content, Nitrogen Oxides, Sulfates, Chloride Content, HCO_3, CO_3, alkalinity, Total Hardness, Turbidity, Total Dissolved Hardness, Electrical Conductivity and pH of water showed a drastic decline. Chellanam and Kumbalangi continue to face GW problems from mostly salt-water intrusion. The southern part of the city, Thevara, Pettah, etc. faces low-quality of ground water. Therefore, the Corporation supplies water treated from the Muvattpuzha River for domestic and small-scale industries in these areas.

Table 7.3 shows the classification of water by the authors based on the water quality criteria by CPCB. The study area consists of 10 groundwater monitoring stations shown first in the Table and 9 surface water collection and monitoring stations from which data for the years of study were collected and analyzed (Fig. 7.2).

Shoreline change studies are crucial for understanding the environmental dynamics and vulnerabilities of coastal areas like Kochi. These studies offer insights into erosion-accretion rates, environmental conditions, biodiversity-related conditions, future shoreline change predictions, safe navigation, hazard mapping, and coastal resource management (Khedher et al. 2022). In the context of Kochi, a city located along the coast, shoreline change assessment is particularly important due to its susceptibility to coastal hazards and sea-level rise. Analyzing shoreline changes using the Digital Shoreline Analysis System (DSAS) in ArcGIS provides a robust methodology for assessing coastal dynamics and understanding the evolution of shorelines over time. The DSAS tool, as an add-in to the Esri ArcGIS desktop software, offers capabilities to monitor, display, and map long-term and short-term shoreline advances and retreat (Mulyadi et al. 2022).

The tool enables researchers to conduct detailed analyses of shoreline evolution, erosion, and accretion patterns, providing valuable insights into coastal processes and vulnerabilities. For the case of Kochi, employing the tool has provided a comprehensive assessment of shoreline changes along the coastal areas. By digitizing historical shoreline positions and analyzing them using DSAS, researchers can quantify

Table 7.3 Water quality assessment of groundwater monitoring stations and surface water collection points in the study area

Tables Name of Water Body Location/Station	CPCB Water Quality Criteria Class of Water		
	2015	2018	2022
E87: Kumbalangi2	C	B	B
E88: Chellanam	C	C	E
GWE-07: Chellanam	C	C	C
E92: Njarakkal	A	A	A
GWE-08: Pallippuram 1	A	A	A
SHE_OW06: Pallippuram1	A	B	A
E85: Kochi Corporation	A	C	E
E86: Kochi Corporation	D	D	B
E89: Kochi Corporation	A	A	A
TW 01: Kochi Corporation	A	B	B
Station- Bhoothathankettu	A	A	A
Station-Kalamassery	B	C	C
Station-Mupathadam	D	E	E
Station- SDP at Aluva	A	A	A
Station- Periyar at Pathalam	D	D	D
Station- Periyar at KWA, Aluva	B	C	B
Station-Kalady	A	A	A
Station-Chitrapuzha	C	C	C
Station- Periyar at Aluva	B	C	C

CPCB Water Quality Criteria

Colour	Criteria
Green	Drinking Water Source without conventional treatment but after disinfection
Yellow	Outdoor bathing (organized)
Red	Drinking water source after conventional treatment and disinfection

Source Ernakulam district disaster management plan 2015

the rate of shoreline change, identify erosion hotspots, and predict future coastal dynamics have been identified. This information shows the rate of coastal decline and therefore is crucial for coastal management and environmental conservation efforts starting from the city level. Zone 1 and 2 shows moderate erosion while zone 3 and 6 shows moderate accretion. The most susceptible areas lie in zone 4 and 5 with high erosion rates as shown in Fig. 7.3. The management of natural resources must take environmental sensitivity or vulnerability into account, particularly when examining connections between humans and ecosystems. Environmental Sensitivity is a methodical way to gather data on the sensitivity of coastal areas, biological resources, and human-used resources. Environmental sensitivity is a rating of coastal and marine areas based on how sensitive they are to harm from other sources of damage, such

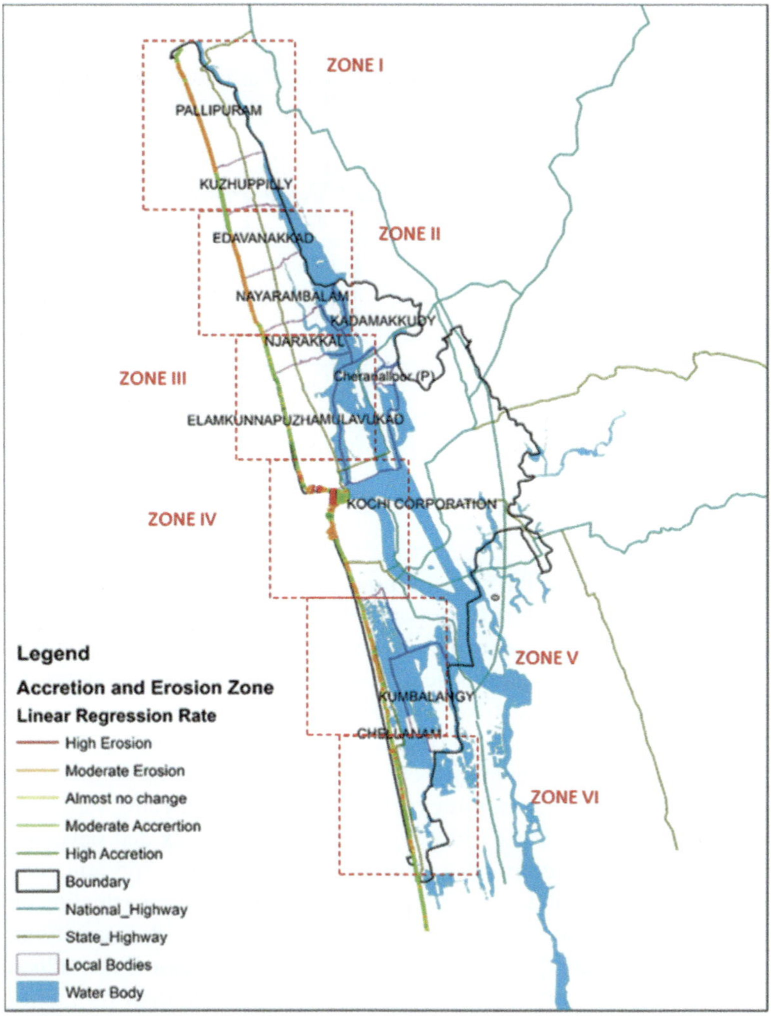

Fig. 7.2 Shoreline change detection for 2023 for zone 1 to 6 of the study region

as oil contamination. Information on coastal and marine areas that need protection, upkeep, and rehabilitation is one benefit of getting a sensitivity map. A map is used to show this information. Environmental sensitivity analysis is crucial for planning and decision-making because it may alert parties to possible land-use conflicts and pinpoint the location and magnitude of any potential negative effects.

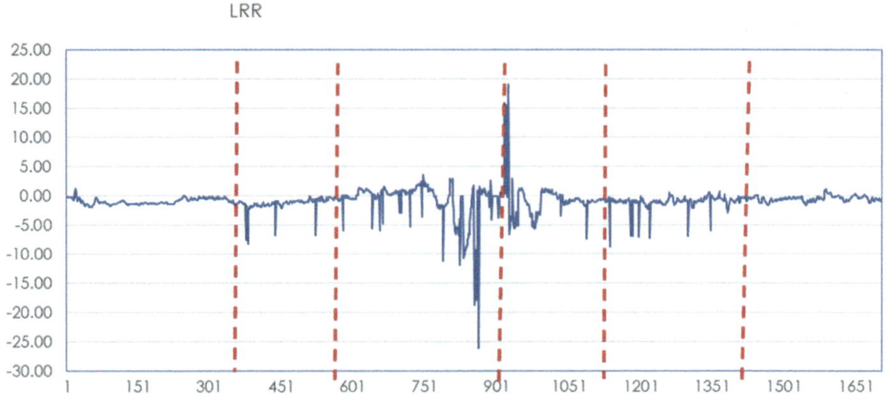

Fig. 7.3 Change in Shoreline 2023

7.4 Analysis and Results

This analysis is conducted through the process of multi-criteria decision-making (MCDM), which seeks to identify the optimal alternative by considering multiple factors. The method offers a variety of tools and techniques that can be used in the case of independent variables in a various field. Multi-Criteria Decision Analysis (MCDA) and AHP, can provide a robust framework for assessing the socio-environmental aspects of vulnerability in urban contexts as seen by Malta and Costa (2021) (Fig. 7.4).

Adopting a multi-hazard approach can significantly improve the assessment of interconnected risks and vulnerabilities in urban areas. By analyzing various hazards and their interactions, planners can identify cascading effects and develop strategies to effectively address complex multi-hazard scenarios. Incorporating the Water Quality Index (WQI) methodology also proves advantageous for evaluating environmental vulnerabilities related to water quality in urban settings. By quantifying water quality parameters and assigning index values, planners can assess the vulnerability of water resources, as explored by Roy and Blaschke (2013). Since different factors may impact the ecosystem in varying ways, it is crucial to reflect this in the calculation process. For this reason, we applied the Analytic Hierarchy Process (AHP) during our analysis, utilizing ArcGIS for the spatial assessment.

7.4.1 Parameters Considered

A list of all the variables was initially compiled that researchers throughout the world frequently utilize to determine which variables are most sensitive to this location. Six effect indices were then considered by taking into account the topography, landscape, biological environment, and development aims of this region. These factors include the Rate of Urbanization/ Urban Sprawl, Proximity to water bodies, Rate of change in

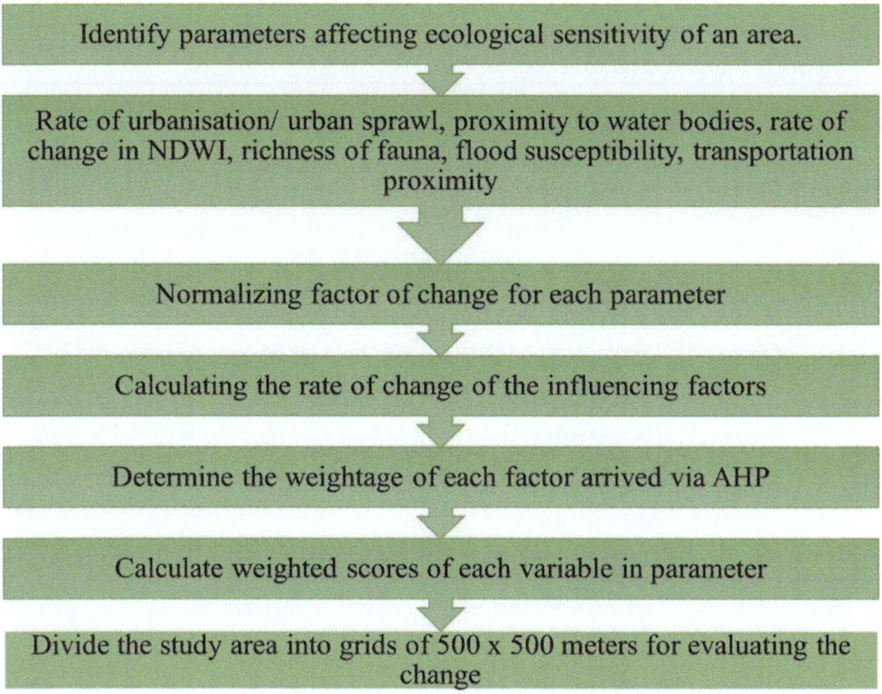

Fig. 7.4 Methodology for environmental vulnerability assessment

NDWI, Rate change in NDVI, Richness of fauna, Flood Susceptibility, Transportation Proximity and Tropical Wetness Index. Table 7.4 explains each index's meaning (Fig. 7.5).

7.5 Results and Discussion

The generated vulnerability map revealed a significant correlation between environmental vulnerability and proximity to water bodies. Areas near wetlands and the coast were found to be particularly affected. Key factors contributing to this heightened vulnerability include: Increased Flooding with regions close to water bodies inherently more prone to inundation during extreme weather events. Urban sprawl further exacerbates this risk by reducing natural drainage pathways and increasing impervious surfaces that impede water infiltration. The deterioration in water quality as seen above due to the rapid urbanization has increased pollution from anthropogenic sources like sewage discharge, industrial effluents, and stormwater runoff that continue to degrade the water in nearby lakes, rivers, and coastal ecosystems. several ongoing unsustainable development practices such as encroachment

Table 7.4 Parameters considered

Parameters considered	Main connotation
Rate of urbanization/ urban sprawl	The amount of settlement may vary, we identify areas where the direction of urban sprawl is increasing as more sensitive
Proximity to water bodies	Distance from water bodies. Areas closer to water bodies are more ecologically sensitive
Rate of change in NDWI	Change of Water Index over a span of 10 years expressed as percentage or decimal. More the change; more is the sensitivity of the area
Rate change in NDVI	Change of the density of vegetation over a span of 10 years expressed as percentage or decimal to see the areas more sensitive to more change
Richness of fauna	Water, as an essential component of the ecological system, supports a rich diversity of life. Rivers and lakes play a crucial role in the existence and growth of cities, as many densely urbanized areas are located near water bodies. These water sources not only provide critical resources but also contribute to the biodiversity that thrives around them
Flood susceptibility	Areas of higher flood susceptibility are more prone to ecological damage
Transportation proximity	Transportation is another significant factor affecting the ecological environment, particularly through infrastructure such as roads and railways. By analyzing the distance from roads, we can assess their impact on ecological spaces using Buffer analysis, which helps to evaluate how transportation infrastructure influences the surrounding environment
Tropical wetness index	High TWI means higher soil moisture therefore more ecologically sensitive

on wetlands and coastal ecosystems shown by increased built area and decrease in area under agriculture or wetlands have been known to be destroying critical habitats for diverse flora and fauna. The environmental vulnerability assessment of Kochi, Kerala, paints a concerning picture. The city faces a multitude of environmental challenges, with areas near water bodies and wetlands exhibiting heightened vulnerability. Unsustainable development practices, rising sea levels, and climate change all contribute to the complex web of threats jeopardizing Kochi's ecological health and long-term sustainability. This study underscores the urgent need for a comprehensive and multi-pronged approach to address these vulnerabilities. By integrating the findings of the vulnerability map with existing knowledge and best practices, Kochi can chart a course toward a more resilient future.

Strengthening Governance and Policy Frameworks:

Effective environmental management hinges on robust governance and clear policy frameworks. The following measures are crucial (Fig. 7.6):

1. *Plans for Urban Development:* Current urban expansion plans need to be critically evaluated. Development projects should prioritize environmental sustainability and strictly avoid encroachment on wetlands and coastal areas.

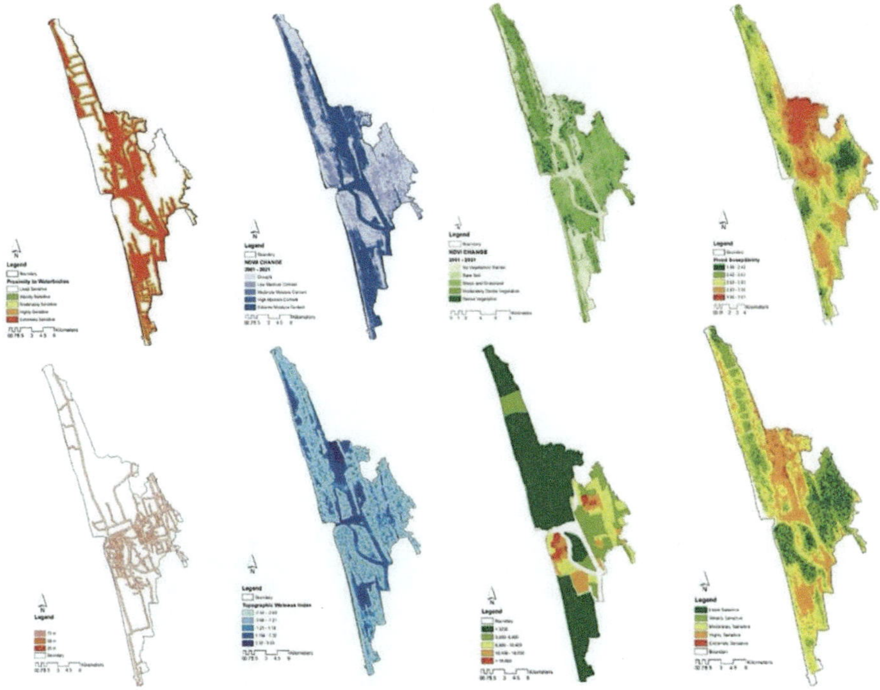

Fig. 7.5 **a–h**: **a** Proximity to water bodies, **b** Change in NDWI from 2001 to 2021, **c** Change in NDVI from 2001 to 2021, **d** Flood susceptibility in the study region, **e** Road buffers, **f** Topographic wetness index, **g** Population density of the region, **h** Derived susceptibility map of the study region

Zoning regulations and environmental impact assessments should be rigorously enforced.

2. *Enhancing Environmental Legislation:* Environmental legislation needs to be strengthened to address contemporary challenges. Stricter regulations on pollution control, waste management, and water resource utilization are essential.
3. *Promoting Inter-departmental Collaboration:* Effective environmental management requires collaboration between various government departments, including urban planning, environment, water resources, and disaster management. A holistic approach that considers the interconnectedness of environmental issues is critical.
4. *Engaging with Stakeholders:* Meaningful stakeholder engagement is vital. Local communities, environmental NGOs, and academic institutions can provide valuable insights and expertise in developing and implementing sustainable solutions.
5. *Investing in Green Infrastructure and Ecosystem Restoration:* Green infrastructure is crucial for reducing environmental risks and strengthening urban resilience. The following are key areas where investment is essential to maximize its benefits.

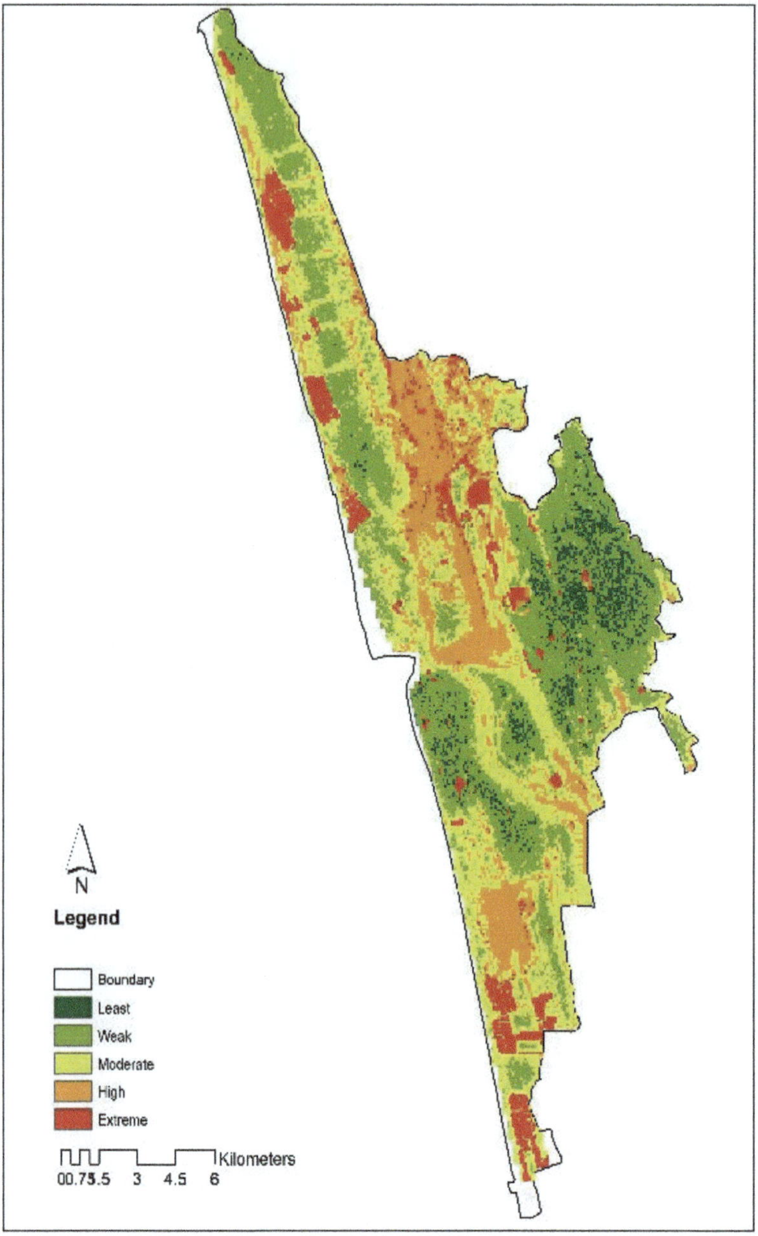

Fig. 7.6 Degree of environmental vulnerability of the study region

6. *Creating Coastal Buffers:* Establishing vegetated buffers along the coastline can help to mitigate the impacts of erosion, storm surges, and saline intrusion. These buffers can be designed to provide additional benefits such as recreational spaces and habitats for native species.
7. *Rehabilitation of Wetlands:* Restoring degraded wetlands is essential for improving water quality, filtering pollutants, and regulating flood flows. These restored wetlands can also provide vital habitat for migratory birds and other species.
8. *Promoting Urban Green Spaces:* Creating parks, green roofs, and urban forests within cities can enhance air quality, reduce the urban heat island effect, and improve stormwater management. These green spaces not only offer recreational opportunities but also contribute to the overall health and well-being of residents.
9. Technological Innovation and Data-Driven Decision Making: Investing in Early Warning Systems: Developing and implementing robust early warning systems for floods, storm surges, and other natural hazards is essential to minimize loss of life and property damage.
10. *Utilizing Remote Sensing and GIS:* Remote sensing and Geographic Information Systems (GIS) can be used to monitor and analyze environmental changes, such as shoreline erosion and changes in vegetation cover. This data can be vital for informing mitigation strategies.
11. *Building Capacity and Public Awareness:* Public awareness and community engagement are vital for building long-term resilience. Educational programs for schools and the general public can raise awareness about environmental challenges and foster a sense of stewardship toward the natural environment.
12. *Community-Based Disaster Preparedness Initiatives:* Engaging local communities in disaster preparedness initiatives empowers them to take ownership of their safety and contribute to building resilient communities.
13. *Capacity Building for Local Governments:* Providing training and resources to local government officials on environmental management best practices can enhance their capacity to implement sustainable solutions.
14. *Public–Private Partnerships:* Developing public–private partnerships can leverage private-sector investment for green infrastructure projects and environmental initiatives.
15. *Climate Change Adaptation Funds:* Exploring and applying for national and international climate change adaptation funds can provide vital financial resources for resilience-building projects.
16. *Green Bonds:* Issuing green bonds can raise capital specifically dedicated to financing environmentally sustainable projects.

Building a more resilient and sustainable Kochi requires a concerted effort from all stakeholders. Collaboration between the government, private sector, civil society, and the local community is essential. The findings presented in this environmental vulnerability assessment provide a crucial starting point for action. By implementing the strategies outlined in this way forward, Kochi can work toward a future where environmental sustainability and economic development go hand-in-hand. The city's continued prosperity hinges.

7.6 Conclusion

In conclusion, the environmental vulnerability assessment of Kochi highlights the critical need for a sustainable and comprehensive approach to address the multifaceted challenges facing the city, particularly in areas near water bodies and wetlands. The assessment reveals that increasing urbanization, unsustainable development practices, and climate change are exacerbating risks such as flooding, water pollution, and habitat destruction. To mitigate these threats, it is essential to strengthen governance frameworks, enhance environmental legislation, and foster inter-departmental collaboration. Investments in green infrastructure, ecosystem restoration, and technological innovations like GIS and early warning systems will be pivotal in building resilience. Furthermore, engaging local communities, and stakeholders, and fostering public–private partnerships can amplify efforts toward sustainable development. By implementing these strategies, Kochi can achieve a balanced integration of environmental sustainability and economic growth, ensuring a resilient and prosperous future for its citizens and ecosystems.

References

Alias J, Wadhwa S (2023) Kochi: provincialising postcolonial metro-cosmopolitan spatialities. Cult Geogr 31:167–182. https://doi.org/10.1177/14744740221147554

Anujan K, Velho N, Kuriakose G, Ebin PJ, Nagendra H (2022a) Beyond the metropolis: street tree densities and resident perceptions on ecosystem services in small urban centers in India. bioRxiv. https://doi.org/10.1101/2022.09.28.509699

Anujan K, Velho N, Kuriakose G, Ebin PJ, Nagendra H (2022b) Beyond the metropolis: street tree densities and resident perceptions on ecosystem services in small urban centers in India. bioRxiv. https://doi.org/10.1101/2022.09.28.509699

Artiningrum T (2023) Determination of Cikapundung River water quality index using IKA-INA method and pollution index. Eduvest J Univ Stud 3(6):1086–1097. https://doi.org/10.59188/eduvest.v3i6.843

Arya LP, Lancelet TS (2021) Urban expansion and rising surface temperature: A multi-temporal assessment in Kochi metropolis, Kerala. Disaster Adv 14(12):48–54. https://doi.org/10.25303/1412da4854

Barros JL, Tavares AO, Santos PP, Freire P (2022) Enhancing a coastal territorial vulnerability index: anticipating the impacts of coastal flooding with a local-scale approach. Coast Manag 50(4):442–468. https://doi.org/10.1080/08920753.2022.2107858

Bindu G, Raj RK, Baiju MA, Farhana CA (2021) The impact of a building implosion on ambient air quality: a case study in an urban coastal city. Passive Remote Study Aerosol Upper Atmos Earth:1–7. https://doi.org/10.47363/jeesr/2021(3)142

Basheer Ahammed KK, Mahendra RS, Pandey AC (2016) Coastal vulnerability assessment for eastern coast of India, Andhra Pradesh by using geo-spatial technique. Geoinform Geostat Overv 4(1), Article 1000146. https://doi.org/10.4172/2327-4581.1000146

Khedher KM, Abu-Taweel GM, Al-Fifi Z, Qoradi MD, Al-Khafaji ZS, Halder B, Bandyopadhyay J, Shahid S, Essaied L, Yaseen ZM (2022) Farasan island of Saudi Arabia confronts the measurable impacts of global warming in 45 years. Sci Rep 12, Article 18225. https://doi.org/10.1038/s41598-022-18225-5

Krishnapriya MG (2020) Mode choice behaviour of students, integrating residential location characteristics: a study from Kochi city, India. Eur Transp/Trasporti Europei 79:1–17. https://doi.org/10.48295/et.2020.79.5

Makki IM, Manii JK (2020) Water quality index of Euphrates river near Al-Musayyab power plant. Iraqi J Sci 61(11):3002–3008. https://doi.org/10.24996/ijs.2020.61.11.21

Malta FS, Costa EMD (2021) Socio-environmental vulnerability index: an application to Rio de Janeiro, Brazil. Int J Public Health 66, Article 584308. https://doi.org/10.3389/ijph.2021.584308

Mulyadi A, Hamidy R, Musrifin M, Efriyeldi E, Jhonnerie R (2022) Tiga dekade laju perubahan garis pantai di kota Dumai. Dinamika Lingkungan Indonesia 9(1):25–31. https://doi.org/10.31258/dli.9.1.p.25-31

Murali RM, Ankita M, Amrita S, Vethamony P (2013) Coastal vulnerability assessment of Puducherry coast, India, using the analytical hierarchical process. Nat Hazards Earth Syst Sci 13(12):3291–3311. https://doi.org/10.5194/nhess-13-3291-2013

Rahman F (2023) Local economic impact of covid-19 on the urban tourism-related services: a perspective of Kochi heritage city, Kerala. Sustainability 15(24), Article 16585. https://doi.org/10.3390/su152416585

Ramamurthy A, Chundeli FA, Chandran AG, Joshi T (2024) Evaluating the impact of urbanisation on climate change: a case of Kochi City, Kerala State, India. In: Schrenk M, Popovich T, Zeile P, Elisei P, Beyer C, Ryser J, Kaufmann HR (eds) Real corp 2024 proceedings, pp 275–284. https://www.corp.at

Roy DC, Blaschke T (2013) Spatial vulnerability assessment of floods in the coastal regions of Bangladesh. Geomat Nat Haz Risk 6(1):21–44. https://doi.org/10.1080/19475705.2013.816785

Ružić I, Jovančević SD, Benac Č, Krvavica N (2019) Assessment of the coastal vulnerability index in an area of complex geological conditions on the KRK Island, Northeast Adriatic Sea. Geosciences 9(5), Article 219. https://doi.org/10.3390/geosciences9050219

Simonyan GS (2021) Systemic-entropic approach for assessing water quality of rivers, reservoirs, and lakes. IntechOpen. https://doi.org/10.5772/intechopen.93220

Tanim AH, Goharian E (2023) Toward an integrated probabilistic coastal vulnerability assessment: a novel copula-based vulnerability index. Water Resour Res 59, Article e2022WR033603. https://doi.org/10.1029/2022WR033603

Torresan S, Critto A, Rizzi J, Marcomini A (2012) Assessment of coastal vulnerability to climate change hazards at the regional scale: the case study of the North Adriatic Sea. Nat Hazards Earth Syst Sci 12(7):2347–2368. https://doi.org/10.5194/nhess-12-2347-2012

Chapter 8
Exploring the Role of Socio-Cultural Factors in Disaster Risk Reduction—A Case of Indigenous Koli Communities of Mumbai

Sanskriti Shukla and Rama Umesh Pandey

Abstract The chapter explores the heightened vulnerability of Mumbai's indigenous Koli fishing communities to climate change-induced disasters, accentuated by rapid urbanization and their historical ties to the region's ecosystem. Despite their rich cultural heritage, the Kolis are facing significant threats to their traditional practices and collective resilience. Current disaster risk reduction (DRR) literature predominantly emphasizes technical and structural approaches, often overlooking the crucial role of cultural dimensions in shaping community resilience. This oversight can undermine the effectiveness of DRR strategies and exacerbate marginalization. Employing a mixed-method approach, this study integrates qualitative ethnographic methods, quantitative indicator-based index approaches, and geospatial vulnerability analysis to investigate the socio-cultural factors affecting DRR in Koli communities. Geographic Information System (GIS) techniques are utilized to map disaster risks within Koliwadas, providing spatial insights into vulnerabilities. Concurrently, household surveys assess socio-cultural resilience factors such as community infrastructure, sense of place, social capital, and cultural practices, leading to the creation of a socio-cultural resilience index. The findings highlight the necessity of incorporating both tangible and intangible cultural aspects into DRR frameworks to enhance community resilience effectively. The research advocates for participatory approaches that empower local communities, ensuring their agency in decision-making processes and the sustainability of DRR initiatives. Mainstreaming cultural considerations into DRR policies and practices can harness the full spectrum of community assets, fostering more resilient and adaptive societies amidst increasing climate uncertainties.

S. Shukla (✉) · R. U. Pandey
School of Planning and Architecture Bhopal, Bhopal, Madhya Pradesh, India
e-mail: 24m0201@iitb.ac.in

R. U. Pandey
e-mail: ramapandey@spabhopal.ac.in

L. K. Dashora et al. (eds.), *Urban Sustainability*, Disaster Risk Reduction,
https://doi.org/10.1007/978-981-96-4890-0_8

Keywords Disaster risk reduction · Socio-cultural resilience index · Indigenous koli communities · Vulnerability mapping

8.1 Introduction

Climate change has caused weather-related disasters to escalate five-fold, over the past 50 years, resulting in 115 fatalities and a staggering daily economic loss of US$ 212 million (WMO 2021). Natural disasters that can affect people, property, and finances include earthquakes, volcanic eruptions, and storms. Individuals' ability to respond to these threats is determined by their vulnerability and resilience (Gaillard 2006). A decade of disasters has resulted in substantial losses, both in terms of human lives lost and money spent; roughly 700,000 people have perished, 1.4 million have been injured, and 23 million have been displaced. There have been 1.5 billion affected people in all, with vulnerable groups being disproportionately impacted (IPCC 2012; Field et al. 2012a, 2012b). The Climate Vulnerability Index (CEEW 2021) states that over 80% of Indians reside in districts that are extremely vulnerable to extreme hydro-meteor disasters, with Assam, Andhra Pradesh, Maharashtra, Karnataka, and Bihar being particularly vulnerable to extreme climate events like floods, droughts, and cyclones. When it comes to its economic and physical vulnerability to climate change, Mumbai is India's most vulnerable coastal district (TERI 1996). Because Mumbai was formerly made up of seven islands and was reclaimed, it is just around 15 m above sea level, or slightly above it in some locations. The natural drainage patterns and ecosystems have become ecologically imbalanced due to rising urbanization and human activity. This has left the area extremely vulnerable to coastal disasters like flooding and cyclones. The indigenous populations that live near the water or in low-lying areas bear a disproportionate share of this burden. The Koli fishing groups and the Worli tribes, who are native to Mumbai, have long been the city's inhabitants. They continue to leave their mark on the city, weaving an indigenous thread through its diverse population. They are stunning signs of how life, livelihoods, festivals, faith, and beliefs are integrated with nature in close-knit, sustainable communities.

In indigenous societies, natural calamities are potent agents of socio-cultural transformation. There are two schools of thought on how indigenous populations should react to natural calamities. First of all, traditional cultures that rely on their environment are weak and unable to handle large-scale, quickly developing environmental disasters on their own. The second theoretical framework, on the other hand, believes that traditional civilizations may bounce back from the effects of natural events on their own (Gaillard 2006). Natural hazard-induced environmental changes compelled these cultures to make only minor adaptations without altering the core principles of their social structure. Therefore, there is a need to investigate the dynamic relationship between indigenous communities' cultural vulnerability to disasters and how cultural ecosystem services can play an instrumental role in enhancing disaster resilience in these communities, given the rising frequency and severity of disasters and their disproportionate impact on these communities, who are the keepers of traditional knowledge and practices.

Socio-cultural factors are frequently said to be overlooked while developing and putting into practice disaster risk reduction measures in the mainstream research on the subject (Kasdan 2016; Wisner et al. 2004; Paton and Johnston 2016). According to Kasdan (2016) and Kulatunga (2010), neglecting to take socio-cultural factors into account may make a community more susceptible to catastrophes and result in the creation of ineffective DRR plans. Therefore, it is crucial to research the socio-cultural aspects of disaster resilience in order to incorporate them with DRR strategies, given the rising frequency and severity of disasters as well as their disproportionate impact on indigenous communities, who are the keepers of traditional knowledge and sustainable practices.

Although most research to date has been on the physical aspects of disaster risk reduction (DRR), little is known about how socio-cultural factors contribute to resilience building. It is noteworthy that climate change vulnerabilities comprise social, economic, and cultural factors in addition to physical exposure. Factors including cultural values, social capital, and resource accessibility are important in determining how well a community can adjust to climate change. Unfortunately, these cultural components are frequently ignored in mainstream DRR literature, which could result in solutions that are ineffective. By exploring the frequently disregarded field of intangible cultural factors in disaster studies, the study aims to fill this gap. Previous research has shown that ignoring these cultural factors might increase community vulnerability and lead to ineffective DRR measures (Paton and Johnston 2016; Wu and Chen 2023). There is still a lack of research demonstrating a spatial relationship between disaster resilience and socio-cultural elements, despite the body of literature on disaster resilience in indigenous communities.

The aim of this paper is to enhance understanding of the intricate interactions between cultural values, environmental systems, and disaster resilience, with a focus on the socio-cultural aspects of disaster resilience within the indigenous Koli community.

The objectives are:

1. To link socio-cultural factors to disaster risk reduction through a literature review and bibliometric analysis.
2. To map coastal vulnerability in Koliwadas.
3. To develop and calculate a socio-cultural resilience index using identified indicators, through mapping, household surveys, and observations in selected Koliwadas.

8.2 Conceptual Framework

"Culture" as a general term refers to a wide range of aspects related to behavior, manner of life, ethnicity, moral standards, beliefs, values, customs, artwork, and intellectual output that influence, originate from, or are shared within a particular community. In a nutshell, culture is a society's expression of humanity. Many cultural dimensions are used to describe different aspects of severe hazards resulting from natural catastrophes or climate change. They include cultural aspects of risk perception, negative cultural components linked to danger, vulnerability, and fear, humanitarian compassion culture, organizational or institutional culture and the reactions to it, and the culture of taking preventive actions to lower risks (Valverde n.d.; International Science Council 2023). This includes using construction materials that are resistant to extreme weather conditions and developing and maintaining a "Risk Management Culture," "Safety Culture," or "Adaptation Culture."

Ethnicity and culture have a significant impact on how well a community adapts to climate change. As an illustration, current study conducted in Northern Burkina Faso reveals that, although coexisting in the same physical area and experiencing climate change together, two diverse ethnic groups have adopted different solutions because of their cultural values and historical contacts. The culture of local risk management, deeply ingrained in communities, is essential. Cultivating a "culture of prevention" is essential for creating local networks that can continuously assess and mitigate hazards at the micro level.

Various indicators within the socio-cultural dimension have been extensively explored in the literature. According to Li and Tan (2019), social capital—which includes social network, trust, and identity factors—has received the most attention. Furthermore, Dube and Munsaka (2018), Ostadtaghizadeh et al. (2015) and Kasdan (2016) research include civic engagement, empowerment, and community participation as key metrics. Other dimensions include the tangible components of the availability and accessibility of essential services, as well as the intangible aspects covered by Paton et al. (2021) and Gaber et al. (2019). Intangible elements include family vitality, self-efficacy, and community efficacy. The feeling of location and community, along with common values and standards, are among the commonly accepted indicators, as noted in the works of Li and Tan (2019).

8.2.1 Identification of Indicators

In conducting a bibliometric analysis for the literature review to understand the relationship between cultural ecosystem services and disaster resilience as seen in Fig. 8.1, the dimensions database was utilized with a refined search string ("socio-cultural factors" OR "cultural influences") AND ("community disaster resilience" OR "risk reduction") AND ("indigenous communities" OR "native populations") AND ("natural disasters"). This search yielded 192 journal papers. Initial analysis

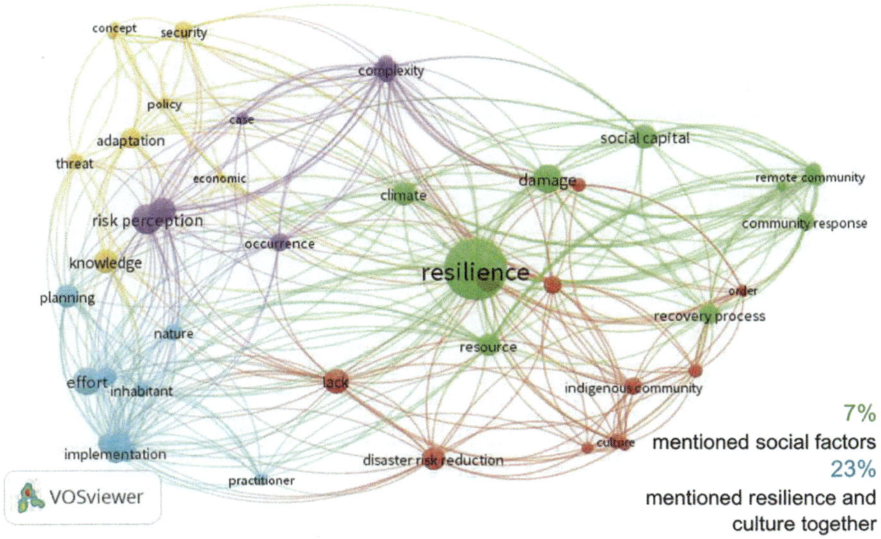

Fig. 8.1 Bibliometric analysis using VOSviewer

revealed that 7% of these papers explicitly mentioned socio-cultural factors, while 23% discussed the interplay between resilience and culture.

Subsequently, 11 papers were further shortlisted for in-depth examination based on criteria including publication after 2015 and maximum citations. From these selected papers, a detailed index framework was developed, comprising indicators and variables pertinent to the study's focus. This bibliometric approach not only facilitated the identification of relevant literature but also laid the groundwork for a comprehensive analysis of socio-cultural dynamics in the context of disaster resilience within indigenous communities.

The papers examined the intricate relationship between community engagement, social capital, essential community infrastructure, culture, awareness, and empowerment in the context of disaster resilience within indigenous communities (Axon 2020; Bourke et al. 2022; Cardona et al. n.d.; Marasco et al. 2022; Gaber et al. 2019; Hu et al. 2021; Kasdan 2016; Kumar n.d.; Li and Tan 2019; Nakamura and Kanemasu 2020; Valverde n.d.). Some studies emphasized the pivotal role of community engagement in fostering collaboration, trust, and collective action, thereby enhancing resilience to natural disasters (Kasdan 2016). Social capital emerged as a critical factor, with research highlighting how strong social networks and cohesive community bonds contribute to effective disaster response and recovery efforts (Marasco et al. 2022). Culture was recognized as a foundational element influencing community resilience, with indigenous knowledge systems and traditional practices often providing valuable insights and strategies for coping with natural hazards (Kumar n.d.; Usman et al. 2022). Awareness and empowerment were identified as essential components in building resilience, with emphasis placed on the importance of

Table 8.1 Indicator scoring matrix

	INDICATORS	1	2	3	4	5	6	7	8	9	10	11	Score
1.	Leadership												3
2.	Community Participation												3
3.	Collective efficacy												2
4.	Empowerment												2
5.	Sense of Place												6
6.	Security												1
7.	Cultural factors (tangible & intangible)												5
8.	Degree of social equity												1
9.	Risk awareness												4
10.	Diversity of work opportunities												3
11.	Shared values												4
12.	Community knowledge												4
13.	Information												3
14.	Essential services												7
15.	Social Capital												8

education, capacity-building initiatives, and inclusive decision-making processes to empower communities and strengthen their ability to mitigate risks and adapt to changing environmental conditions (Tammar et al. 2020; Li and Tan 2019).

From the 11 identified papers in the bibliometric analysis, a matrix was constructed (Table 8.1) to evaluate various dimensions based on their mentions across the papers. Among these dimensions, it was observed that sense of place, social capital, cultural factors, and essential services garnered the highest scores of—6, 8, 5, and 7, respectively. These dimensions were selected to form the foundation of the socio-cultural resilience index, given their significant presence in literature. This index aims to provide a structured framework for assessing the socio-cultural aspects of resilience within indigenous communities facing natural disasters, emphasizing the critical role played by factors such as sense of place, social capital, cultural practices, and access to essential services in enhancing community resilience.

8.2.2 Proposed Framework

The proposed framework as seen in Fig. 8.2 illuminates the pivotal role of socio-cultural factors in shaping a community's resilience to disasters. It underscores the significance of a community's vulnerability and social capital in determining its capacity to withstand and recover from adversity. Positioned at the heart of the framework are elements such as social capital, sense of place, community infrastructure, and culture, which collectively influence disaster resilience. Social capital, characterized by collective action, communication, and mutual support, emerges as a cornerstone in enhancing a community's preparedness, mitigation, and response to disasters. Moreover, the framework emphasizes the importance of a strong sense of

Fig. 8.2 Conceptual framework

place and community identity in bolstering resilience, highlighting the detrimental effects of social divides and loss of identity on vulnerability. By acknowledging how socio-cultural factors shape perceptions of risk and inform disaster preparedness and response behaviors, the framework underscores the necessity of understanding and integrating these dynamics into the development of effective disaster risk reduction strategies.

8.3 Material and Methods

The study begins with a thorough examination of the background context of the primary subject under investigation. As can be observed in Fig. 8.3, the first stage involves a comprehensive review of the literature, with an emphasis on socio-cultural factors and disaster resilience in the context of indigenous groups. The objectives of the study are thus shaped by the findings of this evaluation of the literature. After a strong theoretical base has been established, the research carefully outlines its goals and objectives. An extensive literature analysis that critically assesses previous research efforts in addition to compiling definitions and concepts pertinent to the selected issue follows this phase. In doing so, the study seeks to extract knowledge on useful approaches and markers for the upcoming empirical inquiry.

The stated goals are carefully broken down into four main parts. First, to determine socio-cultural elements and the indicators that go along with them that affect indigenous communities' resilience to disasters, the research does a thorough secondary literature review. This preparatory stage is essential for providing guidance for the research's later empirical stages. The second goal is the coastal vulnerability mapping through geospatial techniques and remote sensing. This entails defining several criteria and scales to define the study area in a sophisticated manner.

Subsequently, third stage adopts a mixed research approach, integrating both qualitative and quantitative methods to comprehensively explore the research domain. The qualitative phase involves in-depth exploration through ethnographic study,

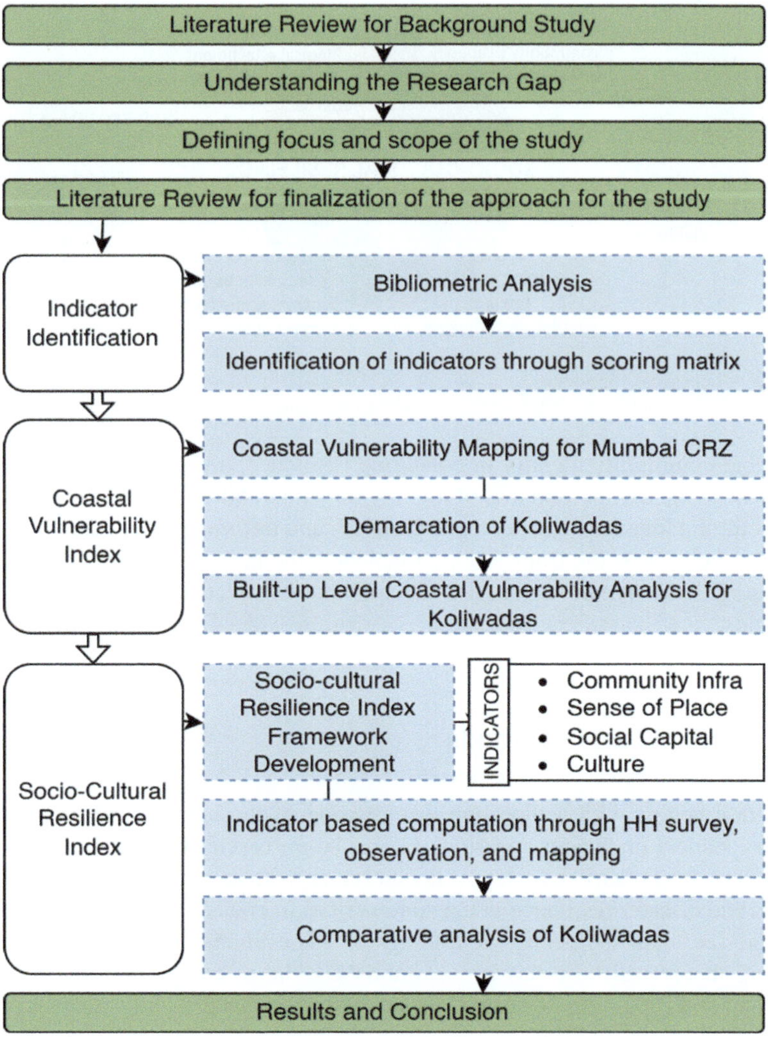

Fig. 8.3 Methodology flowchart

Table 8.2 Sample size

S. no	Location	Sample size	Gender distribution	
			Male	Female
1	Khar Danda Koliwada	65	32	33
2	Worli Koliwada	52	23	29
3	Colaba Koliwada	35	17	18
	Total	152	72	80

employing methods such as observations, semi-structured interviews, and Focus Group Discussions (FGDs). Subsequently, the qualitative data undergoes- structural coding and thematic analysis to derive meaningful insights. Followed by developing a case-based assessment framework to quantify socio-cultural resilience. This involves the application of observations and household surveys to measure resilience across selected dimensions, indicators, and variables. This dual-method approach aims to provide a holistic understanding of socio-cultural resilience within the specified context.

Selection Criteria for Koliwadas

Three Koliwadas were chosen for thorough examination and analysis. The selection criteria used were: presence of Koli community, proximity to the coast, accessibility, absence of language barrier, and presence of socio-cultural factors.

Household Survey Sampling

The number of samples to be collected for the household surveys (Table 8.2) was calculated using random sampling. The sample size for finite population was calculated using Cochran formula (modified for smaller population). During the calculation, 95% of confidence interval and 8% of margin of error was considered.

8.4 Introduction to Study Area

The Kolis, an indigenous fishing community in Mumbai, India, have long resided in the city's coastal districts but are now being pushed to its vulnerable peripheries. Socio-economically, they are highly dependent on fishing, making them susceptible to declines in fish populations due to overfishing, pollution, and environmental degradation, leading to reduced income and food security (Senapati and Gupta 2014). Limited access to alternative employment and education exacerbates their poverty and marginalization. Environmentally, they face severe threats from climate change, including rising sea levels, storm surges, and extreme weather, which damage their homes and coastal ecosystems. The removal of protective mangroves further exposes them to these hazards. Socio-culturally, modernization and urbanization erode their

Fig. 8.4 Vernacular house

traditional fishing practices and cultural identity, as urban infrastructure encroaches on their grounds and globalization draws younger generations away from traditional knowledge, weakening community cohesion (Movik et al. 2023). The subsequent sections provide the community profiles of the four selected Koliwada settlements in Mumbai (Figs. 8.4, 8.5 and 8.6).

8.4.1 Khar Danda Koliwada

Khar Danda, occupying an area of 23 hectares, is home to a population of 12,951, resulting in a density of 563 people per hectare. A significant portion, accounting for 53% of the population, is engaged in fishing, highlighting the area's reliance on this livelihood. Regarding housing typology, 11% of the structures are vernacular in design, as seen in Fig. 8.7. Noteworthy community initiatives include a village beautification project. The community's socio-economic fabric is intricately linked to fishing, with 60% of its populace involved in this occupation. Additionally, the society distributes essential resources such as oil and nets, subsidized by the government. However, environmental concerns persist, with 75% of the catch comprising garbage rather than fish, underscoring the impact of pollution on local marine resources. The community faced adversity during Cyclone Tarkte on May 17, 2021, which damaged 18 boats. Despite receiving no assistance from the coast guard, the community rallied together, employing communal efforts to rescue affected fishermen. Each boat, supporting 25–30 dependents, plays a vital role in the community's livelihood. The cultural identity of Khar Danda is rich and vibrant, with traditions such as Narli

Fig. 8.5 Worship place at the dock

Fig. 8.6 Danda dock

Poornima, where coconuts are offered to the sea for boat protection, and Holi festivities featuring dressing competitions and boat decorations on Rangpanchmi (Figs. 8.8, 8.9 and 8.10).

8.4.2 Worli Koliwada

Worli Koliwada, spanning 24 hectares, accommodates a population of 10,265, resulting in a density of 427 people per hectare. A notable portion, approximately 37%, of the populace is engaged in fishing, highlighting the community's reliance on this traditional livelihood. In terms of housing typology, 27% of the structures are built in vernacular style as seen in Fig. 8.11, showcasing the area's architectural heritage. The community is actively involved in initiatives such as The Heritage Project, led by the RFG Foundation. Structurally, the community is organized into 4–5 societies, with prominent ones like the Nakva Matsya Vavsay Society leading various activities. These activities range from managing a cooperative bank to hosting marriage halls and facilitating access to boat-related resources. The community also convenes annually for a comprehensive society-wide meeting, while 11 trustees hold monthly gatherings. Elections, conducted every five years by the fish department, ensure democratic representation within the community leadership. Additionally, the community receives alerts from local authorities such as the police and coast guard, contributing to disaster preparedness. Architecturally, Worli Koliwada exhibits vernacular elements such as wood, tiles, and Khol, reflecting its cultural heritage. Efforts toward heritage conservation are actively supported by organizations like the RFG Foundation and Locavore, underscoring the community's commitment to preserving its cultural identity and architectural legacy (Fig. 8.12).

8.4.3 Colaba Koliwada

Colaba Koliwada, occupying a modest area of 9 hectares, is home to a population of 6,938, resulting in a remarkably high density of 770 people per hectare. The community's livelihood predominantly revolves around fishing, with a staggering 81% of its residents engaged in this occupation, including 12% in artisan fishing. Unlike other Koliwadas, Colaba Koliwada does not have any vernacular houses, reflecting a departure from traditional architectural styles, visible in Fig. 8.13. Structurally, the community is organized into three societies overseeing the activities of approximately 300 boats. A vast majority, around 90%, of the population are directly involved in fishing activities, including selling and auctioning the catch, which is predominantly supplied to Sasson Dock. However, the community faces challenges, including a significant reduction in catch volume over the past 15 years, attributed to various factors such as increased competition from large commercial LED fishing boats and environmental issues like open dumping in the sea. To mitigate risks, the community

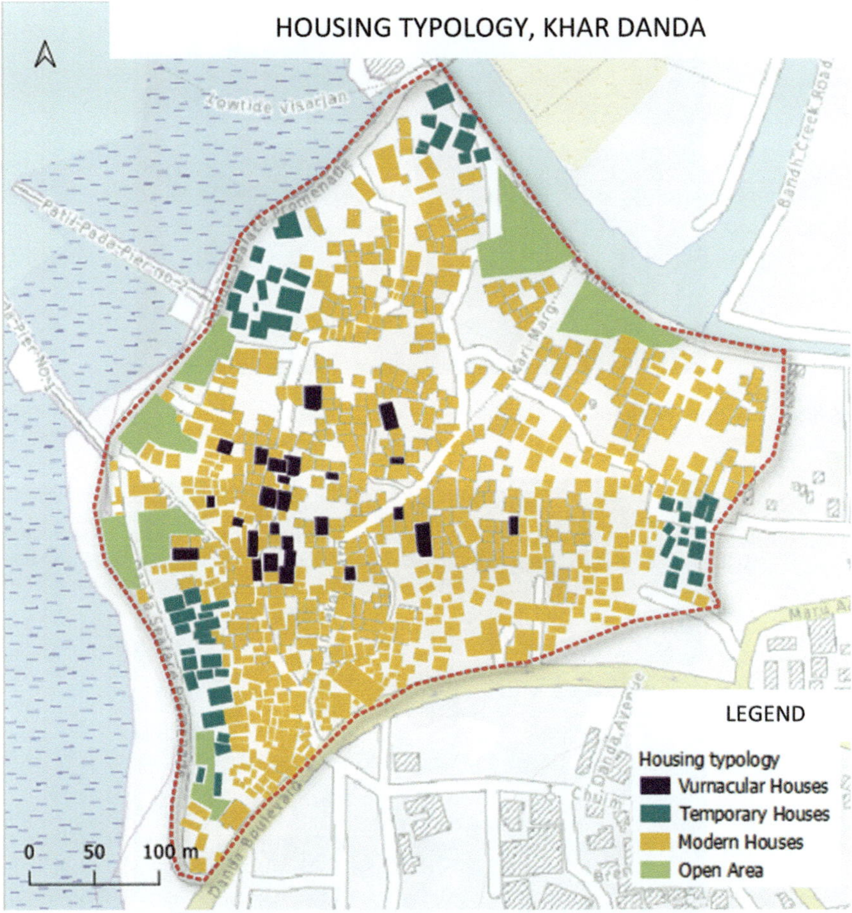

Fig. 8.7 Housing typology map, Khar Danda (*Source* Survey by author)

Fig. 8.8 View from Worli fort

Fig. 8.9 Low cost protective structure

utilizes modern technology like the Vendy app for weather forecasts and prioritizes human safety during high tides. In case of losses, reports are submitted to the fish department, and compensation is received directly from the government into individual accounts. Cultural traditions remain integral to the community, with rituals such as worshiping Khar Devi on Gudi Parva observed. Additionally, the community has undergone displacement from the main Colaba area and is experiencing a shift in occupational dynamics, with only 10% of the new generation engaged in fishing, highlighting ongoing changes within the community's socio-economic fabric.

Fig. 8.10 Vernacular house, Worli

8.5 Results and Discussions

This portion elucidates the Coastal Vulnerability Index (CVI) mapping and its implications for understanding Mumbai's coastal vulnerability, focusing on the Coastal Regulation Zone (CRZ) buffer zone and the building-level vulnerability of the three Koliwadas. GIS-based geospatial analysis sheds light on the complex interplay of geophysical factors shaping Koliwada vulnerability compared to Mumbai's overall coastal vulnerability. Additionally, it introduces the socio-cultural resilience index (SCR Index), a novel framework designed to assess community resilience through socio-cultural factors. It details the methodologies used for data collection, and presents the results of the index calculation, providing valuable insights into the socio-cultural determinants of resilience (Fig. 8.14).

8.5.1 Coastal Vulnerability Index

The Coastal Vulnerability Index (CVI) was computed utilizing remote sensing data from diverse sources and subsequently classified into risk categories according to predefined benchmarks. This process involved the integration of individual layers through ArcGIS weighted overlay analysis tool, culminating in the generation of a composite CVI map. This study employs a methodology based on Gornitz's Coastal Vulnerability Index to assess the vulnerability of Mumbai's coastline within a 500-m Coastal Regulation Zone (CRZ) buffer. Vulnerability, as an internal risk factor, signifies the intrinsic predisposition of a subject or system to damage when exposed to hazards. Six geophysical parameters, as seen in Table 8.3, were utilized to compute

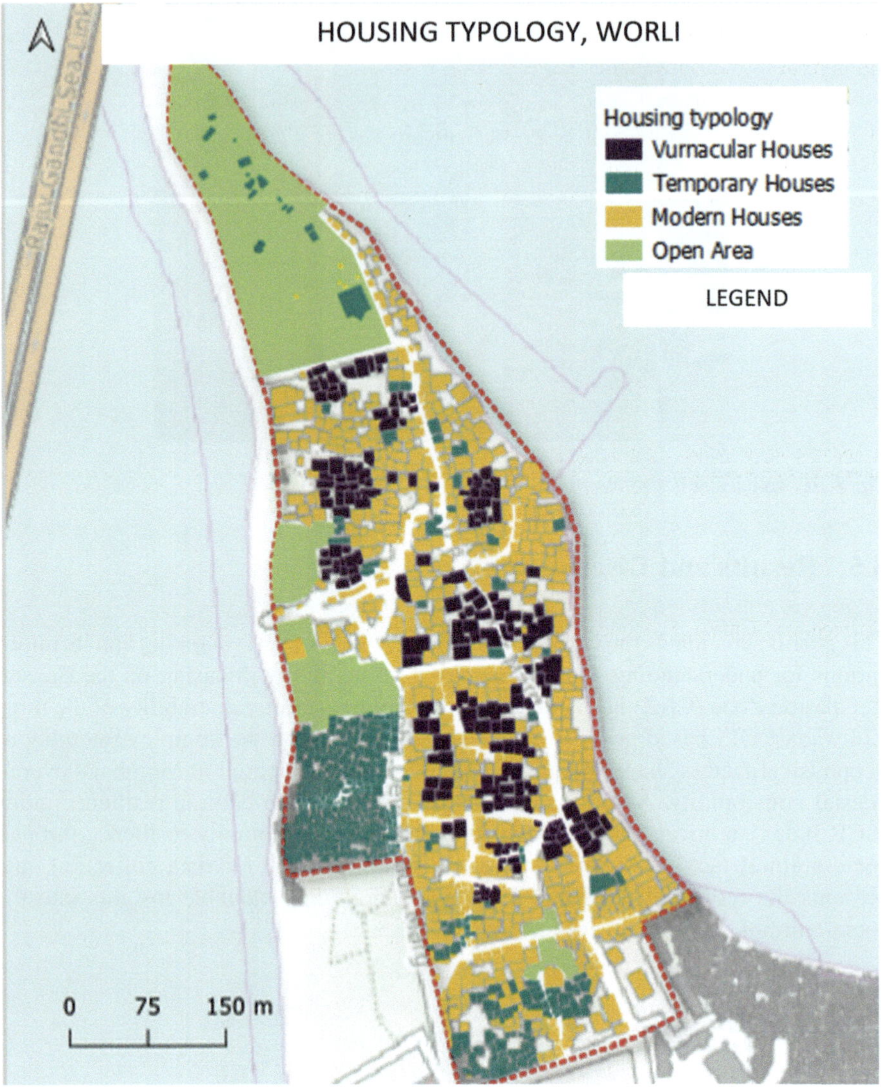

Fig. 8.11 Housing typology map, Worli (*Source* Survey by author)

Fig. 8.12 Open space, Colaba

the Coastal Vulnerability Index (CVI), categorized as low, moderate, or high-risk factors.

The study analyzed various factors influencing coastal vulnerability. Coastal geomorphology insights were derived from 1:250 K maps by the Geological Survey of India (GSI), identifying landforms like dissected plateaus and coastal plains (Fig. 8.18). Slope data, critical for assessing inundation risks from storms and tsunamis, were calculated using GEBCO data with one-minute grid resolution, assigning risk ratings to different areas (Fig. 8.17). Elevation, crucial for assessing susceptibility to sea-level rise, was analyzed using Shuttle Radar Topography Mission (SRTM) data with 30-m resolution (Fig. 8.16). Shoreline change rates were determined using the Digital Shoreline Analysis System (DSAS) with Landsat images, creating over 2000 transects at 100-m intervals for manual digitization in ArcGIS (Fig. 8.15). Tidal range analysis, based on data from WXTide 32 software, calculated maximum and minimum tide amplitudes to categorize coastal regions by susceptibility to tidal fluctuations (Fig. 8.21). Lastly, sea-level change rates, assessed using AVISO altimetry data, provided essential insights into global and localized processes impacting Mumbai's coastline, with sea-level anomaly (SLA) time-series data from 1993 to 2020 (Figs. 8.19 and 8.20).

The Coastal Vulnerability Index (CVI) for Mumbai's coastline is 6.17 out of 10, indicating moderate vulnerability. Worli Koliwada shows a higher CVI of 6.58, while Colaba Koliwada has a significantly elevated CVI of 7.61, and Khar Danda Koliwada records a CVI of 6.12. Spatial analysis reveals high-risk buildings are concentrated along the coastline's edges, suggesting greater susceptibility to coastal hazards in these areas. This assessment highlights the need for targeted interventions, especially in Colaba, to enhance resilience using culturally sensitive approaches

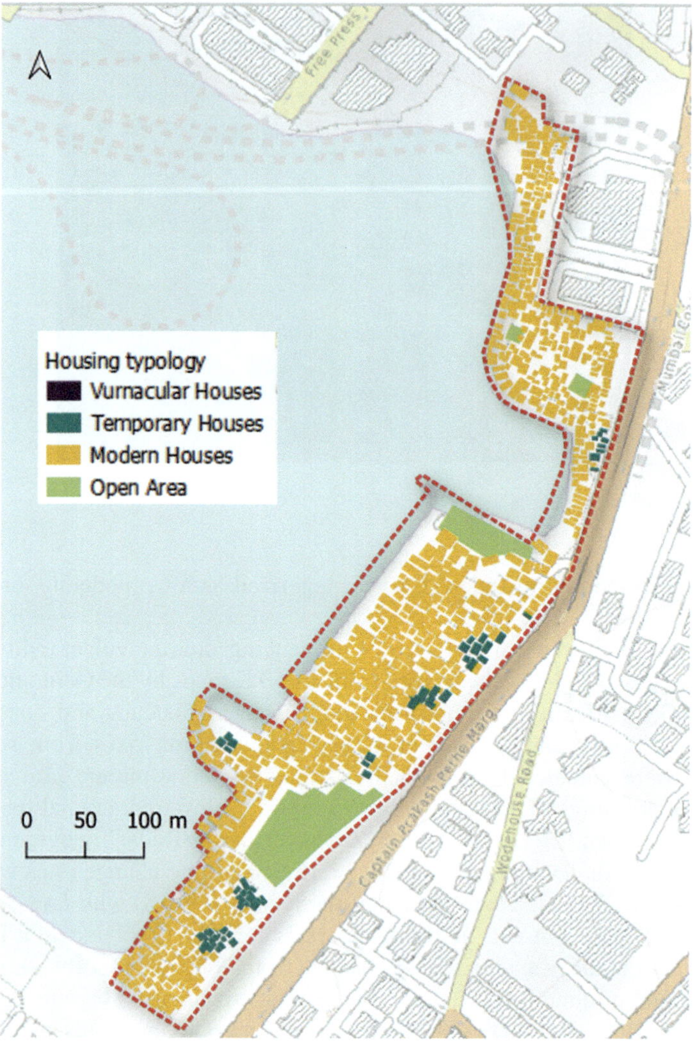

Fig. 8.13 Housing typology map, Colaba (*Source* Survey by author)

Fig. 8.14 Alley, Colaba

Table 8.3 Data sources

Type	Parameter	Source
Geological	1. Geomorphology	Bhukosh, GSI
	2. Slope	NASA USGS
	3. Elevation	NASA USGS
	4. Shoreline change	Landsat 8 (DSAS, ArcGIS)
Physical	5. Tide tange	WX Tide 32
	6. Sea-level-change	GLOSS

Fig. 8.15 Courtyard in Colaba

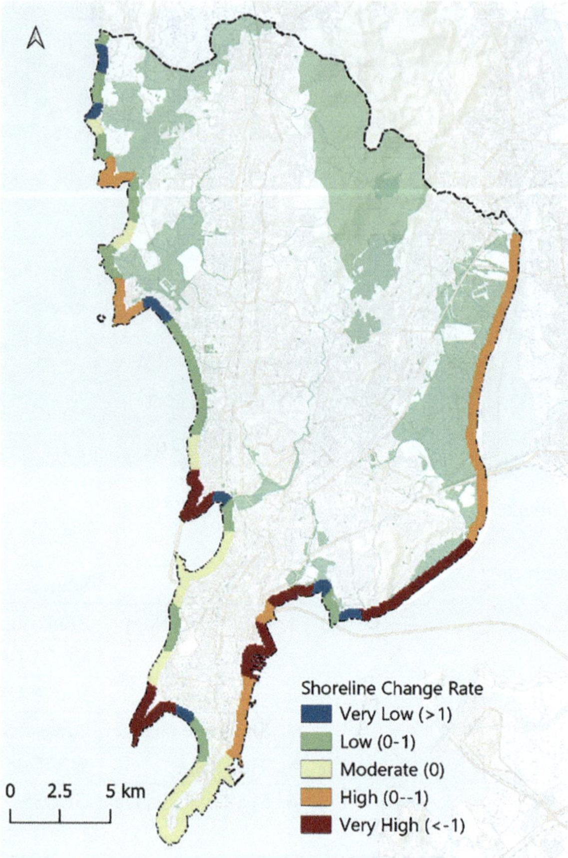

Fig. 8.16 Shoreline change map

Fig. 8.17 Elevation map

Fig. 8.18 Slope map

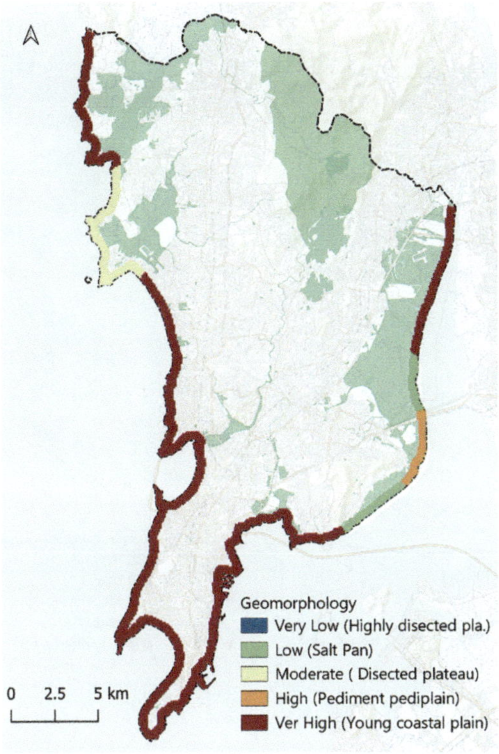

Fig. 8.19 Geomorphology map

and to implement focused mitigation and adaptation measures in high-risk zones to minimize potential impacts (Figs. 8.22 and 8.23).

8.5.2 Socio-Cultural Resilience Index

The socio-cultural resilience index (SCR Index) aims to comprehensively assess the resilience of a community across four dimensions as discussed in Fig. 8.2. It encompasses 15 indicators and 31 variables as can be seen in Fig. 8.24. The methodology involves normalizing actual indicator scores, assigning equal weights to all indicators and dimensions, and aggregating scores to derive an overall SCR Index, scaled out of 100.

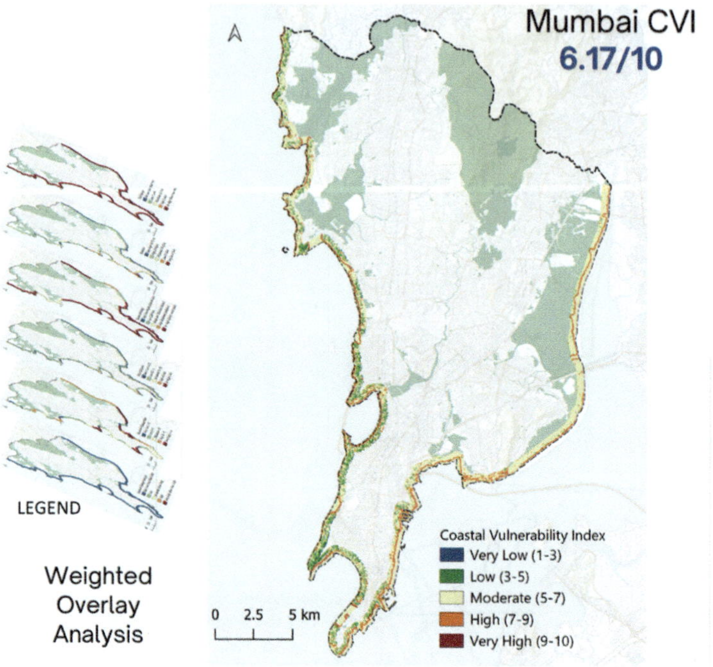

Fig. 8.20 Mumbai coastal vulnerability index

The process of normalization of actual indicator scores involves transforming the scores of each indicator onto a common scale to enable easier comparison and aggregation. Subsequently, the dimension index is calculated by summing the normalized scores of indicators within each dimension, thereby representing the resilience level within that dimension. Further, respective weights are assigned to the four dimensions

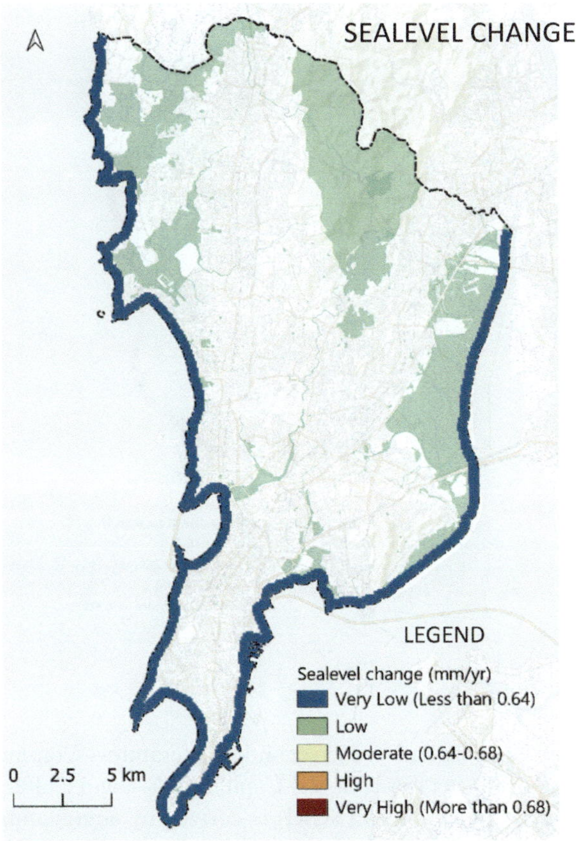

Fig. 8.21 Sealevel change map

Fig. 8.22 Tidal range map

on the basis of expert consultation and secondary literature—Community Infrastructure (35%), Sense of Place (15%), Social Capital (30%), and Culture (20%). Lastly, the socio-cultural resilience index (SCRI) is derived by aggregating the dimension indexes with equal weights allotted to each dimension. To standardize scores across diverse contexts, the maximum possible SCR Index value is capped at 100.

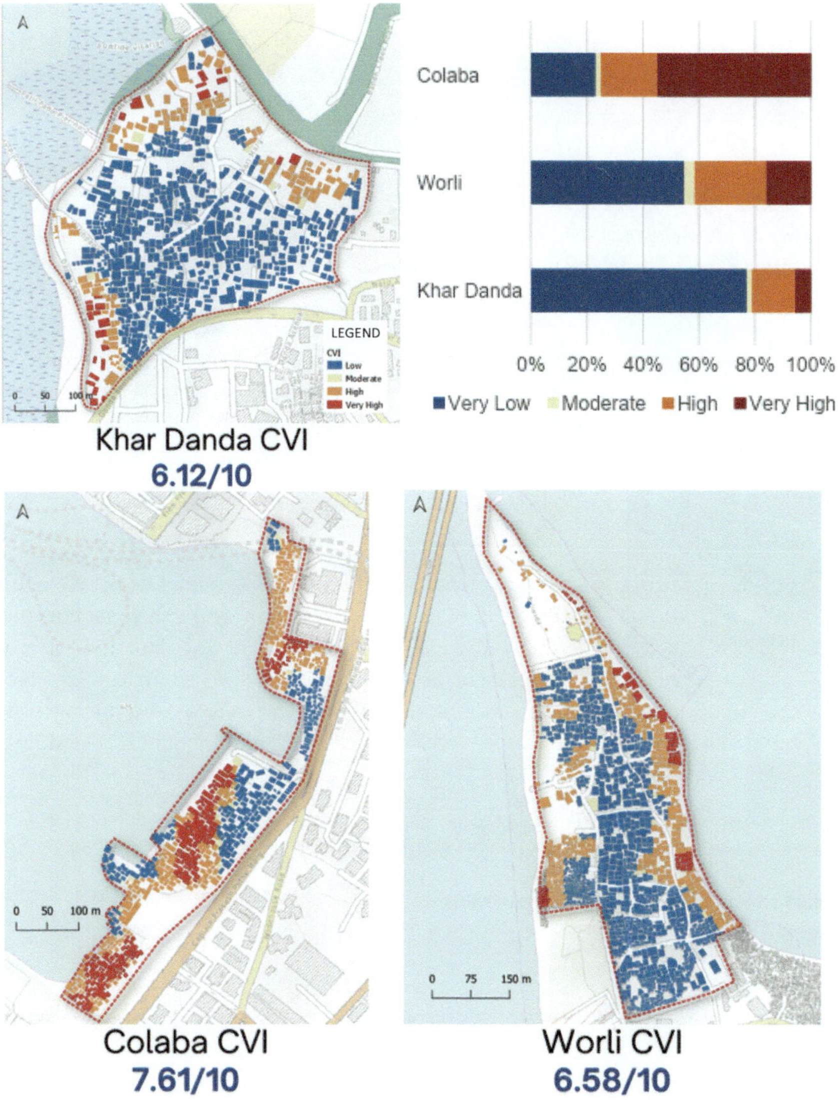

Fig. 8.23 Community wise built-up vulnerability

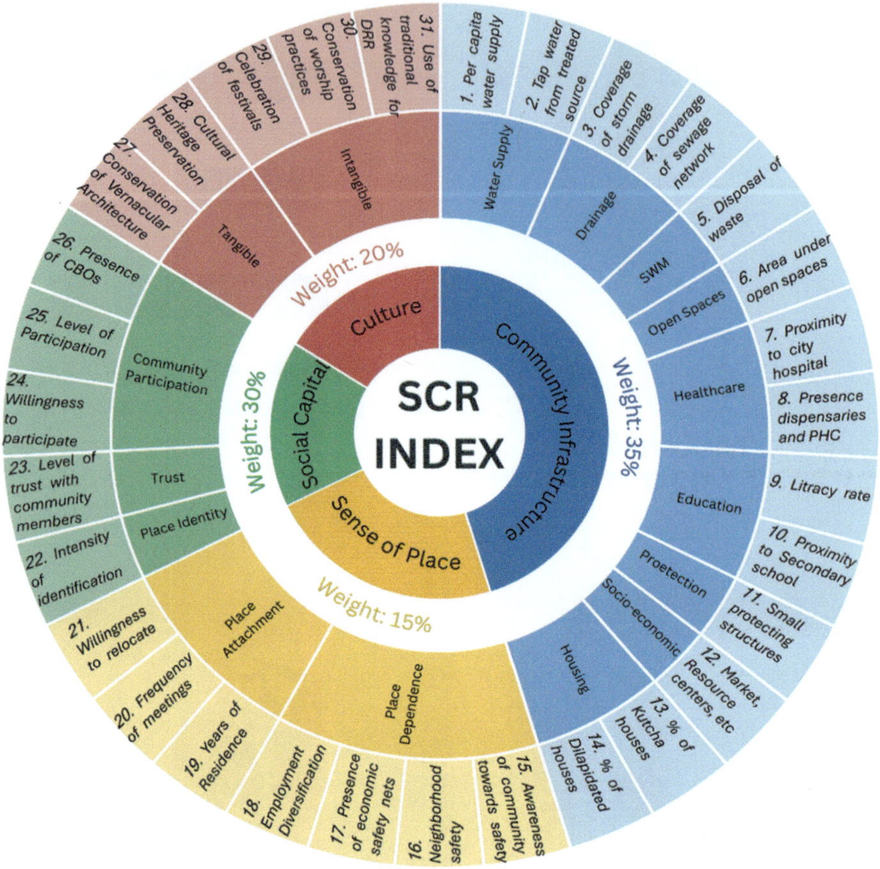

Fig. 8.24 Socio-cultural resilience framework

COMMUNITY INFRASTRUCTURE				
INDICATORS	SUB-INDICATORS	SCORES		
		1/ LOW	3/ MEDIUM	5/ HIGH
1. Water supply	1. Per capita municipal water supply	<60 lpcd	80-100 lpcd	>135lpcd
	2. Tap water from treated source	<30%	30-60%	>60%
2. Drainage system	3. Coverage of storm water drainage system	<20%	40-60%	80-100%
	4. Coverage of sewage network	<20%	40-60%	80-100%
3. SWM	5. Disposal of waste	>80% open dumping	40-60%	<20% open dumping
4. Open spaces	6. Area under parks, playgrounds, open, & forest area	<0-1 sqm/ person	3-5 sqm / person	>8 sqm/ person
5. Healthcare	7. Proximity to city level government hospital	Beyond 5km	Beyond 3km	Within 1km
	8. Presence of dispensaries and PHC	< 1/10000	1/7500 people	1/ 5000 people
6. Education	9. Litracy rate	less than % share at city level	equal to % share at city level	more than % share at city level
	10. Proximity to Secondary school	Beyond 5km	Beyond 3km	Within 1km
7. Protective	11. Small-scale and low-cost protecting structures	Absent	Present but poorly maintained	Well maintained structures
8. Socio-economic	12. Prsesnce of market grounds, resource centers, religious centers	< 1/10000	1/7500 people	1/ 5000 people
9. Housing Condition	13. Percentage of Kutcha houses	>80%	40-60%	<20%
	14. Percentage of Dilapidated houses	>80%	40-60%	<20%

SENSE OF PLACE				
INDICATORS	SUB-INDICATORS	SCORES		
		1/ LOW	3/ MEDIUM	5/ HIGH
9. Place Dependence	15. Awareness of community towards safety	No knowledge of safety apps, kits, shelters	Moderate knowledge	Expert level knowledge
	16. Neighbourhood safety	High crime rates, frequent incidents	Generally safe with some concerns	Extremely low crime rates
	17. Presence of economic safety nets	No economic safety nets in place	Limited social safety nets with gaps	Strong community-based support systems
	18. Employment Diversification	Absence of alternative employment opportunities	Limited presence of SMEs or entrepreneurial ventures	Diverse industries and economic sectors
10. Place attachment	19. Years of Residence	< 1yr	5-25 yrs	Generations
	20. Frequency of meetings	Rarely	Weekly	Every day
	21. Willingness to relocate	Yes	Maybe	No

SENSE OF PLACE				
INDICATORS	SUB-INDICATORS	SCORES		
		1/ LOW	3/ MEDIUM	5/ HIGH
11. Place identity	22. Intensity of identification	Possess knowledge and understanding of culture and heritage, unique symbols	Emotional expression of culture and heritage, unique symbols and language	Active transmission of culture and heritage, unique symbols and language
12. Trust	23. Level of trust with community members	Unwillingness to share information	Moderate information and resource sharing	Unrestricted information sharing
13. Community Participation	24. Willingness to participate	Dedicates minimal time	Regularly contributes a moderate time	Consistently volunteers significant time
	25. Level of Participation	Lack of clear goals, and minimal impact on decision-making	Moderate participation, diverse representation, and some influence	Inclusive participation, empowered community, collaborative decision-making
	26. Presence of CBOs	Very few CBOs only one type of CBO like a religious organization	Moderate number of CBOs with some diversity in focus areas	Very high number of CBOs with excellent diversity

CULTURE				
INDICATORS	SUB-INDICATORS	SCORES		
		1/ LOW	3/ MEDIUM	5/ HIGH
14. Tangible	27. Conservation of Vernacular Architecture	Original materials and techniques largely lost, replaced by modern, incompatible materials	Significant portions of original materials and design remain, with some modifications	Exceptional preservation of original materials, building methods
	28. Cultural Heritage Preservation	Absence of preservation measures, minimal community involvement, severe neglect, and limited capacity	Adequate preservation measures, established management plans, community awareness	Extensive preservation efforts, active community engagement, & well-funded conservation projects
15. Intangible	29. Celebration of festivals	Minimal participation in activities associated with the festival, often celebrated individually	Moderate participation in a variety of activities, possibly involving extended family	Actively leads or initiates activities within the household or community,
	30. Conservation of indigenous worship practices	Indigenous worship practices are rarely or never observed.	Indigenous worship practices are observed occasionally or inconsistently.	Indigenous worship practices are deeply integrated and are observed with regularity
	31. Use of traditional knowledge for DRR	Complete disregard, absence of engagement	Significant efforts to incorporate traditional knowledge	Extensive integration of traditional knowledge

Source: Literature Study by Author

Community Infrastructure

An analysis of community infrastructure reveals significant disparities among various localities. Khar has a regular municipal water supply at 120 L per capita per day, while Worli has the highest water supply coverage at 78%. Conversely, Colaba struggles, with 60% of households lacking trunk infrastructure and the highest rate of open dumping at 51%. Community-led groups in Worli, under SMPA, improve local infrastructure by establishing embankments, raised platforms, and community shelters. Similarly, initiatives in Khar Danda and Worli address specific community needs. However, Colaba faces housing issues, with 45% of homes dilapidated and 23% classified as kutcha houses, leading to a lower infrastructure score. The analysis highlights the need for targeted interventions to improve water supply, waste management, and structural resilience, helping localities like Colaba enhance their infrastructure and promote equitable development (Figs. 8.25, 8.26, 8.27 and 8.28).

Sense of Place

In Worli, three cooperative banks support local financial needs, and 67% of families have deep, multi-generational roots. Khar Danda boasts a diversified economic base, with 53% of residents having supplementary incomes and 13% showing a high willingness to relocate if necessary. Worli's strong tradition of community engagement includes regular meetings of society trustees and members. In contrast, other local societies meet annually at the temple. Worli Koliwada excels in community cohesion and safety, with frequent meetings and gatherings, while Colaba has the lowest score in this domain, highlighting a weaker sense of place and community identity. Addressing these disparities can strengthen community bonds and enhance overall well-being across localities.

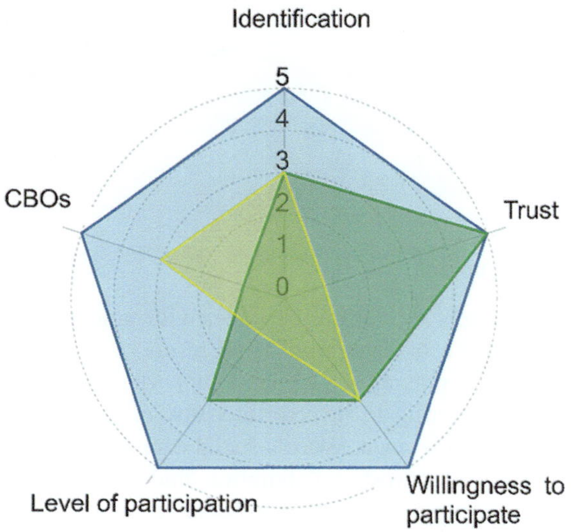

Fig. 8.25 Social capital scores

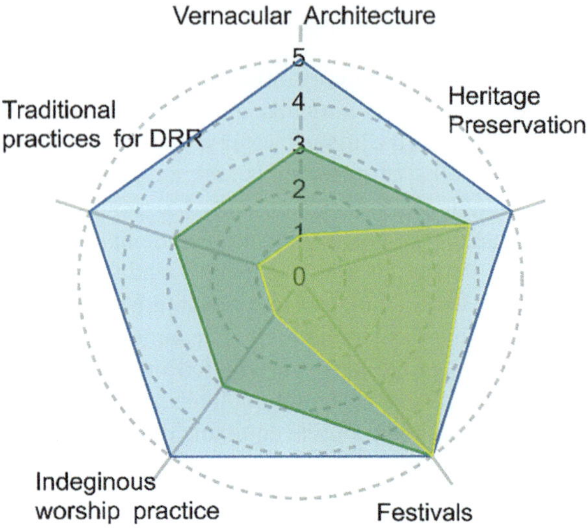

Fig. 8.26 Cultural scores

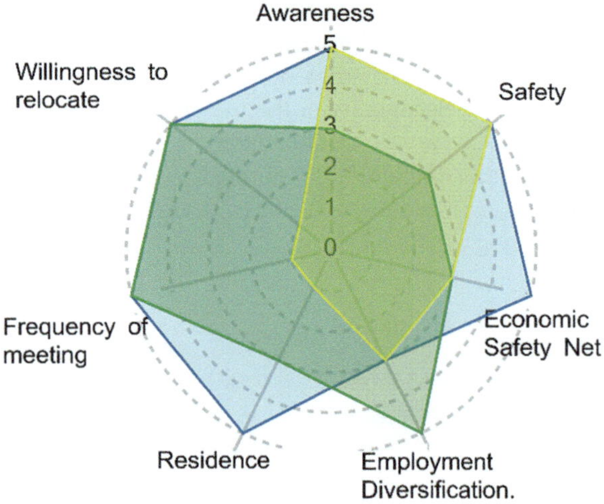

Fig. 8.27 Sense of place scores

Social Capital

Worli excels in cultural heritage, with unique symbols and language fostering a strong sense of belonging. High levels of community participation and influence in decision-making are evident, supported by five community-based organizations (CBOs) catering to diverse interests, including religious groups, fishermen, boat owners, and women. Worli's social capital is robust, scoring a perfect 30 out of 30, reflecting high trust, collaboration, and active community engagement. In contrast, Colaba struggles with lower social capital, highlighting the need to strengthen community bonds and cohesion to enhance overall well-being and resilience.

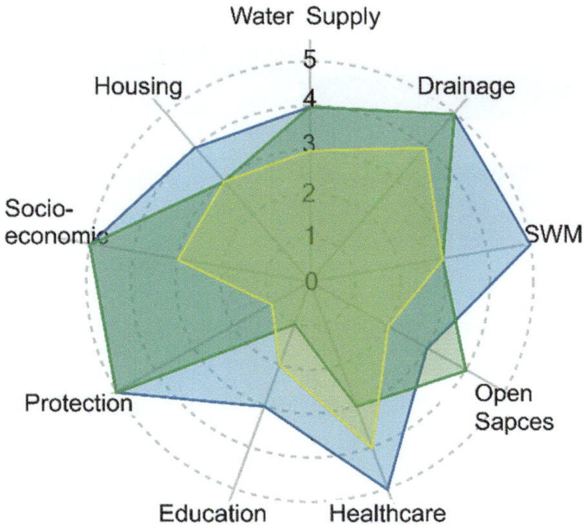

Fig. 8.28 Community infrastructure

Culture

The RPG Foundation's (an NGO working toward sustainable communities) Heritage Project preserves the region's rich culture through heritage walks with food stops and sustainable tourism, while also safeguarding traditional fishing techniques. Worli and Khar Danda host Kul Devi (family deity or clan goddess) temples, central to religious and cultural identity, fostering community cohesion. Traditional festivals like Narli Poornima, Holi, and Gudi Parwa are celebrated across Koliwadas, uniting communities in cultural festivities. In Worli, 30% of fishing is done by artisans using traditional methods, preserving local fishing heritage and sustainability. Khar Danda's mangrove plantations reflect its commitment to environmental conservation. Worli, scoring a perfect 20 out of 20, showcases cultural vibrancy and a strong dedication to preserving both tangible and intangible heritage elements.

In analyzing the socio-cultural factors influencing community disaster risk reduction in Indigenous Koliwadas of Mumbai, it becomes evident that these communities face similar challenges related to natural disasters and environmental issues. However, they also exhibit robust social cohesion, reliance on traditional industries like fishing, and a deep-rooted cultural identity that contributes significantly to their resilience in the face of adversity (Table 8.4). Community wise dimension scores

Infrastructure disparities were noted, particularly in Colaba, contributing to heightened vulnerability. Urgent attention is required to address these deficiencies and enhance disaster resilience. Worli stands out for its well-established physical and social structures, including heritage sites, which foster community pride and commitment to cultural preservation. Strong community-based organizations (CBOs) and proactive NGO initiatives further contribute to disaster resilience in Worli, serving as

Table 8.4 Community wise dimension scores

INDICATOR	KHAR	WORLI	COLABA
Community Infrastructure	25.00	29.00	19.50
Sense of Place	11.57	14.14	8.14
Social Capital	18.00	30.00	13.20
Culture	14.40	20.00	9.60
SCRI	**68.97**	**93.14**	**50.44**

(*Source* Calculated from household survey for each indicator)

platforms for social interaction, mutual support, information sharing, and the preservation of intangible culture. Overall SCRI was highest for Worli with 93.14/100, followed by Khar Danda and lowest for Colaba 50.44.

The findings reveal a moderate level of vulnerability across the entire Mumbai coastline, with specific Koliwadas exhibiting varying degrees of susceptibility. Worli Koliwada emerges as the most vulnerable, followed by Colaba and then Khar Danda. This underscores the necessity of targeted interventions, particularly in Colaba, to enhance resilience with culturally sensitive approaches. Spatial analysis highlights the concentration of high-risk buildings along the coastal periphery, indicating the need for focused mitigation and adaptation measures in these vulnerable zones. Additionally, the socio-cultural factors influencing disaster risk reduction in Indigenous Koliwadas showcase their resilience amidst challenges, with robust social cohesion and reliance on traditional industries like fishing. Infrastructure disparities, especially in Colaba, contribute to heightened vulnerability, emphasizing the urgency of addressing these deficiencies. Notably, Worli stands out for its established physical and social structures, supported by strong community-based organizations and proactive NGO initiatives, contributing to disaster resilience. Integrating socio-cultural factors into disaster risk reduction strategies is crucial, requiring collaborative efforts and culturally sensitive interventions to ensure the sustainability of these coastal communities in the face of increasing environmental challenges. The findings emphasize the importance of integrating socio-cultural factors into disaster risk reduction strategies, acknowledging the unique strengths and vulnerabilities of Indigenous Koliwada communities. Culturally sensitive interventions, community engagement, and collaborative efforts between stakeholders are essential for building resilience and ensuring the sustainability of these coastal communities in the face of increasing environmental challenges.

8.6 Conclusion

In conclusion, this thesis has highlighted the critical importance of socio-cultural factors in disaster risk reduction (DRR) strategies, particularly in coastal and low-lying fishing communities like the indigenous Koliwadas of Mumbai. While much of the existing research on DRR has focused on physical aspects, such as infrastructure and environmental vulnerabilities, this study underscores the significance of cultural values, social cohesion, and community engagement in resilience building.

Climate change vulnerabilities encompass not only physical exposure but also social, economic, and cultural dimensions. Neglecting these cultural components in DRR initiatives could lead to ineffective solutions that fail to address the unique needs and strengths of local communities. By exploring intangible cultural factors and their impact on disaster resilience, this study aims to bridge this gap in the literature and provide valuable insights for more contextually relevant and effective DRR strategies.

The study initiates with an in-depth exploration of socio-cultural factors and disaster resilience within indigenous communities and contributed to the DRR knowledge in three ways. Firstly, through a thorough secondary literature review, socio-cultural elements influencing indigenous communities' resilience to disasters are identified. Secondly, utilizing geospatial techniques and remote sensing, coastal vulnerability mapping is conducted, defining criteria and scales for precise delineation. Thirdly, a mixed research methodology, incorporating qualitative ethnographic studies and quantitative case-based assessments, comprehensively explored the research domain.

This study underscores the importance of integrating socio-cultural factors into DRR strategies to enhance the resilience of coastal communities like the indigenous Koliwadas of Mumbai. Culturally sensitive interventions, community engagement, and collaborative efforts between stakeholders are essential for building resilience and ensuring the sustainability of these communities in the face of increasing environmental challenges.

In conclusion, the study contributes to a more comprehensive understanding of the complex interplay between cultural values, environmental systems, and resilience in the face of disasters. By recognizing and harnessing the strengths of socio-cultural factors, we can develop more effective and sustainable DRR strategies that empower communities to withstand and recover from adversity.

8.7 Recommendation

Future research can expand on the findings of this thesis by refining and validating the socio-cultural resilience index framework across diverse cultural settings. Longitudinal studies could track the dynamics of community resilience over time, offering insights into how socio-economic and environmental factors influence adaptation.

Engaging communities through participatory action research (PAR) will co-create knowledge and build tailored disaster risk reduction (DRR) strategies. Impact evaluations of socio-cultural interventions can assess their effectiveness, while studies on indigenous knowledge integration could enrich formal DRR frameworks. Lastly, interdisciplinary research exploring cross-sectoral collaboration and policy analysis could inform governance mechanisms that strengthen resilience in coastal communities worldwide.

References

Bourke SC, Chapman J, Jones R, Brinckley M-M, Thurber KA, Calabria B, Doery K, Olsen A, Lovett R (2022) Developing aboriginal and Torres strait islander cultural indicators: an overview from Mayi KUWAYU, the national study of Aboriginal and Torres strait islander wellbeing. Int J Equity Health 21(1):109. https://doi.org/10.1186/s12939-022-01710-8

Dube E, Munsaka E (2018) The contribution of indigenous knowledge to disaster risk reduction activities in Zimbabwe: a big call to practitioners. Jàmbá: J Disaster Risk Stud 10(1):493. https://doi.org/10.4102/jamba.v10i1.493

Field CB, Barros V, Stocker TF, Dahe Q (eds) (2012a) Managing the risks of extreme events and disasters to advance climate change adaptation: special report of the IPCC. Cambridge University Press. https://doi.org/10.1017/CBO9781139177245

Field CB, Barros V, Stocker TF, Dahe Q (eds) (2012b) Managing the risks of extreme events and disasters to advance climate change adaptation: special report of the intergovernmental panel on climate change (1st edn). Cambridge University Press. https://doi.org/10.1017/CBO9781139177245

Gaber ARM, El-Kader MHA, Okba EM (2019) Assessment of urban resilience through the development of an urban resilience assessment model (resilience performance index) 12(12)

Gaillard JC (2006) Traditional societies in the face of natural hazards: the 1991 mt. Pinatubo eruption and the Aetas of the Philippines. Int J Mass Emerg Disasters 24(1):5–43. https://doi.org/10.1177/028072700602400101

Grootaert C, van Bastelaer T (n.d.) Understanding and measuring social capital

Hu H, Qiao X, Yang Y, Zhang L (2021) Developing a resilience evaluation index for cultural heritage site: case study of Jiangwan town in China. Asia Pac J Tourism Res 26(1):15–29. https://doi.org/10.1080/10941665.2020.1805476

Intergovernmental Panel on Climate Change (IPCC) (2012) Managing the risks of extreme events and disasters to advance climate change adaptation. In: Field CB, Barros V, Stocker TF, Qin D, Dokken DJ, Ebi KL, Mastrandrea MD, Mach KJ, Plattner G-K, Allen SK, Tignor M, Midgley PM (eds) Cambridge University Press

International Science Council (2023) Report for the mid-term review of the Sendai framework for disaster risk reduction. International Science Council. https://doi.org/10.24948/2023.01

Kasdan DO (2016) Considering socio-cultural factors of disaster risk management. Disaster Prev Manag 25(4):464–477. https://doi.org/10.1108/DPM-03-2016-0055

Khoja L, Schubert R, Joerin J (n.d.) Literature review FRS working paper 2

Kulatunga U (2010) Impact of culture towards disaster risk reduction. Int J Strateg Prop Manag 14(4):304–313. https://doi.org/10.3846/ijspm.2010.23

Li Z, Tan X (2019) Disaster-recovery social capital and community participation in earthquake-stricken Ya'an areas. Sustainability 11(4):993. https://doi.org/10.3390/su11040993

Marasco S, Kammouh O, Cimellaro GP (2022) Disaster resilience quantification of communities: a risk-based approach. Int J Disaster Risk Reduct 70:102778. https://doi.org/10.1016/j.ijdrr.2021.102778

Mohanty A, Wadhawan S (2021) Mapping India's climate vulnerability – A district level assessment. Council on Energy, Environment and Water

Movik S, Adam HN, Alankar A (2023) Claiming space: contested coastal commons in Mumbai. Geoforum 144:103805. https://doi.org/10.1016/j.geoforum.2023.103805

Nakamura N, Kanemasu Y (2020) Traditional knowledge, social capital, and community response to a disaster: resilience of remote communities in Fiji after a severe climatic event. Reg Environ Change 20(1):23. https://doi.org/10.1007/s10113-020-01613-w

Ostadtaghizadeh A, Ardalan A, Paton D, Jabbari H, Khankeh HR (2015) Community disaster resilience: a systematic review on assessment models and tools. PLoS Curr 7

Paton D, Johnston D (2016) Building community resilience to natural hazards: a systematic review of the literature. Nat Hazards 81(1):1–21

Sansare DA, Mhaske SY (2020) Natural hazard assessment and mapping using remote sensing and QGIS tools for Mumbai city, India. Nat Hazards 100(3):1117–1136. https://doi.org/10.1007/s11069-019-03852-5

Senapati S, Gupta V (2014) Climate change, urbanization and livelihood perspective of indigenous fishing communities of Mumbai, India. Present Environ Sustain Dev 8(2):117–129. https://doi.org/10.2478/pesd-2014-0029

Tammar A, Abosuliman SS, Rahaman KR (2020) Social capital and disaster resilience nexus: a study of flash flood recovery in Jeddah City. Sustainability 12(11):4668. https://doi.org/10.3390/su12114668

Usman M, Mario M, Tamrin S, Amandaria R (2022) Socio-cultural intervention as disaster mitigation in Indonesia. JED (Jurnal Etika Demokrasi) 7(3):534–542. https://doi.org/10.26618/jed.v7i3.8179

Wu J-Y, Chen L-C (2023) Traditional indigenous ecological knowledge to enhance community-based disaster resilience: Taiwan mountain area. Nat Hazard Rev 24(1):05022014. https://doi.org/10.1061/NHREFO.NHENG-1673

Chapter 9
Exploring the Risk and Socioeconomic Vulnerability of Riverbank Erosion Hazard Along the Upper Tapi River, India

Devika Chandran, Resmi S. R., Bharath Raghavan M., Jagath S. Ashok, and Kalidas A. K.

Abstract This chapter introduces a thorough geospatial methodology for evaluating riverine erosion hazards and vulnerability in the Upper Tapi River basin. This research combines satellite image analysis, geospatial methods, and socioeconomic indicators to create indices for erosion hazards and vulnerability, establishing a risk assessment framework spanning two distinct periods from 1978 to 2020. Satellite imagery and toposheet maps were used to develop the erosion hazard index, utilizing metrics such as net areal change and percentage of eroded length. The vulnerability assessment incorporated three indices: Exposure, susceptibility, and resilience. Geological, geomorphological, and soil characteristics, along with terrain features like slope, relief, and roughness, were included in the exposure index. Demographic factors, such as population growth, settlement density, and agricultural land use intensity, were considered in the susceptibility index. The resilience index encompassed socioeconomic indicators, including literacy rate, access to banking and medical facilities, and forest cover. These three indices were combined to calculate a composite vulnerability index, providing a comprehensive evaluation of the area's vulnerability to erosion-induced hazards. Furthermore, this chapter proposes a risk assessment framework that merges the erosion hazard index with the composite vulnerability index. This approach enables the identification of high-risk areas susceptible to riverine erosion, facilitating targeted disaster risk reduction strategies along the Upper Tapi River. Given the Upper Tapi River's history of frequent flooding, riverbank erosion is considered one of the most hazardous disasters in the study area. The development and implementation of specific policies in these vulnerable regions may help support affected communities.

Keywords Upper Tapi river · Erosion hazard · Vulnerability · Exposure · Susceptibility · Resilience · Risk

---

D. Chandran · Resmi S. R. (✉) · Bharath Raghavan M. · J. S. Ashok · Kalidas A. K.
National Institute of Technology Calicut, Kozhikode, Kerala, India
e-mail: resmi@nitc.ac.in

L. K. Dashora et al. (eds.), *Urban Sustainability*, Disaster Risk Reduction,
https://doi.org/10.1007/978-981-96-4890-0_9

9.1 Introduction

The course of rivers is in a constant state of flux (Whipple and Tucker 2002). The river's dynamic nature is evident through various processes, including channel migration, channel splitting, bank ruptures, formation of point bars, sediment redistribution, erosion, deposition, and the development of branching channels (Sarif et al. 2024; Sah and Das 2018; Das et al. 2019). River bank collapse and erosion occur when the force of water flow exceeds the strength of the bank material (Bandyopadhyay and De 2015). This process is driven by gravitational pull and the water's erosive capabilities. The erosion of riverbanks can be attributed to natural causes or human activities. While natural erosion can yield beneficial outcomes, such as the formation of fertile floodplains and alluvial terraces, excessive erosion in unstable river systems and beyond normal levels on either side of the river is a matter of significant concern (Mollah and Ferdaush 2015).

The formation of river channels and floodplains is significantly influenced by riverbank erosion (Hooke 1979). Riverbanks are ever-changing environments that respond to various external factors. The gradual removal of soil and sediment from riverbanks due to natural forces such as water flow, wind, and geological processes is known as bank erosion. This recurring geomorphic hazard is associated with floodplains and river systems that meander or braid (Biswas and Anwaruzzaman 2019). Lateral migration, which is the movement of a river channel caused by changes in fluid flow and sediment discharge, is always accompanied by erosion of the stream bed or channel wall due to turbulent flow conditions (Siddik et al. 2017; Yang et al. 1999). This process can lead to significant local or regional changes (Collinson and Lewin 1983). Riverbank erosion causes the displacement of riverine households, resulting in the loss of land, homes, ponds, crops, trees, and other valuable resources (Majumdar et al. 2023; Mondal and Tripathy 2020; Sharma 2013; Hooke 1979). Although river bank erosion results in fewer casualties compared to other natural calamities, its economic and social consequences are remarkably severe (Sarif et al. 2024; Hasan et al. 2018). A holistic approach can be used to incorporate various elements contributing to bank erosion, such as hydrological data, geomorphological properties, and land use patterns. By integrating these elements, a zonation map can be created to categorize riverbanks into different vulnerability levels. This method helps identify high-risk areas that require immediate attention and management (Bandyopadhyay et al. 2014). The socioeconomic impacts of riverbank erosion on communities can be evaluated using a combination of field surveys, geographic information system (GIS) tools, and socioeconomic data (Rahman and Gain 2020; Singha et al. 2020; Biswas and Anwaruzzaman 2019; Fekete 2019). This approach enables the assessment of the extent of vulnerability experienced by local populations. By employing these methods, key factors contributing to vulnerability, such as the loss of agricultural land, displacement, and reduced income opportunities, can be identified (Debnath et al. 2016). Certain areas are susceptible to erosion due to a combination of hydrological, geological, and anthropogenic factors. These vulnerable regions face considerable risks to their infrastructure, livelihoods, and shelters (Hasan et al. 2018).

This chapter employs a comprehensive, integrated approach to evaluate the erosional hazard vulnerability and risk in the Upper Tapi River region, utilizing a socio-spatial methodology. The methodological framework applied in this study views vulnerability as a composite of exposure, sensitivity, and capacity factors (Sarif et al. 2024; Jamshed et al. 2020; Ghosh 2017). In contrast to other natural hazards such as earthquakes, landslides, floods, coastal threats, and climate change vulnerability, there is a scarcity of comprehensive socio-spatial approaches for evaluating vulnerability to river bank erosion. The majority of research on river bank erosion has concentrated on the physical alterations, morphometric changes, etc. (Dabojani et al. 2014; Mandal 2017; Hasanuzzaman et al. 2021). This research aims to shed light on the extent and intensity of erosion risk by analyzing long-term spatial and temporal changes in the riverbanks of the Upper Tapi River in India. The analysis utilizes topographical maps from 1978 and high-resolution satellite imagery from 2020. To assess the overall risk, the study considers the region's specific hydrogeomorphic exposure, socioeconomic vulnerability, and structural and non-structural adaptive capacity.

9.2 Study Area

The study area comprises the Upper Tapi sub-basin, which drains into Hatnur Reservoir in Maharashtra, India (Fig. 9.1). Situated between 21° 00′ 00″ N–22° 00′ 00″ N latitude and 75° 45′ 00″ E–78° 30′ 00″ E longitude, this sub-basin covers a 10,193 sq.km. drainage area, excluding the Purna sub-basin. The elevation descends from 752 m near the river's source at Multai to approximately 207 m at the Hatnur Reservoir outlet. The sub-basin is bordered by hill ranges: the Gwaligarh hills in the south and the Satpura hills in the north, with the Khandesh Plains to the west. The region experiences a climate ranging from subtropical to temperate, with the majority of rainfall occurring between June and September. From its origin to Hatnur Reservoir, the Upper Tapi River flows for about 350 km through varied landscapes including alluvial plains, mountainous areas, and deep ravines. Deccan trap lava flows cover the basin in Madhya Pradesh's Betul and Burhanpur Districts, while valleys and cuts dominate the Upper Tapi River's course in Maharashtra's Amravati District (Resmi et al. 2019). The study area is predominantly composed of clayey and loamy soils, which are susceptible to mild to severe erosion.

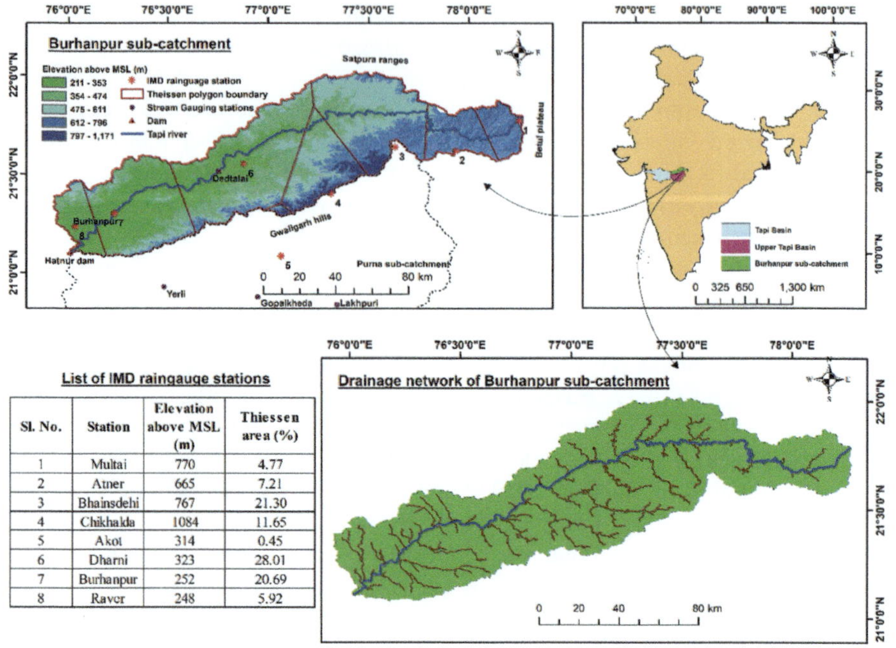

Sl. No.	Station	Elevation above MSL (m)	Thiessen area (%)
1	Multai	770	4.77
2	Atner	665	7.21
3	Bhainsdehi	767	21.30
4	Chikhalda	1084	11.65
5	Akot	314	0.45
6	Dharni	323	28.01
7	Burhanpur	252	20.69
8	Raver	248	5.92

Fig. 9.1 Index map of Upper Tapi river basin. (The figure is generated in ArcGIS 10.8©; Details of primary data source are enlisted in Table 9.1)

9.3 Materials and Method

Details of the data used in this study are presented in Table 9.1. The methodology involved in analyzing erosion hazards, vulnerability, and risk is explained in the following sections.

9.3.1 Identification of Erosion Hazard Areas

A hazard can be defined as the frequency and severity of any extreme event leading to a disaster. Mapping erosion hazards is crucial for comprehending riverine erosion dynamics and potential landscape impacts. By comparing 2020 satellite imagery with 1978 toposheets, changes in eroded areas along the Upper Tapi River banks were measured, enabling the identification of landscape alterations over time. Systematic analysis of bankline shifts and net areal changes allowed for the delineation of areas affected by erosion and deposition. The riverbanks were delineated between 1978 and 2020. The left and right bank lines were manually digitized for the Upper Tapi River in these years using the ArcGIS 10.8©. Toposheets downloaded from the Survey of India's official Web site were used to delineate 1978 banklines. Sentinel-2 satellite

Table 9.1 Details of data collected

Data used	Description	Source
Topographical data toposheets	Scale: 1:50,000 Survey year: 1977–79	Open series map (OSM), Survey of India (SoI)
Digital elevation model	SRTM DEM	USGS earth explorer
Satellite images	Sentinel 2 (2020)—10 m Resolution	USGS earth explorer
Geology geomorphology	Scale: 1:20,00,000 Scale: 1:2,50,000	Geological survey of India
Information from the census of India	Population growth, literacy rate, settlement density, percentage of agricultural land, banking, and medical facility, percentage of forest cover	Census of India 2011
Village boundary	Shapefile (Maharashtra and Madhya Pradesh)	NASA open data portal

images were used to digitize the bank lines for the year 2020. A uniform projection system, UTM 43N was provided for all maps and images. The scale, resolution, and existing features visible in the imagery and toposheet data were carefully considered to ensure the accuracy of the digitization process. These vector-digitized layers of the 1978 and 2020 banklines underwent overlay analysis to obtain the eroded and deposited areas and compute the net areal change. The village boundaries from the NASA open data portal were downloaded, and a buffer area of 10 km from both sides of the Tapi River was created and clipped with the village boundaries (Table 9.2).

Consequently, an erosion hazard assessment involves determining the extent and intensity of erosion in a specific region. This assessment was conducted by calculating the comprehensive erosion hazard index (*CEHI*), which considered two parameters: (a) net areal change (the difference between eroded and deposited areas between 1978 and 2020, with positive values indicating erosion and negative values indicating deposition on riverbanks) and (b) the percentage of eroded riverbanks (the ratio

Table 9.2 Erosion hazard parameters (*Source* Ghosh 2017)

Index	Factors	Parameters	Division standard	Classes
Hazard	Erosion intensity	Net areal change (the gap between the eroded and deposited area between two successive times)	<0	1
			0–0.1 sq. km	2
			>0.1 sq.km	3
		Percentage of eroded river banks (length of eroded stretch to total length of river bank)	0–0.5%	1
			0.5–15%	2
			>1.5%	3

of eroded stretch length to total riverbank length, focusing on the spatial extent of erosion rather than its rate or amount within a given area). All villages were classified into three classes: low (1)-, moderate (2)-, and high (3)—hazard classes for the net areal change and the percentage of eroded riverbanks (Hamid et al. 2017; Laha and Bandyapadhyay 2013; Ghosh and Mukhopadhyay 2016; Bandyopadhyay 1997; Thakur et al. 2012). *CEHI* was calculated by assigning equal weights to the parameters as:

$$CEHI = function\,of\left(\frac{Net\,Areal\,Change + Percentage\,Eroded\,Length}{2}\right) \quad (9.1)$$

9.3.2 Vulnerability Assessment

Vulnerability refers to the anticipated level of damage under specific conditions of exposure, susceptibility, and resilience. It involves identifying areas at risk due to hazardous factors that could potentially result in the loss of lives, property, and other resources. In the coming decades, rivers will become increasingly vulnerable due to human activities like mining and deforestation, as well as ecosystem degradation and climate change. To develop strategies for adaptation and resilience in the face of potentially catastrophic events, vulnerability assessments are necessary. The intensity and severity of hazardous events, such as riverine erosion, vary across different geographical regions. These events often lead to disasters when combined with the vulnerable socioeconomic conditions of the human population (Bandyopadhyay et al. 2014; Bhuiyan et al. 2017; Tha et al. 2024; Debnath et al. 2016; Mollah and Ferdaush 2015).

The susceptibility of a system to be adversely impacted by specific hazards, determined by its position, internal configuration, arrangement, and functions, is known as exposure. Elements of exposure include the lithological makeup, landforms, and hydrological characteristics (Gornitz 1990; Doukakis 2005; Kumar et al. 2010; Feng et al. 2011; Balica et al. 2012; Mani Murali et al. 2013; Barman et al. 2016). The exposure index (*EI*) is composed of six factors that contribute to the risk of biophysical exposure, identified through a thorough review of previous studies. These factors comprise geology, geomorphology, soil composition, terrain roughness, slope, and relief. Each factor is categorized based on its susceptibility to erosion. The area under investigation has a relief range from 26.12 m to 290.64 m, with higher relief values indicating increased vulnerability to riverbank erosion. Terrain roughness is associated with the irregularity of the surface.

$$EI = function\,of\,(Geology + Geomorphology + Soil + Slope + Releif + Roughness\,of\,the\,terrain/6) \quad (9.2)$$

The degree to which a system, population, or environment is vulnerable to adverse effects from exposure to danger is known as susceptibility. This is characterized by components that can potentially trigger a hazard. The susceptibility index (*SI*) was calculated by assigning appropriate rating classes to each parameter and taking their average (Aydın and Uysal 2014; Merlotto et al. 2016). As populations grow, the demand for housing and infrastructure leads to the expansion of urban and rural settlements along riverbanks. This growth often involves clearing vegetation that naturally supports the soil. The loss of trees and plants eliminates root systems that stabilize the soil, increasing the vulnerability of banks to water-induced erosion. This issue is exacerbated by higher concentrations of human habitation, which introduce impervious surfaces like paved roads and buildings, limiting water infiltration into the soil. Consequently, rainfall and surface water runoff enter rivers more rapidly, resulting in increased volume and velocity of water capable of eroding riverbanks. Higher population densities typically lead to increased pollution and waste, potentially including chemicals that degrade soil quality and diminish its erosion resistance. Natural vegetation is often replaced by extensive agricultural land, leading to a higher risk of riverbank erosion during heavy rainfall due to the absence of deep-rooted plants and the disrupted soil structure caused by farming practices involving heavy machinery and irrigation.

$$SI = \mathit{function\ of}\ (\mathit{Population\ growth} + \mathit{settlement\ density} + \mathit{Percentage\ of\ agricultural\ land}/3) \quad (9.3)$$

Adaptive capacity, also referred to as resilience, describes a system's potential to endure and recover from risk-related adverse effects by leveraging its current resources, abilities, and technology. Greater resilience correlates with reduced vulnerability to hazards. The resilience index (*RI*) was calculated using socioeconomic factors, including literacy rates, availability of banking and medical services, and forest cover percentage. The index was determined by assigning rating classes and computing their average. Rural populations with well-equipped healthcare facilities are better prepared to handle health impacts of disasters like floods and landslides, which often accompany riverbank erosion. In these situations, prompt medical attention is vital for treating injuries and preventing disease outbreaks, thus protecting the affected community's health and resilience. Financial institutions contribute to economic stability and enable strategic investments in infrastructure and sustainable practices. Education is crucial for disseminating information about effective land use and management strategies, thereby reducing harmful practices such as deforestation and unsustainable agriculture that exacerbate erosion. Consequently, areas with higher literacy rates often experience more effective community-driven efforts to mitigate riverbank erosion. Areas with a higher concentration of these facilities demonstrated a lower risk of riverbank erosion. Forests stabilize soil and increase its erosion resistance through their extensive root networks. Tree canopies moderate the direct impact of rainfall by decelerating water movement and allowing more time for ground absorption rather than rapid surface runoff. Furthermore, vegetation acts as a physical barrier, trapping sediments and limiting their movement, thus enhancing bank stability.

$$RI = \mathit{function\ of}\ (\mathit{Literacy\ rate} + \mathit{Banking\ facilities} + \mathit{Medical\ facilities} + \mathit{percentage\ of\ forest\ cover}/4) \quad (9.4)$$

Vulnerability in riverine systems is positively influenced by the interplay between exposure and susceptibility, which can have a substantial effect on these ecosystems. Conversely, resilience contributes to mitigating the adverse impacts of hazards and has an inverse relationship with vulnerability. In constructing the indices and assessing risk, all indicators were given equal importance and weighting. Each parameter was evaluated on a qualitative scale ranging from 1 (low) to 3 (high). Table 9.3 outlines the specific rating class criteria used for assessing vulnerability. The composite vulnerability index (*CVI*) is formulated using the following approach:

$$CVI = EI + SI - RI \tag{9.5}$$

9.3.3 Composite Erosion Risk Assessment

Risk denotes the likelihood of experiencing negative consequences, such as injury, loss, or damage, resulting from a particular action, circumstance, or event. It is typically defined as the combination of an event's probability and the magnitude of its impact. Riverine erosion risk pertains to the potential hazards and detrimental effects caused by the erosion of riverbanks and nearby terrain due to flowing river water. This risk is heightened in regions characterized by heavy rainfall, steep river gradients, and friable or unconsolidated soil. Human-induced activities, including deforestation, mining operations, and construction projects, can exacerbate erosion by removing vegetation that helps stabilize the soil (Fuentes et al. 2022; Sarif et al. 2024; Ghosh 2017). A composite risk assessment involves developing a comprehensive risk index that serves as an all-encompassing indicator of the overall erosion risk within the Upper Tapi River basin.

The composite risk index (*CRI*) was derived by integrating *CEHI*, which quantifies the magnitude and spatial distribution of erosion threats, with *CVI*, which encompasses factors influencing the vulnerability and adaptive capacity of communities and ecosystems to erosion effects. This integration enables a thorough assessment of the erosion risk landscape, supporting targeted mitigation strategies and informed decision-making to enhance the study area's resilience against erosion-related dangers. The composite risk was categorized into three levels: low, moderate, and high erosion risk zones. *CRI* was computed as:

$$CRI = CEHI \times CVI \tag{9.6}$$

Table 9.3 Rating class criteria for vulnerability (*Source* Sharma 2013; Ghosh 2017; Bhuiyan et al. 2017; Fuentes et al. 2022)

Index	Parameter	Range	Rating class
Exposure	Geology	Betul gneiss group, BETUL group	1
		Deccan trap, upper gondwana, lower gondwana	2
		Undifferentiated sediments	3
	Geomorphology	Pediment pediplain complex, low dissected structural lower plateau, pond	1
		Moderately dissected structural upper plateau, moderately dissected structural hills and valleys, unclassified waterbodies, older alluvial plain, dam and reservoir	2
		Moderately dissected structural lower plateau, moderately dissected denudational hills and valleys, highly dissected plateau, river, active flood plain, younger alluvial plain	3
	Slope	0°–16.75°	1
		16.75°–33.51°	2
		>33.51°	3
	Relief	29.69–110.74	1
		110.74–191.79	2
		>191.79	3
	Roughness	>0.62	1
		0.37–0.62	2
		0.11–0.37	3
	Soil	Low erosion	1
		Moderate erosion	2
		Severe erosion	3
Index	Parameter	Range	Rating class
Susceptibility	Population growth	<15%	1
		15–50%	2
		>50%	3
	Settlement density	0–10	1
		10–40	2
		>40	3
	Agricultural land	0–10%	1

(continued)

Table 9.3 (continued)

Index	Parameter	Range	Rating class
		10–40%	2
		>40%	3
Resilience	Literacy rate	0–50%	1
		50–70%	2
		70–85%	3
	Banking and credit facility	0	1
		1	2
		>1	3
	Medical facility	0	1
		1	2
		>1	3
	% of forest cover	0–10%	1
		10–50%	2
		50–100%	3

9.4 Results and Discussion

9.4.1 Composite Erosion Hazard Assessment

The Upper Tapi region experienced an increase in riverbank erosion across its entirety over time, especially in its northern stretches. The area's landscape underwent gradual transformations due to both natural and human-induced factors. Erosional processes significantly affected the region, particularly its peripheral areas. Spatiotemporal shifts in banklines between 1978 and 2020 indicated varying levels of erosion hazard intensity in the study area. The eroded and deposited areas revealed intense erosion surpassing deposition, suggesting that the vicinity of the Upper Tapi River is susceptible to erosion. With an average annual rainfall of 833 mm, the Upper Tapi River sub-basin has the potential to cause devastating floods. Historical records show that major floods occurred in the Tapi River in 1998 and 2006 (Timbadiya et al. 2014), contributing to erosion. Regarding net areal change, forests and villages in the central part, including Dawaniya, Harda, and Behda, as well as downstream villages such as Mendhode, Naigaon, and Shemalde, experienced severe erosion with a net areal change exceeding 0.1 sq.km (Fig. 9.2a). These findings align with the research conducted by Resmi et al. (2021), which indicated that the Upper Tapi River exhibited a progressive northward shift between 1978 and 2010.

The extent of riverbank erosion is measured by the proportion of eroded banks. High percentages of eroded riverbanks are observed in forests situated in central and northeastern areas, as well as downstream regions. Several upstream villages,

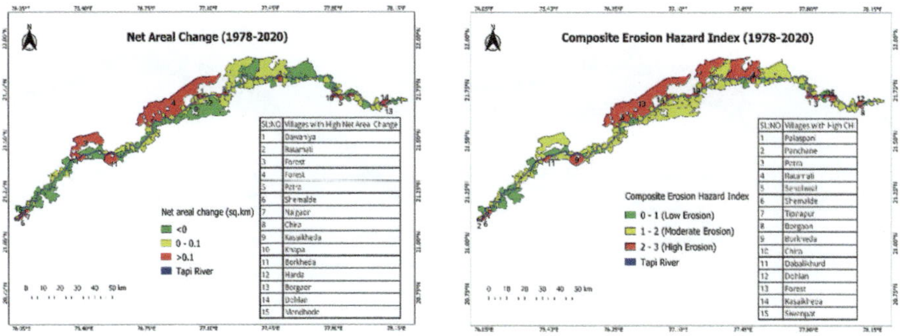

Fig. 9.2 **a** Net areal changes (eroded area)(deposited area) in different villages. **b** Composite erosion hazard map (The figures are generated in ArcGIS 10.8©; Details of primary data source are enlisted in Table 9.1)

including Siwanpat, Tipnapur, Dhanora, Dohlan, and Borgaon, along with downstream villages like Jaisingpura, Dabalikhurd, and Borkheda, are also experiencing considerable erosion. The *CEHI*'s spatial distribution indicated that middle regions, such as Sendhwal, Forest, Kasaikheda, Dawaniya, Harda, Behda, and Chira, were substantially impacted. Moreover, upstream villages like Khapa, Palaspani, Patra, Siwanpat, and Tipnapur, as well as the forest and Ratamati in the northeastern section, display significant erosion hazards. Downstream villages, including Borkheda, Dabalikhurd, Naigaon, Shemalde, and Panchane, are also facing high hazard intensity. This situation is attributed to the reduction in forest cover due to deforestation and the expansion of agricultural land, resulting in diminished vegetation cover (Fig. 9.2b).

9.4.1.1 Composite Vulnerability Assessment

Analysis of exposure indicators consistently showed very high and moderate values across the study area (Fig. 9.3a). The northeastern and southwestern regions, particularly their middle and left portions, displayed very high *EI* values, covering 61% of the total area and including most forested zones. This region is geologically characterized by the Deccan Trap, featuring clayey and loamy soil textures. Geomorphologically, it comprises highly dissected structural upper and lower plateaus, moderately dissected hills and valleys, and moderate slope and relief. The remaining 39% of the area exhibited a moderate *EI*, characterized by undifferentiated sediment, gentle slopes, and relief, encompassing alluvial plains and a pediment-pediplain complex. Forest areas and villages such as Basinda, Bhondilawa, Rangubeli, and Kund demonstrated high *EI* values.

High susceptibility values were observed in 16% of the regions (Fig. 9.3b). This rise was attributed to significant population expansion, resulting in increased settlement concentrations and further transformation of land into farming areas. In Dhulghat, agricultural land reached a peak of 96%. Medium susceptibility values were

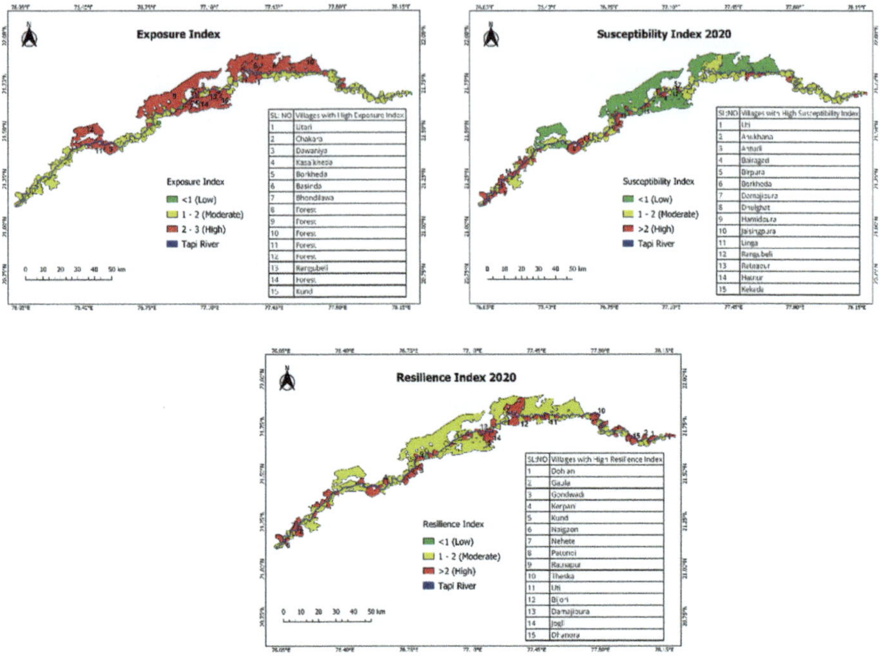

Fig. 9.3 **a** Spatial distribution of exposure index. **b** Spatial distribution of susceptibility index. **c** Spatial distribution of resilience index (The figures are generated in ArcGIS 10.8©; Details of primary data source are enlisted in Table9.1)

found in about 36% of the area, while 48% exhibited low susceptibility values. The highest *SI* values were recorded in villages like Hatnur and Kekada. In contrast, forested regions showed low susceptibility values due to the lack of population growth, settlement density, and agricultural development.

The area with high resilience accounts for 27% due to the excellent literacy rates (70–85%) in these villages (Fig. 9.3c). This high literacy enables residents to be more cognizant of riverbank erosion, prompting them to implement sustainable measures for prevention or mitigation. Improved financial services, enhanced healthcare facilities, and a considerable proportion of forest cover contribute to increased resilience against erosional threats. The forest cover aids in soil stabilization and provides protective vegetation. Upstream villages like Bijori, Damajipura, Jogli, and Dhanora recorded the highest *RI* values. The remaining 72% of the study area exhibited moderate resilience (Fig. 9.4).

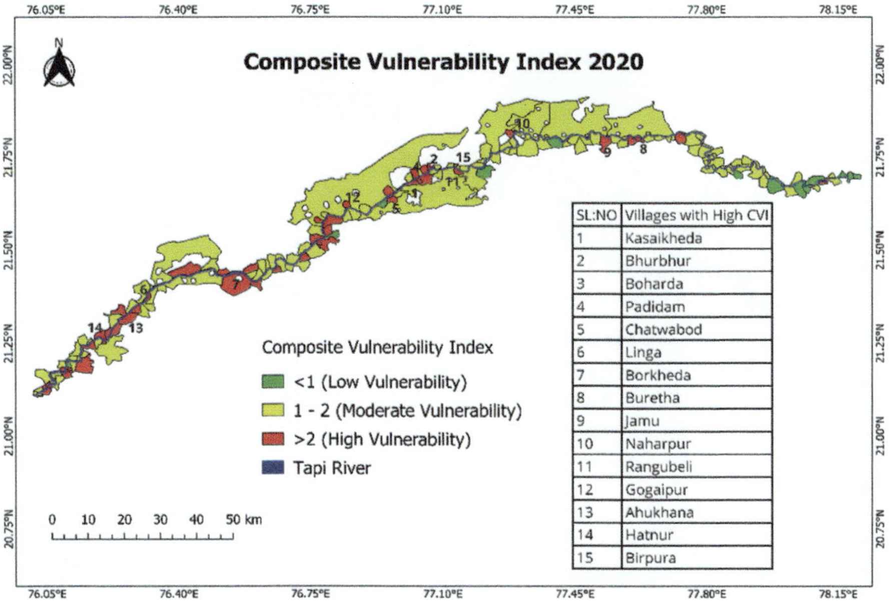

Fig. 9.4 Spatial distribution of composite vulnerability index. (The figure is generated in ArcGIS 10.8©; Details of primary data source are enlisted in Table 9.1)

9.4.2 Spatial Distribution of Risk

The spatial distribution of high-risk values aligned closely with the area most susceptible to erosion and the zones with the highest erosion potential. High-risk regions comprised 63% of the area, suggesting an increasing likelihood of riverbank erosion over time. Among the villages, Dabalikhurd, Borgaon, Kasaikheda, and Borkheda exhibited the greatest risks (Fig. 9.5). Moderate-risk areas accounted for approximately 32%, while low-risk areas made up 5%. The spatial distribution of erosion hazard risk classes revealed that high-risk zones are characterized by high vulnerability and high to moderate erosion intensity.

Geological and landscape analysis showed that areas with the highest risk were composed of undifferentiated sediments and Deccan Trap formations. The upstream geomorphology consisted of moderately dissected structural lower and upper plateaus. Highly dissected structural upper and lower plateaus dominated the middle section, while younger and older alluvial plains were found downstream. These high-risk areas featured clayey and loamy soils, gentle-to-moderate slopes, and high-relief roughness. The combination of these factors contributes to an elevated risk of natural hazards like erosion in these regions.

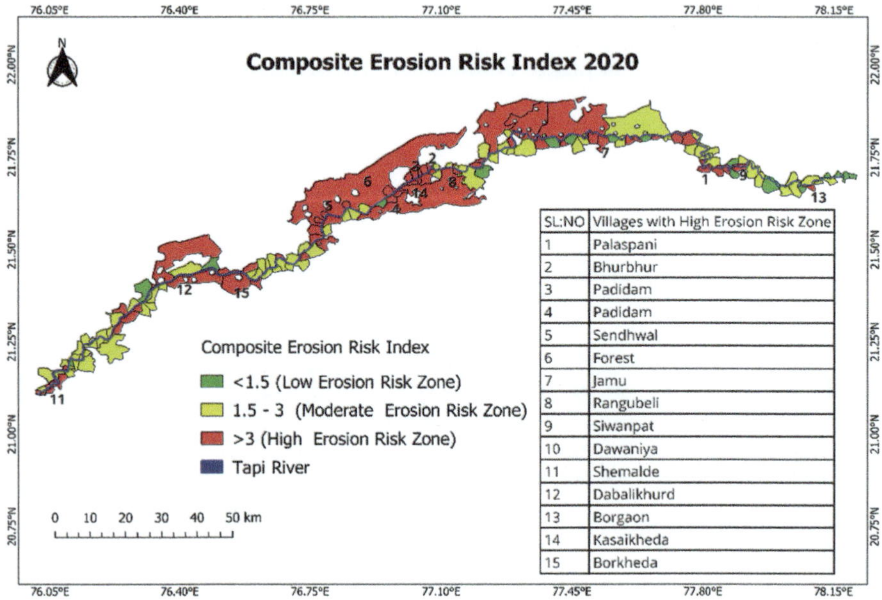

Fig. 9.5 Spatial distribution of erosion risk. (The figure is generated in ArcGIS 10.8©; Details of primary data source are enlisted in Table 9.1)

9.5 Discussions and Recommendations

This study on riverbank erosion risk in the Upper Tapi River, India, from 1978 to 2020, provides in-depth analysis of erosion patterns and their societal and economic effects using satellite data, geographic information systems, and social and economic indicators. This resonates with the past studies that were carried out in Yangtze River delta (Chen et al. 2013), Bhagirathi River (Islam and Guchhait 2017), lower Ganga plain (Talukdar et al. 2021), and Middle Brahmaputra floodplains (Bhuyan et al. 2024). The results indicate a significant rise in riverbank erosion intensity and scope, especially in the central region and various upstream and downstream communities, including Dabalikhurd, Borgaon, Kasaikheda, and Borkheda. This increase is linked to both environmental factors, such as soil composition, and human-induced changes like deforestation and agricultural growth. The vulnerability evaluation showed high exposure in the central, northeastern, and southwestern sections of the study area due to geological and landform characteristics. Erosion susceptibility has grown considerably due to population increase, settlement density, and agricultural expansion, while resilience has improved through enhanced healthcare facilities, higher education rates, and increased forest coverage. Vulnerability is largely influenced by administrative and economic challenges related to infrastructure development and social standards. The analysis suggests that this area is experiencing rapid population growth, along with swift land changes and reclamation of immature land.

To increase the region's resilience, various non-structural measures can be implemented. These include improving literacy rates, initiating afforestation and disaster awareness programs, enhancing medical facilities, implementing scientific land use planning, and establishing Riparian Buffers. The area's socioeconomic growth and infrastructure improvements may bolster its resilience and reduce its vulnerability in the future. Developing new economic opportunities can significantly ease the pressure on agriculture.

The spatial risk distribution revealed 63% high-risk areas, with the most vulnerable zones closely corresponding to regions experiencing severe erosion. To address this escalating risk, the study proposes a multi-pronged approach combining sustainable land use planning, tree planting, disaster awareness initiatives, infrastructure improvements, and scientific structural management techniques. In high-risk areas, scientific structural management techniques such as Riprap, Gabions, Retaining Walls, Geotextiles, and Brush Mattress can be employed to stabilize riverbanks, redirect water flow energy, and prevent erosion. Mitigation strategies and policy initiatives should be tailored to specific locations and adaptable to accommodate variations in the physical parameters of hazards.

9.6 Conclusions

By combining hazard and vulnerability indices, this research offers a comprehensive understanding of riverine erosion risk, enabling the identification of high-risk areas and guiding the development of effective, targeted disaster risk reduction strategies. This study emphasizes the importance of a proactive and integrated approach to managing riverine erosion risks, advocating for a combination of preventive, adaptive, and mitigating strategies to foster a resilient and sustainable future for the Upper Tapi River region. Implementing these measures can reduce vulnerability, enhance resilience, and ensure sustainable development in the face of ongoing environmental challenges. The study of riverbank erosion in the Upper Tapi region highlights several areas requiring further research and management. It is essential to investigate the potential impacts of climate change on riverbank erosion and develop adaptive management strategies to increase resilience. The findings presented here are not definitive but are flexible in terms of indicator selection. Consequently, future research projects can be enhanced by adding or removing indicators based on observational characteristics. This study only considered the lateral migration of the riverbank as an erosional hazard indicator. Cross-sectional aspects such as aggradation and degradation of the river bed were not examined due to data unavailability.

Funding The author(s) declare that financial support was received for the research and/or publication of this article. This research was supported by National Institute of Technology Calicut, under the Faculty Research Grant (FRG) [Grant number: NITC/R&C/2023-2024/FRG/Phase III/(50)].

References

Aydın M, Uysal M (2014) Risk assessment of coastal erosion of Karasu coast in Black Sea. J Coast Conserv 18(6). https://doi.org/10.1007/s11852-014-0343-y

Balica SF, Wright NG, van der Meulen F (2012) A flood vulnerability index for coastal cities and its use in assessing climate change impacts. Nat Hazards 64(1). https://doi.org/10.1007/s11069-012-0234-1

Bandyopadhyay S, De SK (2015) Estimation of bank erosion of the Haora river and proposition of bank erosion vulnerability zonation model. Adv Asian Hum Environ Res (9783319410173). https://doi.org/10.1007/978-3-319-41018-0_9

Bandyopadhyay S, Ghosh K, De SK (2014) A proposed method of bank erosion vulnerability zonation and its application on the River Haora, Tripura, India. Geomorphology 224. https://doi.org/10.1016/j.geomorph.2014.07.018

Bandyopadhyay S (1997) Natural environmental hazards and their management: a case study of Sagar island, India. Singap J Trop Geogr 18(1). https://doi.org/10.1111/1467-9493.00003

Barman NK, Chatterjee S, Paul AK (2016) Estimate the coastal vulnerability in the Balasore Coast of India: a statistical approach. Model Earth Syst Environ 2(1). https://doi.org/10.1007/s40808-015-0074-6

Bhuiyan MAH, Islam SMDU, Azam G (2017) Exploring impacts and livelihood vulnerability of riverbank erosion hazard among rural household along the river Padma of Bangladesh. Environ Syst Res 6(1). https://doi.org/10.1186/s40068-017-0102-9

Bhuyan N, Sajjad H, Saha TK, Sharma Y, Masroor M, Rahaman MH, Ahmed R (2024) Assessing landscape ecological vulnerability to riverbank erosion in the middle Brahmaputra floodplains of Assam, India using machine learning algorithms. Catena 234. https://doi.org/10.1016/j.catena.2023.107581

Biswas R, Anwaruzzaman AKM (2019) Measuring hazard vulnerability by bank erosion of the Ganga river in Malda district using PAR model. J Geogr Environ Earth Sci Int. https://doi.org/10.9734/jgeesi/2019/v22i130136

Chen S, Zha X, Bai Y, Wang L (2019) Evaluation of soil erosion vulnerability on the basis of exposure, sensitivity, and adaptive capacity: a case study in the Zhuxi watershed, Changting, Fujian Province, Southern China. CATENA 177:57–69

Collinson JD, Lewin J (1983) Modern and ancient fluvial systems. Modern Ancient Fluvial Syst. https://doi.org/10.1016/0037-0738(84)90031-9

Dabojani D, Mithun D, Kanti KK (2014) River change detection and bankline erosion recognition using remote sensing and GIS. Forum Geografic XIII(1). https://doi.org/10.5775/fg.2067-4635.2014.038.i

Das VK, Roy S, Barman K, Debnath K, Chaudhuri S, Mazumder BS (2019) Investigations on undercutting processes of cohesive river bank. Eng Geol 252. https://doi.org/10.1016/j.enggeo.2019.03.004

Debnath J, Das N, Ahmed I, Bhowmik M (2016) Vulnerability assessment of socio economic condition due to river bank erosion: a case study on Khowai river, Tripura, India. IOSR J Humanit Soc Sci (IOSR-JHSS) 21(5)

Doukakis E (2005) Coastal vulnerability and risk parameters. Eur Water 11(12)

Fekete A (2019) Social vulnerability change assessment: monitoring longitudinal demographic indicators of disaster risk in Germany from 2005 to 2015. Nat Hazards 95(3). https://doi.org/10.1007/s11069-018-3506-6

Feng TS, Chien LK, Hsu CF, Chiu SY (2011) Establishment and application of coastal vulnerability index in Taiwan nearshore area. In: Proceedings of the international offshore and polar engineering conference

Fuentes JE, Olaya RA, Garcia CE (2022) Evaluation of coastal erosion in the watersheds of municipality of Buenaventura, Colombia: using geospatial techniques and the composite vulnerability index. ISPRS Int J Geo-Inf 11(11). https://doi.org/10.3390/ijgi11110568

Ghosh A, Mukhopadhyay S (2016) Quantitative study on shoreline changes and Erosion Hazard assessment: case study in Muriganga–Saptamukhi interfluve, Sundarban, India. Model Earth Syst Environ 2(2). https://doi.org/10.1007/s40808-016-0130-x

Ghosh A (2017) Quantitative approach on erosion hazard, vulnerability and risk assessment: case study of Muriganga–Saptamukhi interfluve, Sundarban, India. Nat Hazards 87(3). https://doi.org/10.1007/s11069-017-2844-0

Gornitz V (1990) Vulnerability of the east coast, U.S.A. to future sea level rise. J Coast Res 9

Hamid ChM, Ashraf M, Hamid Q, Sarwar SM, Saqib ZA (2017) Geospatial techniques for assessment of bank erosion and accretion in the Marala Alexandria reach of the River Chenab, Pakistan. Sains Malays 46(3). https://doi.org/10.17576/jsm-2017-4603-08

Hasan M, Quamruzzaman C, Rahim A, Hasan I, Methela NJ, Imran SA (2018) Determination of river bank erosion probability: vulnerability and risk in southern shoreline of Bangladesh. Int J Energy Sustain Dev 3(3)

Hasanuzzaman M, Bera B, Islam A, Shit PK (2021) Estimation and prediction of riverbank erosion susceptibility and shifting rate using DSAS, BEHI, and REBVI models: evidence from the lower Ganga river in India. https://doi.org/10.21203/rs.3.rs-1097959/v1

Hooke JM (1979) An analysis of the processes of river bank erosion. J Hydrol 42(1–2). https://doi.org/10.1016/0022-1694(79)90005-2

Islam A, Guchhait SK (2017) Search for social justice for the victims of erosion hazard along the banks of river Bhagirathi by hydraulic control: a case study of West Bengal, India. Environ Dev Sustain 19(2). https://doi.org/10.1007/s10668-015-9739-6

Jamshed A, Birkmann J, Feldmeyer D, Rana IA (2020) A conceptual framework to understand the dynamics of rural-urban linkages for rural flood vulnerability. Sustainability (Switzerland) 12(7). https://doi.org/10.3390/su12072894

Kumar TS, Mahendra RS, Nayak S, Radhakrishnan K, Sahu KC (2010) Coastal vulnerability assessment for Orissa State, East Coast of India. J Coast Res 26(3). https://doi.org/10.2112/09-1186.1

Laha C, Bandyopadhyay S (2013) Analysis of the changing morphometry of river Ganga, shift monitoring and vulnerability analysis using space- borne techniques : a statistical approach. Int J Sci Res Publ 3(7)

Majumdar S, Das A, Mandal S (2023) River bank erosion and livelihood vulnerability of the local population at Manikchak block in West Bengal, India. Environ Dev Sustain 25(1). https://doi.org/10.1007/s10668-021-02046-z

Mandal S (2017) Assessing the instability and shifting character of the river bank Ganga In Manikchak Diara of Malda district, West Bengal using bank erosion hazard index (BEHI), RS & GIS. Eur J Geogr 8(4)

Mani Murali R, Ankita M, Amrita S, Vethamony P (2013) Coastal vulnerability assessment of Puducherry coast, India, using the analytical hierarchical process. Nat Hazards Earth Syst Sci 13(12). https://doi.org/10.5194/nhess-13-3291-2013

Merlotto A, Bértola GR, Piccolo MC (2016) Hazard, vulnerability and coastal erosion risk assessment in Necochea Municipality, Buenos Aires Province, Argentina. J Coast Conserv 20(5). https://doi.org/10.1007/s11852-016-0447-7

Mollah TH, Ferdaush J (2015) Riverbank erosion, population migration and rural vulnerability in Bangladesh (a case study on Kazipur Upazila at Sirajgonj District). Environ Ecol Res 3(5). https://doi.org/10.13189/eer.2015.030502

Mondal T, Tripathy B (2020) River bank erosion and environmental degradation. Int J Res 7(7)

Rahman MS, Gain A (2020) Adaptation to river bank erosion induced displacement in Koyra Upazila of Bangladesh. Prog Disaster Sci 5. https://doi.org/10.1016/j.pdisas.2019.100055

Resmi SR, Patel PL, Timbadiya PV (2019) Morphological study of upper Tapi river using remote sensing and GIS techniques. ISH J Hydraul Eng 25(3). https://doi.org/10.1080/09715010.2017.1409089

Resmi SR, Patel PL, Timbadiya PV (2021) Key morphological changes and their linkages with stream power and land-use changes in the upper Tapi river basin, India. Int J Sediment Res 36(5). https://doi.org/10.1016/j.ijsrc.2021.03.003

Sah RK, Das AK (2018) Morphological dynamics of the rivers of brahmaputra. J Geol Soc India 92(4). https://doi.org/10.1007/s12594-018-1039-y

Sarif MN, Siddiqui L, Nasrin T, Bera S, Ramana GV (2024) Vulnerability of riparian inhabitants to river bank erosion hazard in lower Ganga: an integrated approach. Nat Hazards:1–24

Sharma N (2013) Hazard, vulnerability and risk on the Brahmaputra basin: a case study of river bank erosion. Open Hydrol J 4(1). https://doi.org/10.2174/1874378101004010211

Siddik MdA, Zaman AKMM, Islam MdR, Hridoy SK, Akhtar MtP (2017) Socio-economic impacts of river bank erosion: a case study on coastal island of Bangladesh. J NOAMI 34(2)

Singha P, Das P, Talukdar S, Pal S (2020) Modeling livelihood vulnerability in erosion and flooding induced river island in Ganges riparian corridor. India. Ecol Indic 119. https://doi.org/10.1016/j.ecolind.2020.106825

Talukdar S, Pal S, Singha P (2021) Proposing artificial intelligence-based livelihood vulnerability index in river islands. J Clean Prod 284. https://doi.org/10.1016/j.jclepro.2020.124707

Tha T, Piman T, Kittipongvises S, Ruangrassamee P (2024) Riverbank erosion vulnerability assessment and coping strategies: a case study of the riparian communities in the Mekong river basin in Cambodia. Heliyon 10(3). https://doi.org/10.1016/j.heliyon.2024.e25418

Thakur PK, Laha C, Aggarwal SP (2012) River bank erosion hazard study of river Ganga, upstream of Farakka barrage using remote sensing and GIS. Nat Hazards 61(3). https://doi.org/10.1007/s11069-011-9944-z

Timbadiya PV, Patel PL, Porey PD (2014) One-dimensional hydrodynamic modelling of flooding and stage hydrographs in the Lower Tapi river in India. Curr Sci 106(5)

Whipple KX, Tucker GE (2002) Implications of sediment-flux-dependent river incision models for landscape evolution. J Geophys Res Solid Earth 107(B2). https://doi.org/10.1029/2000jb000044

Yang X, Damen MCJ, van Zuidam RA (1999) Satellite remote sensing and GIS for the analysis of channel migration changes in the active Yellow River Delta, China. ITC J 1(2). https://doi.org/10.1016/S0303-2434(99)85007-7

Chapter 10
Nature-Based Solutions for Mitigating Zoonotic Disease Spillover Risk in Urban Areas

Anupama K. and Saikat Kumar Paul

Abstract Emerging infectious diseases (EIDs) are an increasing concern due to their potential to transform into epidemics and global pandemics. The inherent risks associated with these diseases warrant their classification as biological hazards. About 75% of all the EIDs belong to the category of zoonotic diseases. Most studies primarily focus on the downstream aspects following a disease spillover and its progression into an outbreak. However, there is a significant gap in mitigating the issue by adopting an upstream approach to understand the risk factors contributing to the increased incidence of zoonotic diseases and the possible ways to mitigate it. In this chapter, the urban stressors and its related risk factors that influence the zoonotic spillover risk are identified from the literature review. Based on systems thinking, causal loop diagrams are developed to understand the different pathways of spillover risk in urban environments. The findings indicate that land use changes due to anthropogenic activities act as a major driver that increases contact between humans and zoonotic agents. Nature-based solutions (NBSs) as effective countermeasures to contribute towards mitigating spillover risks are identified as intervention points in the causal loop diagram. NBSs are then categorized based on how they influence the spillover dynamics and various measures are discussed. These solutions address multiple drivers of zoonotic spillover, leading to sustainable urban environments.

Keywords Zoonotic diseases · Spillover risk · Nature-based solutions · Sustainable urban environment

Anupama K. (✉) · S. K. Paul
Indian Institute of Technology Kharagpur, Kharagpur, West Bengal, India
e-mail: anupamaniroop16@gmail.com

S. K. Paul
e-mail: skpaul@arp.iitkgp.ac.in

L. K. Dashora et al. (eds.), *Urban Sustainability*, Disaster Risk Reduction,
https://doi.org/10.1007/978-981-96-4890-0_10

10.1 Introduction

Emerging infectious diseases (EIDs) pose a significant threat to global health. The problem with EID is how it affects people, as pandemics and epidemics posing risks to human health and international stability (McArthur 2019; Murugesan 2020). The twenty-first century has already witnessed five major pandemics COVID, Ebola, MERS, Sine Flu, and SARS, and the number of new infections and outbreaks are increasing over the years (Eltarabily and Elghezanwy 2020). The inherent risks associated with these diseases has led to their classification as biological hazards, as highlighted in The Sendai Framework for Disaster Risk Reduction and the Health EDRM framework, which the World Health Organization adopted in 2019. About 75% of EIDs, are zoonotic in nature and spread from animals to human populations.

The world is becoming increasingly urban and the twentieth century has seen rapid urbanization in the world's population (Aliyu and Amadu 2017). It is predicted that by 2050, nearly 68% of the world's population will dwell in cities and about 90% of this growth is expected from Asia and Africa (United Nations 2018). The rising urbanization coupled with climate change has also led to increased occurrences of disease in cities (Mukherjee 2017). It has also been emphasized that post-pandemic planning decisions of urban areas should be centred around health and should reconnect the urban planning and public heath domains (Gouveia and Kanai 2020). Various studies have highlighted that those cities which have been subjected to rapid urbanization and unplanned growth were extensively affected by EIDs (Kim 2021; Neiderud 2015; Wang et al. 2021). As the emergence of zoonoses are considered as a result of complex interactions between various factors, it becomes important that we decipher the interconnected natural and urban systems and their interactions in urban environments (Finucane et al. 2014; Kim 2021).

Nature-based solutions (NBS) which contribute towards effectively leveraging natural processes and ecosystem services are gaining importance to address different societal challenges including disease mitigation (Eggermont et al. 2015). By integrating NBS into urban planning, cities can enhance their resilience to different environmental stressors (Nesshöver et al. 2017). NBS approaches include green infrastructure, ecological restoration, and sustainable water management practices, which can mitigate the adverse effects of urbanization and climate change on public health (Cohen-Shacham et al. 2016; Kabisch et al. 2017). There is growing evidence that NBS can reduce zoonotic disease spillovers by improving environmental conditions and reducing human-wildlife contact (Mills et al. 2017; Vanhove et al. 2021). Most of the response with regard to zoonotic disease focus on the downstream aspects such as outbreak control and treatment. There is a need to shift and give equal emphasis to upstream approaches to understand and mitigate stressors and risk factors that lead to such zoonotic spillovers and ways to effectively address the same with different interventions like NBS.

In this chapter, we focus on understanding the possible intervention points for NBS in the interaction pathway of urban stressors and risk factors that contribute to zoonotic spillovers, based on an upstream approach. By examining the role of

NBS in mitigating zoonotic spillover risks using causal loop diagrams based on systems thinking, this chapter aims to contribute to the growing body of knowledge on sustainable urban planning and public health.

10.2 Understanding the Role of Nature-Based Solutions in Mitigating Zoonotic Spillover Risk

Urbanization and its associated environmental changes have significantly increased the risk of zoonotic spillover. The role of nature-based solutions (NBS) in mitigating these risks is increasingly being recognized. This section aims to explore how NBS can be utilized to address the drivers and risk factors of zoonotic spillover in urban areas, supported by a system thinking approach using Causal Loop Diagrams (CLDs).

10.2.1 Drivers and Risk Factors of Zoonotic Spillover in Urban Areas

The likelihood of zoonotic spillover is established by the interplay of multiple factors, such as pathogen exposure, disease dynamics in the reservoir host, and human factors that impact susceptibility to infections (Plowright et al. 2017). Urbanization is one of the key stressors that increases the risk of spillover directly or indirectly (Mayer 2000; Mackenstedt et al. 2015; Jones et al. 2017; Liu et al. 2018; Kim 2021). It influences the environmental and social factors which in the process leads to increasing the spillover risk to EIDs. Deforestation, a prominent driver of habitat loss, sets in motion a chain of events where the reduced habitat availability increases the proximity of wildlife to human settlements and domestic animals by increasing risky interfaces (Gottdenker et al. 2014; Mackenstedt et al. 2015; Weiss and McMichael 2004). Additionally, habitat loss leads to altered ecosystems, potentially favouring infectious disease reservoir hosts and elevating exposure risk (Hassell et al. 2017). Altered ecosystems can lead to development of risky interfaces which increases wildlife–human–animal contact and thus the spillover risk.

Climate change, a global phenomenon, indirectly impacts spillover risk by altering disease vectors that carry pathogens by influencing its demography and geographic spread (Brisbois and Ali 2010; Coker et al. 2011; Jones et al. 2017; Baker et al. 2021). These changes can lead to diseases encroaching closer to human populations, increasing spillover risk. In parallel, livestock density can serve as an amplifier for diseases as they are intermediate host for various EIDs and lack of effective management of livestock aids in disease transmission risk (Miao et al. 2020; Jiang et al. 2022).

The increasing impacts of globalization is another risk factor as is evident in the changes that have come in the food habits leading to an increase in the demand for animal products (Field 2009; Jones et al. 2017). Industrialization of animal farming is most often associated with over use of antibiotics and growth hormones leading to antibiotic resistance in pathogens exacerbating disease challenge and elevating the spillover risk (Mayer 2000; Wilcox et al. 2008; Coker et al. 2011).

The changes in host dimension play a crucial role—high population density facilitates increased contact among humans, wildlife, and domestic animals, further enhancing transmission risk (Jones et al. 2008; Jones et al. 2017; Kim 2021). The interaction among these groups increases the likelihood of pathogen transmission, contributing to spillover risk. Land use changes influences habitat fragmentation, which subsequently alters the ecosystems which in turn influences the disease dynamics thus contributing to spillover risk (Jones et al. 2008; Gottdenker et al. 2014; Guo et al. 2019; Connolly et al. 2021; Kim 2021). Changes in landuse, deforestation, and habitat fragmentation have larger implications; as they lead to loss of foraging spaces for reservoir hosts (e.g. bats), resulting in allostatic overload and weakened immunity, leading to increased viral shedding (Plowright et al. 2024). The landscape changes leads to changes in abiotic conditions in urban areas and results in changes in temperature, hydrology, increased light pollution, noise pollution, heavy metal pollution thus influencing the biotic conditions of vector and reservoir host and eventually contributing to spillover dynamics (Alberti 2005; Combs et al. 2022).

Overall, the urbanization, climate change, globalization, and biodiversity loss are important stressors that influence landuse changes which plays a crucial role in spillover risk in an urban context. The environmental changes and risk factors that influence the zoonotic spillover risk in urban areas highlight the significance of exploring how NBS can address these challenges.

10.2.2 Understanding the Role of Nature-Based Solutions for Mitigating Environmental and Public Health Challenges

NBS are approaches that effectively leverage natural processes and ecosystem services to address societal challenges. The International Union for Conservation of Nature (IUCN) classifies NBS into several categories: 'ecosystem restoration approaches, issue-specific ecosystem-related approaches, infrastructure-related approaches, ecosystem-based management approaches, and ecosystem protection approaches' (Cohen-Shacham et al. 2016). These classifications help in understanding the broad scope and potential applications of NBS in various contexts.

Several studies have highlighted the benefits of NBS in addressing environmental and public health issues. Alvarado et al. (2023) explored the relationship between street trees and mental health, underscoring the multifaceted benefits of urban green infrastructure. Kabisch et al. (2016) provide an overview on effectiveness of NBS on improving air and water quality, reducing urban heat islands and enhancing biodiversity, all of which contribute to better health outcomes. The public health benefits of urban green spaces help combat urban heat island effects and pollution (Braubach et al. 2017). There are certain studies that indicate towards the possibility that increase in green spaces in urban areas without careful planning, implementation and proper management would lead to increase in population of disease carrying vectors (Wood et al. 2014; Lõhmus and Balbus 2015). Despite these concerns, the significance of studying NBS in the context of zoonotic disease mitigation remains high. Understanding how NBS can be optimized to reduce zoonotic spillover risks is crucial, given the increasing incidence of EIDs and their profound impacts on global health (Keesing and Ostfeld 2021).

This chapter aims to explore how NBS can influence the dynamics of zoonotic spillover by focussing on the potential of NBS to create healthy ecosystems that reduce the risk of pathogen transmission.

10.2.3 Systems Thinking: Causal Loop Diagram as a Tool

Zoonoses occur as a result of interaction between agent, host, and environment, and this interaction is represented in the epidemiological triad. (Bhopal 2009; Kim 2021) have discussed this approach in understanding the factors associated with host, agent, and environment that influence EIDs. Since the context of this study is the urban environment, it has become important to assess the urban influences on host and agent. A system's thinking approach offers numerous advantages in assessing spillover risk for zoonotic diseases in urban areas and understanding the leverage points within where NBS intervention can help in altering feedback loops or reduce risk factor impacts. It is a very useful approach to get a holistic perspective and understand the interconnections among the different variables and the pathways of influence.

System thinking helps in understanding complex systems by considering the internal feedback loops, time delays, and interactions that influence the behaviour of the system. Over the years, there has been a surge in interest in using the systems thinking to understand urban and population health problems (Owrangi et al. 2015; Roux 2015; Langellier et al. 2019; Diemer and Nedelciu 2020; Rubin et al. 2021). The systems thinking is much required to explore the linkages between urbanization, environmental change, and human health owing to the complexity of interactions involved (Bai et al. 2012). Since there are factors at multiple levels such as individual, neighbourhood, and city that influence the population health outcomes, it is important to consider the feedback loops, their relationships, and influences to address the problems of the complex adaptive system (Bai et al. 2012; Roux 2015; Langellier et al. 2019).

Causal loop diagrams (CLDs) are a popular and widely used tool of systems approach to comprehend the relation between multiple variables that can affect the behaviour of the system (Kim 2000; Ram and Irfan 2021). A CLD visually represents the relationships between various variables and elements that influence the behaviour of a system. In a CLD, arrows connect the variables, indicating causal dependencies. The direction of the connection signifies whether the variables change in the same direction (positive relationship, +) or in opposite directions (negative relationship, −). These positive and negative causal links can create either reinforcing (R) or balancing (B) feedback loops, which help explain the dynamics of the system (Sterman 2004). CLDs are mainly used to understand and describe the complex set of interconnection, feedback loops and linkages by assimilative qualitative and quantitative insights (Coletta et al. 2021; Alvarado et al. 2023). Various studies have used CLD to understand role of NBS in diverse context. For instance, Coletta et al. (2021) utilized CLD with regard to supporting different decisions related to NBS benefits and trade-offs in wetland ecosystem. Alvarado et al. (2023) used CLDs to understand the relationship between street trees and mental health, highlighting the complex interactions and benefits of urban green infrastructure.

The following section will detail the materials and methods used in constructing and analysing the CLD for understanding the role of NBS in mitigating zoonotic spill over risks.

10.3 Materials and Methods

10.3.1 Overview of the Approach

To understand the various stressors and their allied risk factors that contribute to zoonotic spillover, along with the possible NBS solutions, a comprehensive methodology was followed consisting of four key steps. The first step focussed on conducting an in-depth literature review aimed at identifying the stressors and risk factors in urban areas that contribute to zoonotic spillover and understanding how NBS influences the risk of spillover. The review led to the synthesis from literature list of common stressors and risk factors in urban contexts linked to zoonotic spillover and the possible NBS interventions that address these urban stressors. Based on systems thinking, causal loop diagrams are used to visualize the interaction among the key urban stressors and their risk factors to decipher how they contribute to zoonotic spillover. The process of constructing a causal lop diagram involved listing the variables and identifying the feedback loops and interactions. The causal loop diagram was constructed using Venism software. This was followed by two rounds of iterations after consulting with experts in the fields of public health and urban planning. The next step focussed on identifying the key intervention points for NBS. The strategy involved a detailed analysis of the CLD and feedback loops to pinpoint areas and possible zones of influence where nature-based solutions can effectively contribute to reducing spillover risks. Specific leverage points were identified, and the CLD was reconstructed to incorporate and assess the influence pathway after the introduction of key NBS to address the risk factors. The final step involved proposing and categorizing the NBS interventions based on the IUCN framework to group the possible actions to reduce zoonotic spillover risk.

10.3.2 Variable Selection

The selection of variables for the causal loop diagrams was a critical aspect of the methodology. The identification of variables was based on the literature review and expert consultations. Key variables were chosen to represent stressors, risk factors, and ecological processes that contribute to zoonotic spillover. The variables were classified into three main categories:

1. **Urban Stressors**: These variables included factors like urbanization, climate change, and globalization. These are the external pressures in urban environments that exacerbate the conditions for zoonotic diseases to spread.
2. **Risk Factors**: Variables related to human activities and environmental changes that directly increase zoonotic spillover risk were considered, such as increased human–wildlife contact, altered ecosystem, biodiversity loss, vector proliferation and creation of risky interfaces.
3. **NBS Interventions**: Variables such as habitat restoration, wildlife corridors, green infrastructure, and urban forest were included to understand their impact on mitigating risk factors and urban stressors.

The identification of feedback loops and the relationships between these variables was fundamental to mapping the systemic causes of zoonotic spillover and developing intervention strategies. Each of these variables was carefully evaluated for its causal significance, relevance to urban zoonotic spillover dynamics, and its potential for being influenced by NBS interventions.

10.4 Results

10.4.1 CLD Construction: Role of Urban Stressors and Risk Factors in Influencing Zoonotic Spillover

Using an integrated landscape lens to understand the urban stressors and risk factors at varying scales, ranging from local to continental, and systems approach to understand their linkages, feedback loops and nonlinear influences, causal loop diagram (CLD) is built using Venism software. The main stressors in urban context considered in the system are 'urbanization', 'climate change' and 'globalization' which play a key role in influencing the key risk factors leading to zoonotic spillover. The key risk factor variables are 'land use change', 'altered ecosystem', 'biodiversity loss' and creation of 'risky interfaces'. The arrows from the stressors in the diagram to the risk

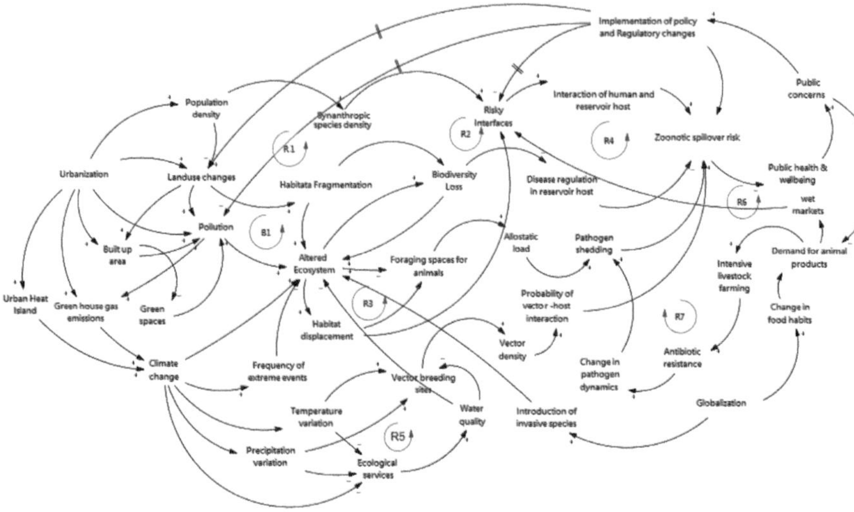

Fig. 10.1 CLD to describe the urban stressors and risk factors linked to Zoonotic spillover

factors represent causal dependencies between variables, creating either reinforcing (R) or balancing (B) loops which drive system behaviour. The CLD addressing the interactions between urban stressors and zoonotic spillover risk factors and feedback loops as shown in Fig. 10.1. The analysis of feedback loops related to the zoonotic spillover are in Table 10.1. The key feedback loops related to the urban stressors are discussed here.

'Urbanization' is a primary driver of 'land use changes' variable which fragment natural habitats. The 'fragmentation of natural habitats' due to urbanization has a detrimental effect on biodiversity and ecosystem stability. 'Biodiversity loss' in urbanized landscapes diminishes the resilience of ecosystems, alters pathogen dynamics and 'disease regulation in reservoir host'. 'Reduced disease regulation in reservoir host' can disrupt the stabilizing effects, making the ecosystem more vulnerable to 'zoonotic spillover' as shown in the reinforcing loop R1 in Fig. 10.1. Changes in 'land use' and increase of 'built-up area' lead to reduction in 'green spaces' and increase in 'pollution'. These pollutants further degrade habitats and 'alters ecosystem'. These changes in biotic and abiotic conditions of ecosystem influence the habitat of reservoir host and vectors of zoonoses leading to 'reduced foraging spaces' and also reservoir host displacement from their natural habitats. The reduction in foraging spaces increases the physiological stress on reservoir hosts increasing the 'allostatic load', which leads to increase viral 'pathogen shedding'. This increases the chances of direct contact with the pathogen thus increasing chances of 'zoonotic spillover' (loop B1). Urbanization which leads to 'population density' increase also has an impact on the increase on the 'synanthropic species density' in urban areas. These species are reservoir host to various zoonoses and thus the proximity in shared urban spaces also contributes to creation of 'risky interfaces. Another

Table 10.1 Main feedback loops with internal and external variables and related dynamics

Loop	Dynamics	Internal variables	External variables
R1	Land use change	Land use changes Habitat fragmentation Biodiversity loss Disease regulation in reservoir host Zoonotic spillover risk Public health and wellbeing Public concerns Implementation of policy and Regulatory changes	Urbanization Population density
B1	Land use change	Landuse changes Built up area Green spaces Pollution Altered ecosystem Foraging spaces for animals Allostatic load Pathogen shedding Zoonotic spillover risk Public health and wellbeing Public concerns Implementation of policy and Regulatory changes	Urbanization Climate change Habitat fragmentation
R2	Land use change	Land use changes Habitat Fragmentation Altered Ecosystem Habitat displacement Risky Interfaces Interaction of human and reservoir host Zoonotic spillover risk Public health and wellbeing Public concerns Implementation of policy and Regulatory changes	Urbanization Synanthropic species denisty Population density
R3	Altered ecosystem	Altered ecosystem Biodiversity loss Disease regulation in reservoir host Zoonotic spillover risk Public health and wellbeing Public concerns Implementation of policy and Regulatory changes Pollution	Climate change Frequency of extreme events Habitat fragmentation

(continued)

Table 10.1 (continued)

Loop	Dynamics	Internal variables	External variables
R4	Vector breeding sites	Climate change Precipitation variation Vector breeding sites Vector density Probability of vector-host interaction Zoonotic spillover risk Public health and wellbeing Public concerns Implementation of policy and Regulatory changes Pollution Green house gas emissions	Temperature variation Water quality
R5	Ecological services	Climate change Ecological services Water quality Altered ecosystem Biodiversity loss Disease regulation in reservoir host Zoonotic spillover risk Public health and wellbeing Public concerns Implementation of policy and Regulatory changes Pollution Green house gas emissions	Temperature variation Precipitation variation
R6	Demand for animal products	Demand for animal products wet markets Risky Interfaces Interaction of human and reservoir host Zoonotic spillover risk Public health and wellbeing Public concerns	Globalization Change in food habits
R7	Demand for animal products	Demand for animal products Intensive livestock farming Antibiotic resistance Change in pathogen dynamics Pathogen shedding Zoonotic spillover risk Public health and wellbeing Public concerns	Globalization Change in food habits

main influence pathway that leads to zoonotic spillover is through creation of 'risky interfaces' which increases the chances of interaction between human–reservoir host and vectors as shown in loop R2.

'Climate change' impacts zoonotic spillover risk pathway by influencing the precipitation and temperature patterns, increasing the 'vector density' and the 'probability of vector host interaction', leading to zoonotic spillover risk (loop R4). 'Globalization' has influenced human lifestyles, economies, and environments, leading to important implications for zoonotic spillover risk as the enhanced interconnectivity has 'altered food habits', 'increased the demand for animal products', intensified livestock production which has increased 'antibiotic resistance' and thus 'pathogen dynamics' leading to spillover risk (loop R6). Globalization has also influenced the consumption of wild animals and the operation of 'wet markets', creating 'risky interfaces' for zoonotic spillover through increased chances of 'interaction of human-wildlife- animals' (loop R7).

10.4.2 Revised CLD Including the Effects of NBS

A revised version of CLD presented in Fig. 10.1 is built by taking into account the multiple NBS intervention points. These interventions focus on pathways that would lead to addressing the risk factors to spillover risk as shown in Fig. 10.2. The NBS interventions influence multiple feedback loops. For instance, the intervention strategy of 'restoring habitats' through reforestation and rehabilitating degraded lands, enhances ecosystem resilience and stability, thus reducing the effects of 'altered ecosystem' resulting from 'urbanisation' led 'land use changes' and 'habitat fragmentation' (influencing loops R1, R2, R3, R5, B1). 'Habitat restoration' reduces species displacement, promotes 'biodiversity', and mitigates zoonotic spillover by maintaining a balanced ecological function. 'Sustainable livestock production' as a NBS intervention strategy influences the feedback loop R7 by reducing the disease transmission risk among livestock and thus reduction in spillover risk by influencing 'pathogen dynamics'. 'Coastal and riverine ecosystem management', including maintaining mangroves, restoring coastal dunes, and managing floodplains, buffers against increase in natural disasters due to 'climate change' and preserves 'ecological stability' of the region (loop R3).

Increase in 'Green infrastructures' such as green roofs, urban parks, and green walls reducing 'UHI effects', green infrastructure lowers 'local temperatures', 'improves air quality' and helps reduce 'pollution' and in turn the pollution-driven 'altered ecosystem' and contributes to reduced zoonotic spillover (loop R4 and B1). 'Urban wetland restoration' improves 'ecological services' enhancing the 'water quality' by removing pollutants and pathogens from stormwater and wastewater. This stabilizes the conditions contributing to vector and host populations reducing the prevalence of waterborne zoonotic pathogens. This address 'water quality' and 'pathogen dynamics' by managing urban water pollution and controlling vector populations (loop R5).

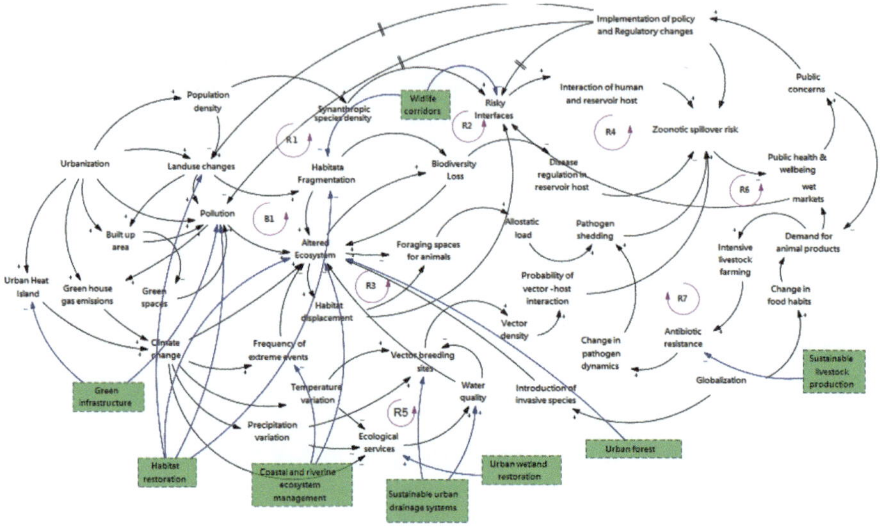

Fig. 10.2 CLD including the effects of NBS

10.4.3 Influence Pathways of Nature-Based Solutions (NBS) in Mitigating Zoonotic Spillover Risk

This section discusses the role of NBS in influencing the multiple pathways leading to zoonotic spillover. Based on the inferences from the revised CLD, it can be seen that NBS can contribute significantly in mitigating zoonotic spillover risk by addressing the complex environmental and ecological pathways that facilitate pathogen transmission. These pathways include direct human contact with pathogens, interactions with reservoir hosts and vectors, and increased human-wildlife interaction in risky interfaces.

10.4.3.1 Role of NBS in Influencing Pathogen Dynamics

Pathogen dynamics, encompassing the survival, replication, and transmission of pathogens, are influenced by environmental conditions and ecological interactions. NBS can play a role in stabilizing these dynamics by enhancing ecosystem services and reducing environmental degradation, thereby limiting the prevalence and spread of pathogens in urban areas. One of the main pathways that impacts zoonotic spillover is owing to the loss of biodiversity due to degraded natural habitats as it leads to dilution effect and thus influences pathogen dynamics leading to dominance and increased disease prevalence (Keesing and Ostfeld 2021). NBS solutions, such as habitat restoration of wetlands and forests near urban areas, can help in this regard by promoting ecological resilience (Xie and Bulkeley 2020; Babí Almenar et al.

2021; Dunlop et al. 2024). Urban green spaces, such as parks and urban forests, act as microhabitats for various species, enhancing local biodiversity and ecosystem services. Emphasizing on climate resilient landscaping, reforestation, and afforestation measures also aids in stabilizing microclimates and in the process influencing urban heat island effects. Improving water quality can limit the survival and propagation of waterborne pathogens thus mitigating zoonotic spill over from contaminated water sources. NBS like constructed wetlands for waste treatment and storm water management can help filter and thus reduce pathogen load and contaminants (Matthew et al. 2022; Matter and Gado 2024). Rain gardens and bioswales are also useful to manage urban runoff and improve the water filtration. These water management measures can improve the water quality and thus reduce breeding grounds for pathogens.

10.4.3.2 Role of NBS in Influencing Reservoir Host and Vectors

Reservoir hosts and vectors play a key role in the spillover and transmission of zoonotic diseases. NBS can be used as an ecological countermeasure to manage the habitats of these species which will aid in managing the reservoir host/vector density and also interactions with humans, thereby contributing towards mitigating zoonotic spillover risks. Habitat restoration is an important measure that provides alternative foraging and nesting sites, thus reducing the dependency of reservoir hosts on core urban areas (Plowright et al. 2024). This minimizes human–wildlife interactions and decreases the concentration of reservoir hosts near human populations. NBS measures like wetland restoration, forest restoration (Prist et al. 2023) helps keep in check the disease vector population of mosquitoes, ticks, etc., by supporting the natural habitats of birds, amphibians, fishes, etc. Since livestock act as intermediatory host to zoonotic spillover risk, sustainable livestock production and farming practices need to be carried out (Coker et al. 2011; Ahmed et al. 2019).

10.4.3.3 Role of NBS in Influencing Risky Interfaces to Reduce Human–Wildlife Animal Interaction

Risky interfaces between humans, wildlife, and domestic animals facilitate zoonotic spillover. Focussing on urban planning approaches that addresses such risks by integrating green buffers and transition zones in urban planning to separate human settlements from wildlife habitats will act as physical barriers that reduce direct contact between humans and wildlife and thus reduces risky interfaces (Khor et al. 2022). Providing green buffers along rural–urban transition zones also aid in reducing the risky interfaces. NBS that help combat habitat fragmentation is creation of urban green corridors which act as wildlife corridors to connect natural habitats and facilitate safe wildlife movement (Glidden et al. 2021). This helps in minimizing conflict, habitat fragmentation, and the need to venture into human dominated landscapes and in the process avoids creation of risky interfaces for enhanced contact.

By influencing the root causes of zoonotic spillover risk, NBSs represent a powerful tool for reducing the incidence of zoonotic diseases. However, their successful implementation requires a coordinated approach across multiple sectors, including urban planning, public health, and environmental management. Though the findings of the study clearly point towards the significant role NBS have with regard to mitigating zoonotic spillover, further studies can focus on quantitatively assessing the influence over time.

10.5 Recommendations

The integration of NBS with urban planning and management strategies helps build ecosystem resilience, restore natural habitats, and reduce human–animal interactions that lead to zoonotic spillover risk. These NBS interventions can influence the different pathways of spillover risk by mitigating the risk factors linked to pathogen dynamics, interaction of reservoir host and vectors, and by reducing risky interfaces leading to human–wildlife animal interaction. Based on the CLD analysis and insights into various stressors leading to zoonotic spillover, the following categories of NBS interventions are recommended, aligning with the IUCN classification as shown in Table 10.2.

10.6 Conclusion

This chapter addresses the possible NBS intervention strategies that could reduce the risk by influencing the pathways contributing to zoonotic spillover risks in urban areas by focussing on an upstream approach. A systems thinking approach, using causal loop diagrams, was used to decipher the interactions and feedback loops of urban stressors and its risk factors leading to zoonotic spillover. Subsequently, potential NBS intervention points in reducing the risk were explored by incorporating it in the CLD to understand the leverage points and their influence in mitigating the risk by influencing casual links. Urbanization, climate change, biodiversity loss, globalization, and its influence of food production practices are key stressors in urban areas that create feedback loops that increase the likelihood of zoonotic spillover. These factors tend to increase risk of zoonotic spillover by either increasing the chances of direct contact with the pathogen or by facilitating increased interaction between human and reservoir host. Land use change driven by anthropogenic activities play an important role in influencing multiple pathways that increases zoonotic spillover risk. These changes cause disruption to natural ecosystem that had played a key role in acting as barriers to human exposure to wildlife and linked pathogens. Incorporating NBS into these systems can help in risk reduction by influencing the feedback loops by addressing the key pathways to spillover risk. The NBS help mitigate the risk by influencing pathogen dynamics, interaction of reservoir host and vectors, and

Table 10.2 NBS categories and possible intervention solutions

NBS category	Intervention point	Action
Ecosystem restoration approaches	Rehabilitate degraded ecosystems to restore natural barriers and enhance ecosystem resilience	**Reforestation of degraded urban fringes**: Restore degraded land, plant native trees and vegetation on the periphery of urban areas, and restore the connectivity between fragmented forest patches **Wetland rehabilitation projects**: Restore wetlands by reintroducing native plant species, enhancing water quality, and restoring natural structure **Constructed wetlands**: Create habitats within urban areas by building wetlands to manage stormwater and wastewater
Issue-specific approaches	Address specific challenges related to ecosystem services	**Ecosystem-based adaptation**: Measures such as installing green roofs and increase green cover to reduce the impacts of UHI **Ecosystem-based mitigation**: Measures such as vegetative buffers and afforestation help reduce risky interfaces **Ecosystem-based disaster risk reduction**: Reduce the impact of urban flooding and extreme weather events by the use of natural barriers like mangroves and dunes **Sustainable livestock production**: Undertake sustainable farming practices, such as rotational grazing to reduce disease transmission and environmental impacts

(continued)

Table 10.2 (continued)

NBS category	Intervention point	Action
Infrastructure-related approaches	Develop and implement natural and green infrastructure to support ecosystem functions	**Natural infrastructure**: Manage flood risks and enhance biodiversity by creating and restoring natural ecosystems like wetlands and riparian zones **Green Infrastructure**: Implement features like green roofs, rain gardens, etc., to improve stormwater management, reduce urban heat island effects **Sustainable urban drainage systems**: Use permeable pavements, rain gardens, and detention basins to manage stormwater, reduce flooding, and enhance urban water quality **Bioswales**: create bioswales to capture and filter runoff, reducing pollution and enhancing urban green spaces

(continued)

Table 10.2 (continued)

NBS category	Intervention point	Action
Ecosystem-based management approaches	Manage ecosystems to address broader ecological and societal issues	**Integrated coastal zone management**: Balance development and conservation in coastal areas to prevent habitat loss and reduce human-wildlife conflict **Integrated water resources management**: Implement sustainable water use and management practices that support aquatic ecosystems and reduce the risk of waterborne zoonotic pathogens **Creation of wildlife corridors: Help minimize conflicts, reduce habitat fragmentation** Create corridors that connect fragmented habitats thus reduce risky human–wildlife interfaces
Ecosystem protection approaches	Safeguard existing natural habitats to preserve biodiversity and prevent habitat fragmentation	**Conservation of urban forests and green spaces**: Measures to protect urban forests and green spaces, maintain natural habitats and biodiversity **Area-based conservation approaches**: Establish protected areas and buffer zones around criticaSl habitats like wetlands to reduce human encroachment and thus maintain ecosystem integrity **Riparian buffers**: Establish vegetated areas along streams and rivers to filter runoff, reduce erosion, and create habitats for wildlife

by influencing the risky interfaces to reduce human–wildlife animal interaction. The study grouped potential NBS interventions according to the IUCN framework, which includes strategies such as habitat restoration, biodiversity conservation, ecosystem management, and the development of green infrastructure. While this chapter highlights the potential for NBS to mitigate zoonotic spillover risks, further research can be carried out to assess the long-term impact of such interventions. Future studies could focus on developing quantitative models to predict how specific NBS interventions influence zoonotic transmission dynamics over time. In conclusion, integrating nature-based solutions into urban and public health planning presents a proactive and multifaceted approach to reducing zoonotic spillover risks. As cities continue to expand and evolve, NBS will play a pivotal role in creating resilient, healthy, and sustainable urban environments that can withstand the challenges posed by emerging infectious diseases.

References

Ahmed S et al (2019) Does urbanization make emergence of zoonosis more likely? Evidence, myths and gaps. Environ Urban 31(2). https://doi.org/10.1177/0956247819866124

Alberti M (2005) The effects of urban patterns on ecosystem function. Int Reg Sci Rev. https://doi.org/10.1177/0160017605275160

Aliyu A, Amadu L (2017) Urbanization, cities, and health: the challenges to Nigeria—a review. Ann Afr Med. https://doi.org/10.4103/aam.aam_1_17

Alvarado MR et al (2023) Street trees and mental health: developing systems thinking-informed hypotheses using causal loop diagramming. Ecol Soc 28(2). https://doi.org/10.5751/ES-14013-280201

Babí Almenar J et al (2021) Nexus between nature-based solutions, ecosystem services and urban challenges. Land Use Policy 100. https://doi.org/10.1016/j.landusepol.2020.104898

Bai X et al (2012) Health and wellbeing in the changing urban environment: complex challenges, scientific responses, and the way forward. Curr Opin Environ Sustain. https://doi.org/10.1016/j.cosust.2012.09.009

Baker RE et al (2021) Infectious disease in an era of global change. Nat Rev Microbiol. https://doi.org/10.1038/s41579-021-00639-z

Bhopal RS (2009) Re: book review of "concepts of epidemiology: integrating the ideas, theories, principles and methods of epidemiology, 2nd edn". Am J Epidemiol 169(11). https://doi.org/10.1093/aje/kwp093

Braubach M et al (2017) Effects of urban green space on environmental health. Equity Resilience. https://doi.org/10.1007/978-3-319-56091-5_11

Brisbois BW, Ali SH (2010) Climate change, vector-borne disease and interdisciplinary research: social science perspectives on an environment and health controversy. EcoHealth 7(4). https://doi.org/10.1007/s10393-010-0354-6

Cohen-Shacham E et al (2016) Nature-based solutions to address global societal challenges. IUCN, Gland, Switzerland. xiii + 97pp

Coker RJ et al (2011) Emerging infectious diseases in southeast Asia: regional challenges to control. The Lancet. https://doi.org/10.1016/S0140-6736(10)62004-1

Coletta VR et al (2021) Causal loop diagrams for supporting nature based solutions participatory design and performance assessment. J Environ Manage 280. https://doi.org/10.1016/j.jenvman.2020.111668

Combs MA et al (2022) Socio-ecological drivers of multiple zoonotic hazards in highly urbanized cities. Glob Change Biol. John Wiley and Sons Inc, pp 1705–1724. https://doi.org/10.1111/gcb.16033

Connolly C, Keil R, Ali SH (2021) Extended urbanisation and the spatialities of infectious disease: demographic change, infrastructure and governance. Urban Stud 58(2). https://doi.org/10.1177/0042098020910873

Diemer A, Nedelciu CE (2020) System dynamics for sustainable urban planning. https://doi.org/10.1007/978-3-319-95717-3_115

Dunlop T et al (2024) The evolution and future of research on Nature-based Solutions to address societal challenges. Commun Earth Environ 5(1). https://doi.org/10.1038/s43247-024-01308-8

Eggermont H et al (2015) Nature-based solutions: New influence for environmental management and research in Europe. GAIA Ecol Perspect Sci Soc. https://doi.org/10.14512/gaia.24.4.9

Eltarabily S, Elghezanwy D (2020) Post-pandemic cities—the impact of covid-19 on cities and urban design. Architect Res 10(3)

Field HE (2009) Bats and emerging zoonoses: henipaviruses and SARS. Zoonoses Public Health. https://doi.org/10.1111/j.1863-2378.2008.01218.x

Finucane ML et al (2014) A conceptual framework for analyzing social-ecological models of emerging infectious diseases. In: Understanding society and natural resources: forging new strands of integration across the social sciences. https://doi.org/10.1007/978-94-017-8959-2_5

Glidden CK et al (2021) Human-mediated impacts on biodiversity and the consequences for zoonotic disease spillover. Curr Biol. https://doi.org/10.1016/j.cub.2021.08.070

Gottdenker NL et al (2014) Anthropogenic land use change and infectious diseases: a review of the evidence. EcoHealth. Springer Science and Business Media, LLC, pp 619–632. https://doi.org/10.1007/s10393-014-0941-z

Gouveia N, Kanai C (2020) Pandemics, cities and public health. Ambiente e Sociedade 23. https://doi.org/10.1590/1809-4422ASOC20200120VU2020L3ID

Guo F, Bonebrake TC, Gibson L (2019) Land-use change alters host and vector communities and may elevate disease risk. EcoHealth 16(4). https://doi.org/10.1007/s10393-018-1336-3

Hassell JM et al (2017) Urbanization and disease emergence: dynamics at the wildlife–livestock–human interface. Trends Ecol Evol. https://doi.org/10.1016/j.tree.2016.09.012

Jiang X et al (2022) Factors associated with severe fever with thrombocytopenia syndrome in endemic areas of China. Front Public Health 10. https://doi.org/10.3389/fpubh.2022.844220

Jones BA, Betson M, Pfeiffer DU (2017) Eco-social processes influencing infectious disease emergence and spread. Parasitology. https://doi.org/10.1017/S0031182016001414

Jones KE et al (2008) Global trends in emerging infectious diseases. Nature 451(7181). https://doi.org/10.1038/nature06536.

Kabisch N et al (2016) Nature-based solutions to climate change mitigation and adaptation in urban areas: perspectives on indicators, knowledge gaps, barriers, and opportunities for action. Ecol Soc 21(2). https://doi.org/10.5751/ES-08373-210239

Kabisch N et al (2017) Nature-based solutions to climate change adaptation in urban areas—linkages between science. Policy Pract. https://doi.org/10.1007/978-3-319-56091-5_1

Keesing F, Ostfeld RS (2021) Impacts of biodiversity and biodiversity loss on zoonotic diseases. Proc Natl Acad Sci USA. https://doi.org/10.1073/PNAS.2023540118

Khor N et al (2022) Envisaging the future of cities

Kim DH (2000) Systems thinking tools: a user's reference guide systems archetypes I: diagnosing systemic issues and designing high-leverage interventions systems, pegasus communications

Kim D (2021) Exploratory study on the spatial relationship between emerging infectious diseases and urban characteristics: cases from Korea. Sustain Cities Soc 66. https://doi.org/10.1016/j.scs.2020.102672

Langellier BA et al (2019) Using community-based system dynamics modeling to understand the complex systems that influence health in cities: the Salurbal study. Health Place 60. https://doi.org/10.1016/j.healthplace.2019.102215

Liu Q et al (2018) Landscape of emerging and re-emerging infectious diseases in China: impact of ecology, climate, and behavior. Front Med. https://doi.org/10.1007/s11684-017-0605-9

Lõhmus M, Balbus J (2015) Making green infrastructure healthier infrastructure. Infect Ecol Epidemiol. https://doi.org/10.3402/IEE.V5.30082

Mackenstedt U, Jenkins D, Romig T (2015) The role of wildlife in the transmission of parasitic zoonoses in peri-urban and urban areas. Int J Parasitol Parasites Wildlife. https://doi.org/10.1016/j.ijppaw.2015.01.006

Matter NM, Gado NG (2024) Constructed wetlands nature-based solutions to enhance urban resilience in Egyptian cities. HBRC J 20(1). https://doi.org/10.1080/16874048.2024.2311521

Matthew R et al (2022) Research note: climate change, peri-urban space and emerging infectious disease. Landsc Urban Plan 218. https://doi.org/10.1016/j.landurbplan.2021.104298

Mayer JD (2000) Geography, ecology and emerging infectious diseases. Soc Sci Med. https://doi.org/10.1016/S0277-9536(99)00346-9

McArthur DB (2019) Emerging infectious diseases. Nurs Clin North Am. https://doi.org/10.1016/j.cnur.2019.02.006

Miao D et al (2020) Mapping the global potential transmission hotspots for severe fever with thrombocytopenia syndrome by machine learning methods. Emerg Microbes Infect 9(1). https://doi.org/10.1080/22221751.2020.1748521

Mills JG et al (2017) Urban habitat restoration provides a human health benefit through microbiome rewilding: the microbiome rewilding hypothesis. Restor Ecol 25(6). https://doi.org/10.1111/rec.12610

Mukherjee S (2017) Emerging infectious diseases: epidemiological perspective. Indian J Dermatol. https://doi.org/10.4103/ijd.IJD_379_17

Murugesan V (2020) How can urban planners address emerging zoonoses? A scoping review of recommended interventions. Urbana Urban Aff Public Policy. https://doi.org/10.47785/urbana.4.2020

Neiderud CJ (2015) How urbanization affects the epidemiology of emerging infectious diseases. Afr J Disabil 5(1). https://doi.org/10.3402/iee.v5.27060.

Nesshöver C et al (2017) The science, policy and practice of nature-based solutions: an interdisciplinary perspective. Sci Total Environ. https://doi.org/10.1016/j.scitotenv.2016.11.106

Owrangi AM, Lannigan R, Simonovic SP (2015) Mapping climate change-caused health risk for integrated city resilience modelling. Nat Hazards 77(1). https://doi.org/10.1007/s11069-014-1582-9

Plowright RK et al (2017) Pathways to zoonotic spillover. Nat Rev Microbiol. https://doi.org/10.1038/nrmicro.2017.45

Plowright RK et al (2024) Ecological countermeasures to prevent pathogen spillover and subsequent pandemics. Nat Commun 15(1). https://doi.org/10.1038/s41467-024-46151-9

Prist PR et al (2023) Promoting landscapes with a low zoonotic disease risk through forest restoration: the need for comprehensive guidelines. J Appl Ecol. https://doi.org/10.1111/1365-2664.14442

Ram SA, Irfan ZB (2021) Application of system thinking causal loop modelling in understanding water crisis in India: a case for sustainable integrated water resources management across sectors. HydroResearch 4. https://doi.org/10.1016/j.hydres.2021.02.001

Roux AVD (2015) Health in cities: is a systems approach needed? Cadernos de Saúde Pública 31(suppl 1). https://doi.org/10.1590/0102-311xde01s115

Rubin DM et al (2021) Facilitating understanding, modeling and simulation of infectious disease epidemics in the age of covid-19. Front Public Health 9. https://doi.org/10.3389/fpubh.2021.593417

Sterman JD (2004) Systems thinking and modeling for a complex world, interfaces

United Nations (2018) World urbanization prospects 2018. Department of Economic and Social Affairs. World Population Prospects 2018

Vanhove MPM et al (2021) Global change increases zoonotic risk, covid-19 changes risk perceptions: a plea for urban nature connectedness. Cities Health 5(sup1). https://doi.org/10.1080/23748834.2020.1805282

Wang X et al (2021) Using a one health approach to prioritize zoonotic diseases in China, 2019. Plos One 16 (11 November). https://doi.org/10.1371/journal.pone.0259706

Weiss RA, McMichael AJ (2004) Social and environmental risk factors in the emergence of infectious diseases. Nat Med. https://doi.org/10.1038/nm1150

Wilcox BA, Gubler DJ, Pizer HF (2008) Urbanization and the social ecology of emerging infectious diseases. In: The social ecology of infectious diseases. https://doi.org/10.1016/B978-012370466-5.50009-1

Wood CL et al (2014) Does biodiversity protect humans against infectious disease? Ecology 95(4). https://doi.org/10.1890/13-1041.1

Xie L, Bulkeley H (2020) Nature-based solutions for urban biodiversity governance. Environ Sci Policy 110. https://doi.org/10.1016/j.envsci.2020.04.002

Chapter 11
Re-envisioning India's Silicon Valley as a Water-Sensitive City: Bengaluru

Varun Prasad, Chithra K., and Bimal P.

Abstract India's urban area is home to over 500 million people (36% of the total population) and is growing rapidly. During the past five decades, the Bengaluru region has attracted 13 million people, leading to the rapid transformation of its LULC pattern and becoming one of the fastest-growing cities in the world. Due to this, there has been a change in drainage patterns, an increase in impervious surfaces, and encroachment on catchment areas. Improper management and the continued trend in the city's growth will lead to higher demand, causing water scarcity, inadequate drainage systems, frequent floods, etc. Water-sensitive city provides various ecological, social, and economic benefits. It includes water security by efficiently using diverse resources, mitigating flood risk and damages, and improving and protecting the health of wetlands, river basins, etc. This chapter focuses on the city's long-term viability, sustainability, liveability, and resilience. A detailed analysis of the city's existing situation, including the existing demand and supply, urban floods, water drainage network, urban sprawl, etc. The aspects of water availability, accessibility, safety, efficiency, and affordability are assessed in the study. A prediction of urban growth for two different scenarios was carried out to identify the rate of encroachment and the relation between urban sprawl and waterbodies. The prediction was carried out for three periods, 2031, 2041, and 2047, using the CA-Markov method in IDRISI software. With the help of SWOT, a water-sensitive master plan has been created to resolve the current challenges and the future demand for resources and infrastructure. The recommendations contain plans to effectively manage its water resources, enhance water quality, and promote resilience in changing hydrological conditions while maintaining a high quality of life for its residents to make Bengaluru a water-sensitive city by 2047.

Keywords Water-sensitive city · Resilient city · Sustainable city · Urban water management · Sustainable urban water management

V. Prasad (✉) · Chithra K. · Bimal P.
National Institute of Technology Calicut, Kozhikode, Kerala, India
e-mail: varunap192@gmail.com

L. K. Dashora et al. (eds.), *Urban Sustainability*, Disaster Risk Reduction,
https://doi.org/10.1007/978-981-96-4890-0_11

11.1 Introduction

Urbanisation has progressed rapidly during the twentieth and twenty-first centuries. The global urban population proportion increased from 13% in 1900 to 29% in 1950, climbing to 49% in 2005 and currently at 55%. Projections indicate that this trend will persist, with an estimated 70% of the world's population residing in cities within a few decades (United Nations Department of Economic and Social Affairs and Pollution Division 2018). The rapid pace of urbanisation introduces significant challenges, particularly in water management. Competition for water resources arises from different sectors, such as agriculture and industry, complicating management efforts. Additionally, climate variability and change, including rising temperatures and shifting weather patterns, further exacerbate these pressures on water resources (Flörke et al. 2018; Field et al. 2014a). As cities expand, water demand increases, straining supply systems and degrading water quality. Over the past six decades, domestic water consumption has quadrupled due to increased access to drinking water infrastructure, population growth, and rising incomes (Flörke et al. 2013). By 2050, global domestic water withdrawal is projected to increase by 50–250% (Wada et al. 2016). Metropolitan regions are at greater risk from climate change, which is also likely to make the future's major water concerns worse rather than better (Hallegatte et al. 2013; McDonald et al. 2011; Field et al. 2014b). Rapid urbanisation, ageing infrastructure, rigid institutions, and interconnected global economies have intensified extreme weather events like floods, droughts, and heat waves (Chester et al. 2021). Climate change-induced uncertainties, including irregular rainfall and frequent extreme weather events, further heighten urban water insecurity.

A water-sensitive city (WSC) adopts integrated approaches to address urban water challenges, enhance resource efficiency, prevent floods, and adapt to changes in water supply. Key strategies include using green and blue infrastructure, permeable pavements, urban wetlands, water recycling, stormwater harvesting, and ecosystem restoration, as successfully implemented in cities like Melbourne (City of Melbourne WSUD Guidelines) and Singapore (Cui et al. 2021). Historically, urban development has prioritised economic growth over sustainable water management, resulting in issues such as inadequate stormwater systems, water contamination, and groundwater depletion, which are intensifying due to climate change. WSCs aim to integrate water management practices that reduce demand, maximise recycling, and build resilience to climate impacts while fostering environmental sustainability and addressing social inequalities related to water scarcity.

India is a South Asian country with over 1. 4 billion people, occupying only 2.4% of the world's land mass with an urban population of 36% and rising (India Population 2024). Bengaluru is one of India's fastest-growing cities and the 'Silicon Valley of India' because of the booming IT industry in the last decades (Kar 2024). The city has witnessed a rapid and unprecedented urban sprawl and development due to recent economic developments, with an urban growth rate of 925% (1973 to 2013) across four decades (Asulabha et al. 2016). The number of lakes in the city reduced from nearly 285 to 194 as of 2006. Over the past four decades, water bodies have experienced a 79% reduction (Shivamurthy 2017), contributing to increased flood frequency and heightened levels of pollution. A recent study identified Bengaluru as one of the cities with the lowest water availability in the Global South, with access limited to an average of two to three hours every three days (Mitlin et al. 2019). Climate change has also exacerbated challenges in managing water resources and infrastructure due to outdated systems and the absence of long-term, visionary strategic planning (Pantelis Adamopoulou et al. 2023). Despite ongoing efforts to enhance urban water management in Bengaluru, several critical research gaps persist:

1. **Lack of Integrated Water Management Approaches**: Urban planning often needs a comprehensive framework incorporating water-sensitive urban design, green infrastructure, and integrated planning strategies.
2. **Inadequate Climate Change Adaptation**: Existing water management practices are insufficient to address the impacts of climate change on water resources.
3. **Limited Implementation of Water-Sensitive Urban Design**: Although global guidelines for water-sensitive cities exist, their application in Bengaluru has been constrained by institutional, financial, and public engagement challenges.
4. **Insufficient Community Engagement**: Public awareness and participation in water conservation and sustainable use remain inadequate.

This chapter aims to analyse Bengaluru's current water management practices by addressing gaps in the city and proposing strategic recommendations to transform it into a water-sensitive city (WSC). The objectives of the chapter include:

1. Analysing water management challenges causing water stress, floods, and poor water quality in Bengaluru.
2. Identify sustainable water management practices that are economically viable for Bengaluru's environment and socioeconomic fabric, specifically green and blue infrastructure.
3. Policy and technological interventions for improving water-integrated urban planning and community participation.

The chapter explores the concept of incorporating water-sensitive cities into urban development to mitigate the negative effects of uncontrolled urbanisation and climate change, enhance water sustainability, and improve the quality of life. It provides practical guidelines for strategic water management in challenging urban environments and contributes to the discussion on building resilient, sustainable cities.

Water-Sensitive City (WSC) Concept—Background and Evolution

A water-sensitive city (WSC) is an innovative urban development framework that integrates water management with sustainability, emphasising the entire water cycle, including water supply, stormwater, and wastewater (Rogers et al. 2020; Realising the vision of a water sensitive city—The source 2024). It promotes decentralised systems focused on water conservation, rainwater harvesting, and wastewater reclamation, enhancing urban resilience. The WSC concept evolved in response to the limitations of traditional, centralised water systems, which struggled to meet the demands of urbanisation, climate change, and water scarcity. WSCs aim to secure water supply by integrating green infrastructure and ecosystem renewal into urban planning. They address global challenges such as climate change, urban population growth, and rising water demand while reducing flood risks, improving water quality, and mitigating the urban heat island effect (Rohr 2019). Using sustainable infrastructure like rain gardens, permeable pavements, and wetlands, WSCs promote ecological balance, reduce reliance on traditional water supply systems, and enhance urban livability through green spaces and biodiversity.

Case Study 1: Melbourne

Melbourne, Australia, faced critical challenges, including drought, water shortages, and increasing urban water demand (Land, Water and Planning 2016; Melbourne Water System Strategy 2017), prompting the city to adopt a water-sensitive city (WSC) approach for improved resilience. Through water-sensitive urban design (WSUD), Melbourne integrated natural water management into urban planning by implementing key initiatives, including incentivising rainwater harvesting, establishing systems to collect, treat, and store stormwater for non-potable uses, and utilising green infrastructure such as rain gardens, permeable pavements, and living roofs(Low et al. 2015). Public awareness campaigns further promoted water conservation and the use of water-efficient fixtures. The WSC approach in Melbourne successfully reduced water consumption, improved water quality, controlled urban flooding, mitigated the urban heat island effect, and enhanced ecological and recreational spaces, positioning the city as a global model for water-sensitive urban design.

Case Study 2: Singapore

Singapore, facing limited water resources and a dense population, developed a sustainable water policy centred on integrating new technologies and water-sensitive principles (Cecilia et al. 2013). The city-state adopted the ‘Four National Taps’ strategy, which includes water from local catchments, imported water, desalinated water, and NEWater (a system that treats sewage water using advanced membrane technology and ultraviolet disinfection, supplying 40% of the nation’s water demand). Singapore has also expanded rainwater harvesting, turning two-thirds of its land into water catchments (Irvine et al. 2014). This water-sensitive approach has secured a reliable water supply, enhanced flood management, cooled urban temperatures, and created new public spaces, positioning Singapore as a model for cities with similar water challenges.

Case Study 3: Toronto

Toronto, Canada, faces water management challenges, including flooding, ageing water and sanitation infrastructure, and the impacts of climate change, such as more frequent and intense storms (D’Andrea 2020). The city has implemented a broad water-sensitive strategy to enhance sustainability and urban resilience with crucial measures, including green roof programs (Carter and Fowler 2008), urban parks integrating natural hydrological systems, and low impact development (LID) techniques such as permeable pavements and bioswales, which manage stormwater and improve groundwater recharge (Wang and Sandberg 2014). Rainwater harvesting also reduces reliance on potable water for non-essential uses. These initiatives have reduced flood risks, mitigated the urban heat island effect, and improved urban livability (Wang and Sandberg 2014). Educational programs and incentives have further increased community participation and awareness. Overall, Toronto's proactive approach has strengthened the city's preparedness for climate change and extreme weather events, promoting long-term sustainability.

11.2 Study Area

The Greater Bengaluru city in Karnataka, India, spans 741 square kilometres (km^2) (Fig. 11.1). It is situated between the latitudes 12°39′00" to 13°13′00" N and longitudes 77°22′00" to 77°52′00" E, located on the Deccan plateau at an altitude of 900m mean sea level (MSL). Bengaluru is recognised as one of the world's fastest-growing megacities, with a population of approximately 13 million (Bangalore, India Metro Area Population 1950–2023|MacroTrends 2023). Over the years, the city, originally known as Garden City, has gradually evolved into an industrial and software hub, becoming India's Silicon Valley.

Bengaluru is the second largest metropolis in India and has expanded spatially more than ten times since 1949. In 1949, the city's total area was approximately 69 square kilometres, and the area has increased to 741 square kilometres. The city's population has grown significantly from 6.5 million in 2001 to 9.5 million in 2011,

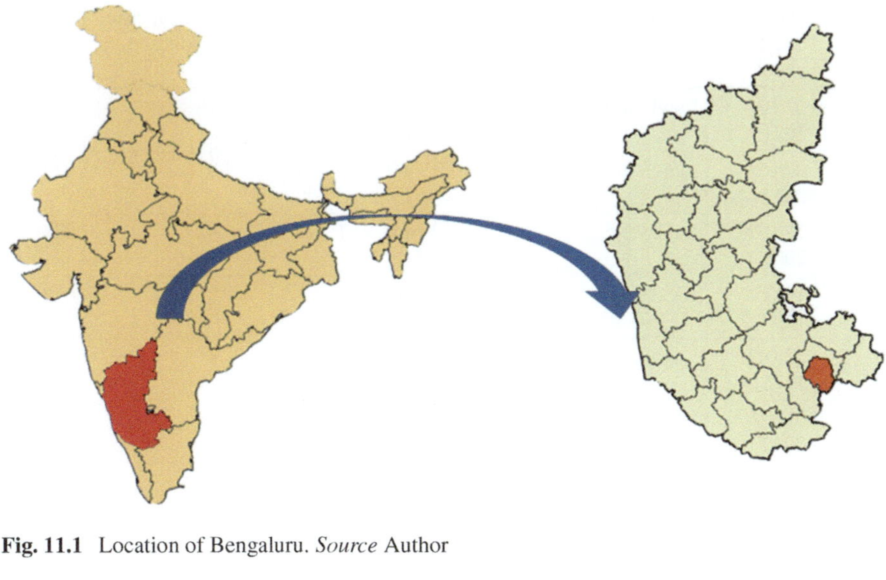

Fig. 11.1 Location of Bengaluru. *Source* Author

representing a growth of 46.68% over the decade. The city has a gross domestic product (GDP) of $110 billion, which accounts for over 87% of Karnataka's total GDP. The Bengaluru IT hub contributes the largest share, accounting for 34% to 40% of India's IT exports ('Bengaluru's IT Dominance 2023). The Bruhat Bengaluru Mahanagara Palike (BBMP) is the administrative body overseeing civic amenities and certain infrastructural assets within the Greater Bengaluru metropolitan area. The Bengaluru Water Supply and Sewerage Board (BWSSB) manages and maintains the city's water supply and sewage infrastructure.

11.3 Materials and Methods

11.3.1 Materials

Landsat Images for five years—1972, 1991, 2001, 2011, and 2021—were downloaded from the U.S. Geological Survey (USGS) Earth Explorer to map the temporal Land Use and Land Cover (LULC) (Table 11.1). Google Earth Engine was used to create flood-vulnerable areas. The digital elevation map has also been downloaded for stream order classification.

Ancillary Data were used: Road & Rail network, Metro & Rail Station, Boundary, and Restricted areas for prediction of LULC for the years 2031, 2041, and 2047 (Table 11.2).

Table 11.1 Imageries attributes

SI no	Date of acquired maps	Data type	Spatial resolution (m)	Path/ row	Number of bands	Radiometric resolution	Cloud cover
1	02/03/ 2021	LandSat-8	30	144/ 51	11	16	0.00
2	07/03/ 2011	LandSat-5	30	144/ 51	7	8	0.00
3	27/03/ 2001	LandSat-5	30	144/ 51	7	8	0.00
4	16/03/ 1991	LandSat-5	30	144/ 51	7	8	0.00
5	22/01/ 1972	LandSat-1	30	144/ 51	7	8	0.00
6	01/1-/ 2012	DEM-SRTM	30	–	–	–	–

Table 11.2 List of secondary data

SI no	Data type	Source type	Source
1	Road network	Secondary	Open street map
2	Rail network	Secondary	Open street map
3	Metro station	Secondary	Google-my maps
4	Railway station	Secondary	Google-my maps
5	Boundary	Secondary	Karnataka-GIS
6	Restricted area	Secondary	BDA-masterplan-2015
7	Precipitation data	Secondary	Karnataka state natural disaster monitoring centre (KSNDMC)
8	Flood vulnerable points	Secondary	Karnataka state natural disaster monitoring centre (KSNDMC)

11.3.2 Methods

Figure 11.2 depicts the flow of methodology used for the study. The study addresses global and regional urban water scarcity, resilience, and sustainability concerns. Based on these considerations, the study formulates its aims and objectives and a literature review exploring how cities can transition into water-sensitive urban areas, drawing insights from relevant case examples. The existing scenario analysis assesses various aspects of water management, including drainage systems, water supply, urban floods, livability, sustainability practices, and land use analysis. Surveys and research analysis revealed the water availability, accessibility, and affordability. To predict future trends, the study employs a CA–Markov model. This model projects regional growth and urban sprawl for specific years (2031, 2041, and 2047) under two scenarios: one without constraints (unplanned development) and another considering waterbody restrictions (planned development). Additionally, flood vulnerability assessment is conducted using Google Earth Engine code. By mapping flood-prone areas for multiple years, the study identifies challenges and gaps that need addressing for cities to become water-sensitive. The results were overlayed and analysed, leading to a strength, weakness, opportunities, and threats (SWOT) analysis highlighting the city's strengths, weaknesses, opportunities, and threats to water resources. The study concludes by proposing an action plan that includes spatial planning, policies, and recommendations, aiming to transform the city into a water-sensitive urban area by 2047.

11.3.2.1 Support Vector Machine Classification

Support vector machine (SVM) offers a principled approach owing to its statistical learning theory-based mathematical underpinning. It finds applications across a broad

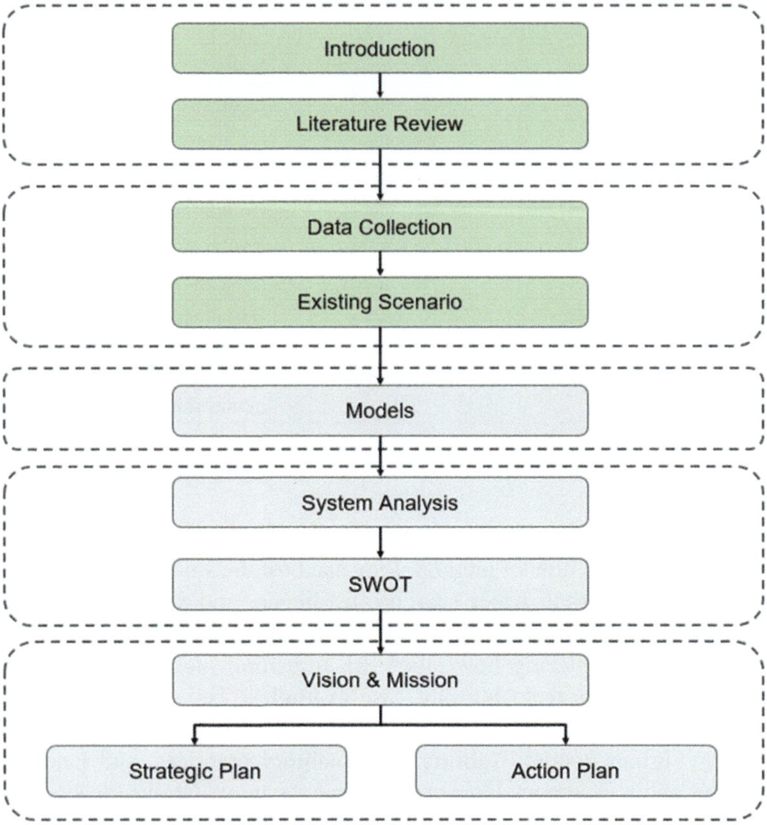

Fig. 11.2 Overall methodology. *Source* Author

spectrum, including classification, regression, novelty detection, and feature reduction problems. The support vector machine (SVM) algorithm is designed to establish an optimal decision boundary, known as a hyperplane, that separates various classes within an n-dimensional space. This facilitates the rapid classification of new data points. SVM identifies the support vectors, critical data points that help define the hyperplane (Pantelis Adamopoulou et al. 2023). Figure 11.3 depicts the use of a hyperplane to classify two distinct categories. Support vectors represent the data points or vectors closest to the hyperplane, wielding the most substantial influence on its placement.

Figure 11.4 represents the methodology for preparing the LULC map. The Landsat Images were downloaded from the USGS Website. Radiometric corrections (it reduces the impact of atmospheric, sensor, and lighting conditions to increase the precision and quality of remote sensing images) were done on each band and were

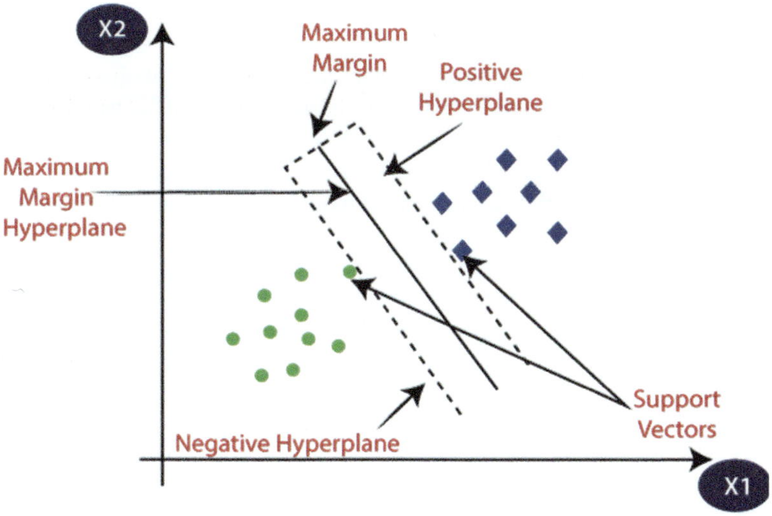

Fig. 11.3 Support vector machine example. *Source* https://www.analyticsvidhya.com/blog/2021/10/support-vector-machinessvm-a-complete-guide-for-beginners/

combined to create a composite band. A false colour composite was used for supervised classification, and samples were classified into five classes: Built-up, waterbodies, dried waterbodies, vegetation, and barren land. These samples were further used to train SVM and classify the raster using the composite band and the trained sample, resulting in a LULC map.

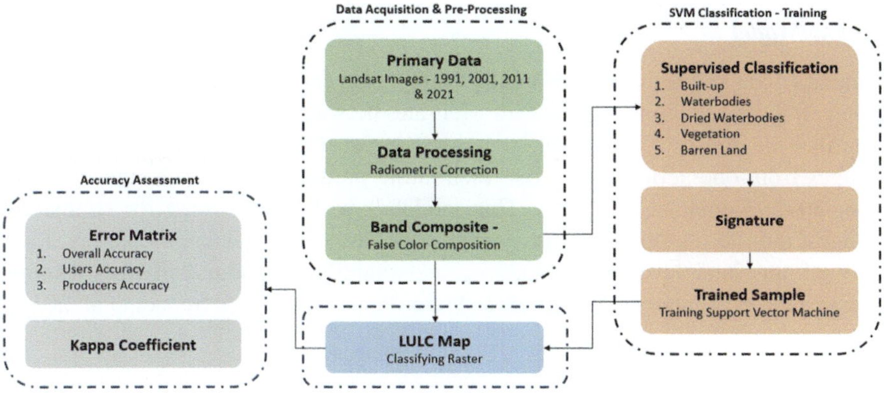

Fig. 11.4 Methodology adopted for LULC map production. *Source* Author

11.3.2.2 Accuracy Assessment

Accuracy assessment involves comparing the classified image with a more reliable data source. In the case of SVM classification, the accuracy was assessed for five years using the error matrix method and the Kappa coefficient. The error matrix is one commonly used method for assessing the accuracy of thematic maps. It involves sampling a specific class from a categorised map and using the field to validate the actual class.

Overall accuracy is calculated by dividing the total number of correctly classified sample points by the total number of sample points (Mondal et al. 2012). Kappa coefficient—The discrete multivariate approach known as the kappa coefficient is used to quantify the precision of maps produced by remote sensing and error matrices. These methods are thought to be more suitable than continuous data.

11.3.2.3 LULC Prediction Simulation Using CA–Markov Model

CA–Markov model is one of the planning support tools for the analysis of temporal changes and spatial distribution of LULC (Hua 2017). It is one of the most widely used, among many LULC modelling tools and techniques, which models both spatial and temporal changes (Regmi et al. n.d.; Weng 2002). CA–Markov model combines cellular automata and the Markov chain to predict the Land Use & Land Cover Change (LULCC) trends and characteristics over time (Nouri et al. 2014). The dynamics of LULC, forest cover, urban development, plant growth, and watershed management modelling are all characterised by this model. Understanding the interconnections between people and the environment from a long-term perspective requires research of the historical LULCC (Yang et al. 2014).

Markov Model

Figure 11.5 depicts a flow of the methodology used for the prediction. The Markov model effectively demonstrates the transfer rates between different land use types and quantifies the transitions among these types (Sang et al. 2011). It characterises LULC changes from one period to another, facilitating predicting future changes (Behera et al. 2012; Kumar et al. 2014). This model is widely employed for simulating changes, monitoring purposes, analysing LULC trends, ecological modelling, forecasting future land development stability, and quantifying land use changes within a specific region of interest (Weng 2002, Amini Parsa et al. 2016, Subedi et al. 2013).

CA-Markov

The CA–Markov model integrates cellular automata with a transition probability matrix derived from cross-tabulating two distinct images, making it a robust method for spatiotemporal dynamic modelling (Singh et al. 2015; Wang and Zhang 2001). The CA–Markov model effectively combines geographic information systems (GIS) and remote sensing (RS) and offers an accurate method for spatial and temporal dynamic modelling of land use changes.(Kamusoko et al. 2009). CA-Model can be

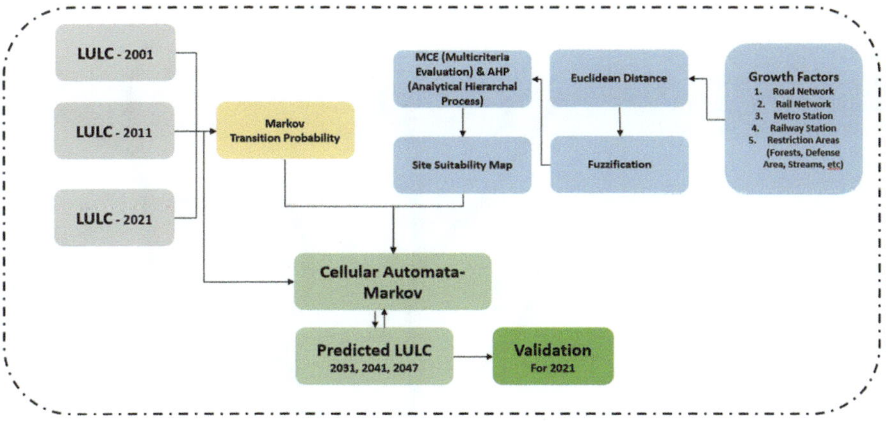

Fig. 11.5 Methodology adopted for prediction model. *Source* Author

expressed by Sang et al. (2011); Subedi et al. 2013).

$$S(t,\ t+1) = f(S(t),\ N)$$

S(t + 1) represents the system status at time t + 1, determined by the state probability at any given time N. For validation, three indicators were employed: Kappa for no ability (K_{no}), Kappa for location ($K_{location}$), and Kappa for quantity ($K_{quantity}$). These indicators validated the CA–Markov model's land use and land cover (LULC) changes predictions. K_{no} provides the overall accuracy of a simulation run, while $K_{location}$ and $K_{quantity}$ assess the agreement in location and quantity between the actual and simulated maps (Zaman and Sayemuzzaman 2014; Pontius and Malanson 2007; Pontius and Schneider 2001). According to Eastman, a Kappa value of 0.80 is considered an acceptable accuracy rate for making reliable future predictions (Eastman 2006).

IDRISI software was used to predict the LULC change for 2031, 2041, and 2047. The 2021 LULC was predicted first to validate the model using 2001 and 2011 data to create the transition probability maps using Markov. Growth factors like road networks, rail networks, metro stations, and railway stations were used to develop Euclidean distance. These Euclidean Distance Raster files were standardised by fuzzified into monotonically increasing and decreasing functions. The fuzzified growth factors and constraints (defence area and forest area) were used for a multi-criteria evaluation through an analytical hierarchical process (AHP) to create site suitability maps. Through the integration of multi-criteria evaluation (MCE), Markov, and cellular automata (CA), the LULC change was predicted. This was further validated with the real/produced 2021 LULC map. Following the same methods and changes in LULC maps from 2001–2011 to 2011–2021, the prediction maps were generated for 2031, 2041, and 2047.

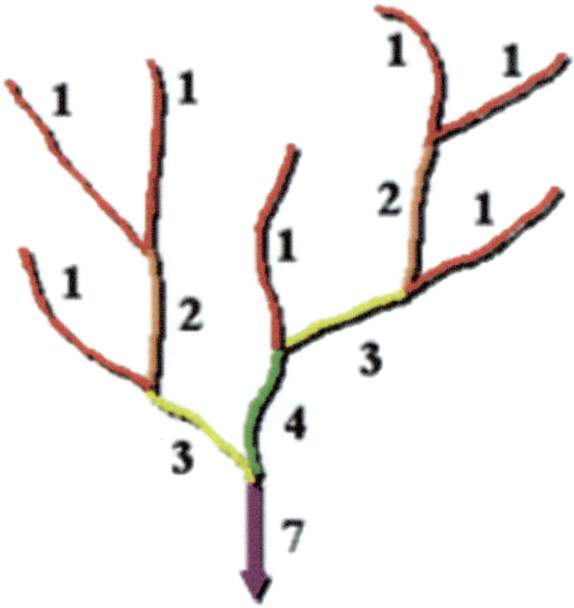

Fig. 11.6 Example for stream order. *Source* https://pro.arcgis.com/en/pro-app/latest/tool-reference/spatial-analyst/how-stream-order-works.htm

11.3.2.4 Stream Order

It is a method for assigning a numerical order to links within a stream network and is categorised into various stream types.(Tarboton et al. 1991). The minuscule numerals represent the minor tributary, and the highest numerical represents a stream or the biggest tributary in that area, as shown in Fig. 11.6.

Stream order is extracted from a digital elevation map (DEM) using ArcGIS software. The DEM file is first filled to remove sinks from the image. The corrected DEM file is used to create a flow direction map, followed by a flow accumulation map. The flow accumulation tool calculates the weight of all cells flowing into each downslope cell in the DEM raster image used to create a stream order map.

11.3.2.5 Flood Vulnerable Areas—Google Earth Engine

Flood-vulnerable areas in Bengaluru were mapped for five years (2017, 2018, 2020, 2021, and 2022) using the Sentinel-1 Synthetic Aperture Radar (SAR) Data in Google Earth Engine to determine the extent of significant floods. SAR can operate in all weather and through cloud cover while providing accurate and crucial information (Recommended Practice 2023). The map was prepared during the city's rainy period, between July and September. The flood maps code was obtained from the United Nations (UN) spider (Fig. 11.7).

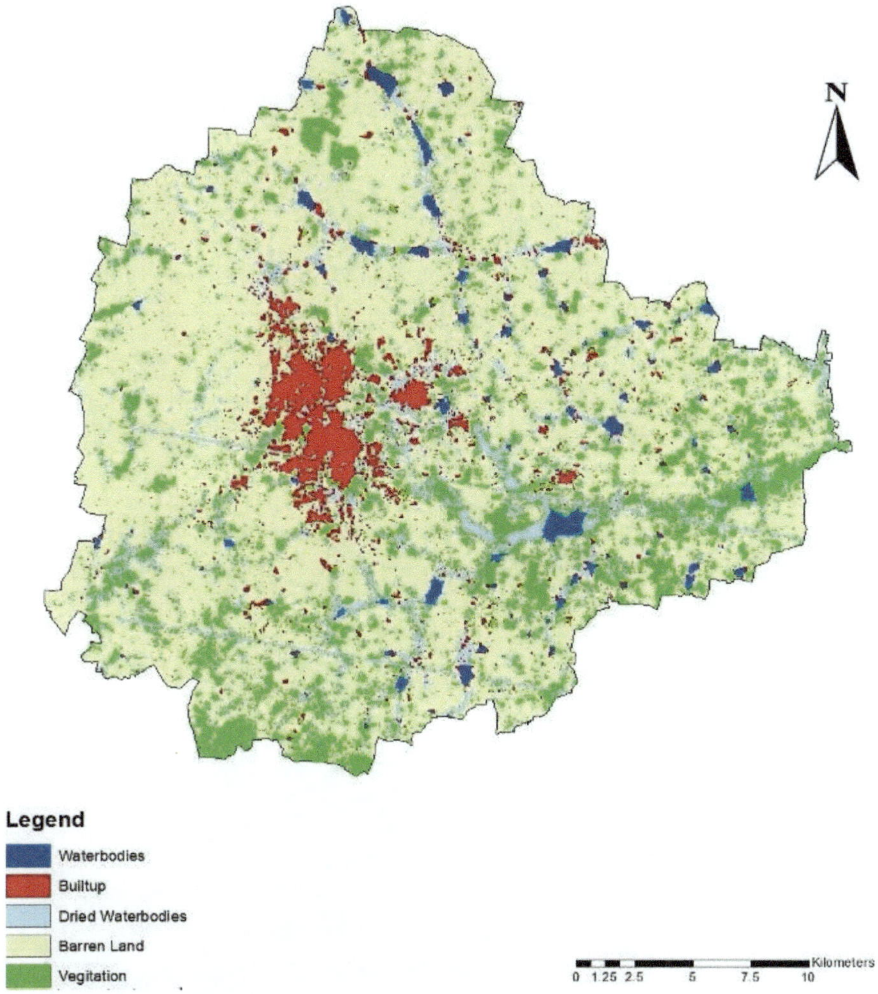

Fig. 11.7 Bengaluru LULC map for 1972. *Source* Author

11.4 Result and Discussion

11.4.1 Spatio-Temporal Distribution of LULC

The Land Use Land Cover (LULC) analysis for the 1991, 2001, 2011, and 2021 periods was carried out through a supervised classification—support vector machine. This technique is used for remote sensing over stable surfaces and enables mapping change between LC classes such as built-up areas, vegetation, barren lands, and water. The accuracy assessment for the maps was done by randomly placing 40

Table 11.3 Accuracy assessment

Year	Kappa coefficient (%)	Overall accuracy
2021	92.7	95.0
2011	86.6	90.0
2001	96.8	97.5
1991	93.5	95.0

Source Author

points on the map and validating it through high-resolution Google Earth Images. The validation was performed through the Kappa coefficient method, with results indicating an accuracy exceeding 80% for all years evaluated (Table 11.3).

The LULC maps for all four years are depicted in Fig. 11.8. The most significant changes in all three time periods (1991–2001, 2001–2011, 2011–2021) are the transformation from vegetation and barren land to built-up. The 1972 LULC was prepared to understand the spatial area of waterbodies, natural drainage, and the sprawl of urbanisation, as shown in Figs. 11.9, 11.10 and 11.11 (Note: The generated maps have a few errors, including misclassification of the dried waterbodies as built up by the SVM classification).

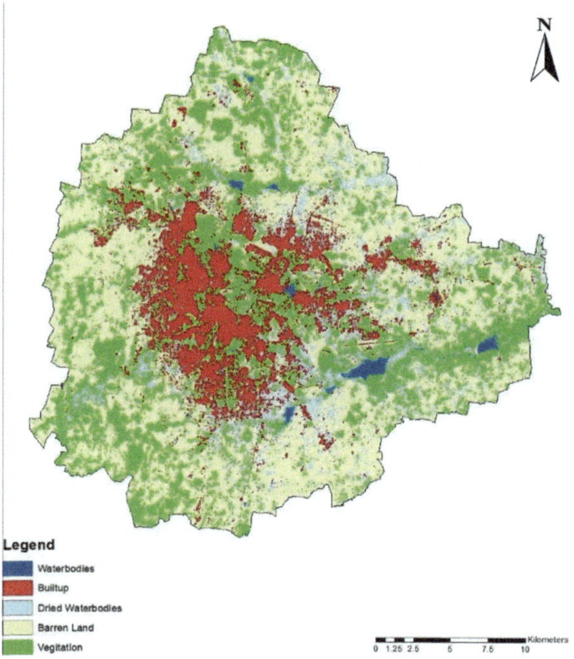

Fig. 11.8 Bengaluru LULC map for 1991. *Source* Author

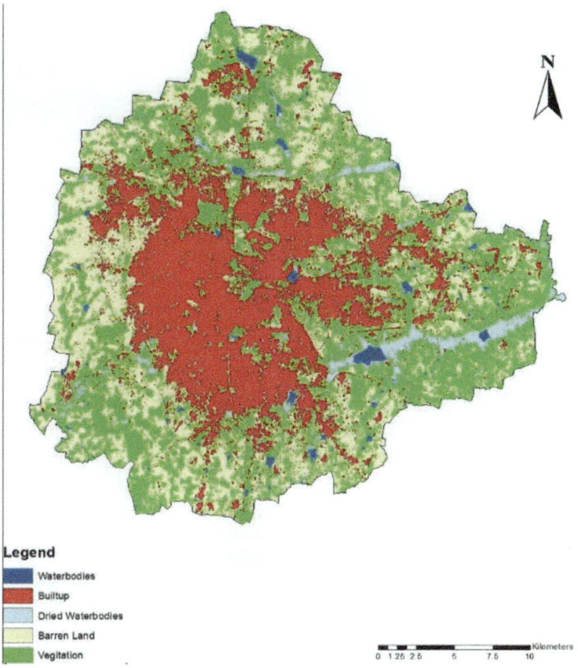

Fig. 11.9 Bengaluru LULC map for 2001. *Source* Author

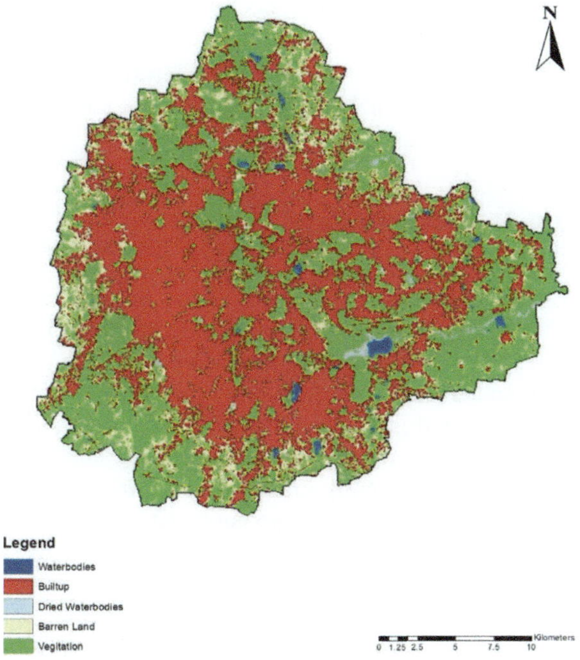

Fig. 11.10 Bengaluru LULC map for 2011. *Source* Author

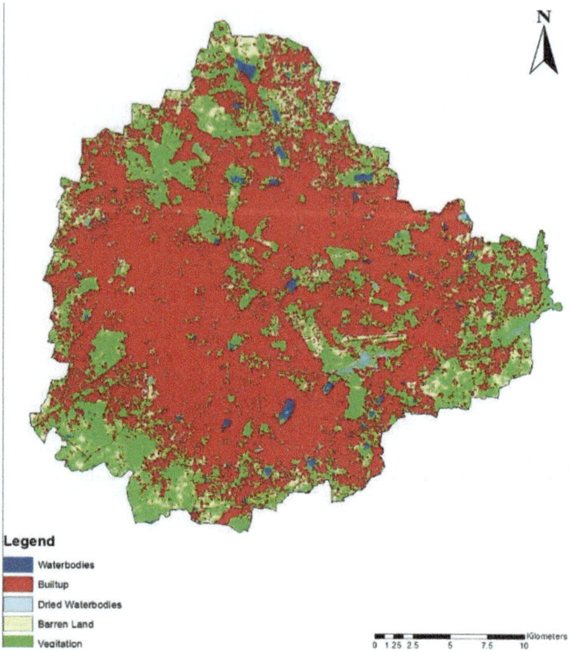

Fig. 11.11 Bengaluru LULC map for 2021. *Source* Author

Trends over the past three decades have evidenced that urban population agglomeration has significant implications for land use planning and natural resource management. Continuously monitoring and analysing the factors driving urban growth while ensuring the sustainable and rational use of natural resources is essential.

11.4.2 Dynamics of LULCC

From 1972 to 2021, Bengaluru's built-up area expanded dramatically, increasing from 36.7 km^2 (5% of the city) to 434.3 km^2 (58.6% of the total land), marking a 1,082% growth over 48 years. This expansion primarily replaced vegetation and open land, significantly impacting the environment, spatially visible in Fig. 11.13. The most rapid growth occurred between 1991 and 2001, driven by infrastructure development, increased population density, and IT sector expansion. Natural land cover has been severely infringed upon, with barren land decreasing from 68% in 1972 to 12% in 2021 and vegetation and water bodies shrinking due to urbanisation (Fig. 11.12). Urban sprawl followed distinct patterns: core-centric expansion, radial growth along transport corridors after 1991, densification from 2011 to 2021, and increased peri-urban development, with critical drivers like changes in LULC, population growth, economic development, policy shifts, and the expansion of transport infrastructure (Table 11.4).

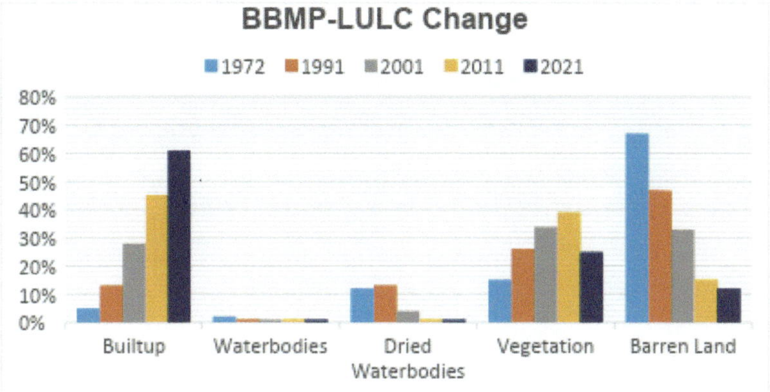

Fig. 11.12 LULC change from 1972–2021. *Source* Author

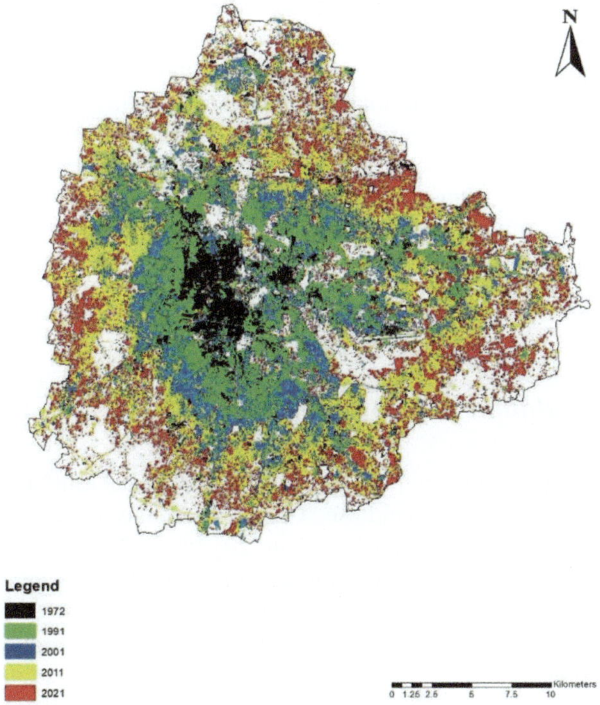

Fig. 11.13 Bengaluru urban growth from 1972–2021. *Source* Author

Table 11.4 Urban growth rate

Built-up		
Year	Area (Sqkm)	Urban growth (%)
1972	36.78	
1991	93.8	255
2001	200.3	214
2011	322.6	161
2021	434.3	135

Source Author

11.4.3 Basin Identification

A basin or valley is a region where water travels from higher altitudes via the river/ stream system to lower elevations while reshaping the channel formations, playing a critical role in water resource management, flood control, and land use planning. Using ArcGIS and a high-resolution digital elevation map (DEM), five major basins were delineated in the study area (Fig. 11.14): the Arkavathi Valley (North), historically used for agriculture but now prone to flooding due to urbanisation altering the drainage system; the Vrishabhavathi Valley (South-West); the Suvarnamukhi and Koramangala Basin (South) have experienced reduced river banks and increased pollution as a result of urbanisation; and the Hebbal Nagavara Valley (East), experiencing moderate urbanisation that has disrupted drainage and increased water management demands.

11.4.4 Precipitation Analysis

Precipitation patterns are crucial for understanding regional rainfall disparities and their effects on hydrology and land use. This section analyses spatial and temporal variations in precipitation in Bengaluru from 2013 to 2022 (Fig. 11.15), revealing fluctuations in rainfall levels, with notable deviations from the mean in certain years. 2015, 2017, 2020, 2021, and 2022 saw above-average rainfall, with 2022 recording the highest at 1,450 mm. Figure 11.18 illustrates the district precipitation data with Fig. 11.17 showing the region's hobli boundaries. Precipitation maps illustrate annual rainfall distribution across the district, with 2022 showing exceptionally high levels in Marathahalli Hobli, registering 1,822 mm, impacting local hydrology. The analysis indicates irregular rainfall trends, likely driven by climate change and urbanisation, pose challenges for urban and rural areas. In contrast, urban regions face drainage and flooding issues, and rural areas experience difficulties with agricultural and water management. High-rainfall areas like Marathahalli Hobli should be better equipped to manage flooding and improve drainage systems (Fig. 11.16).

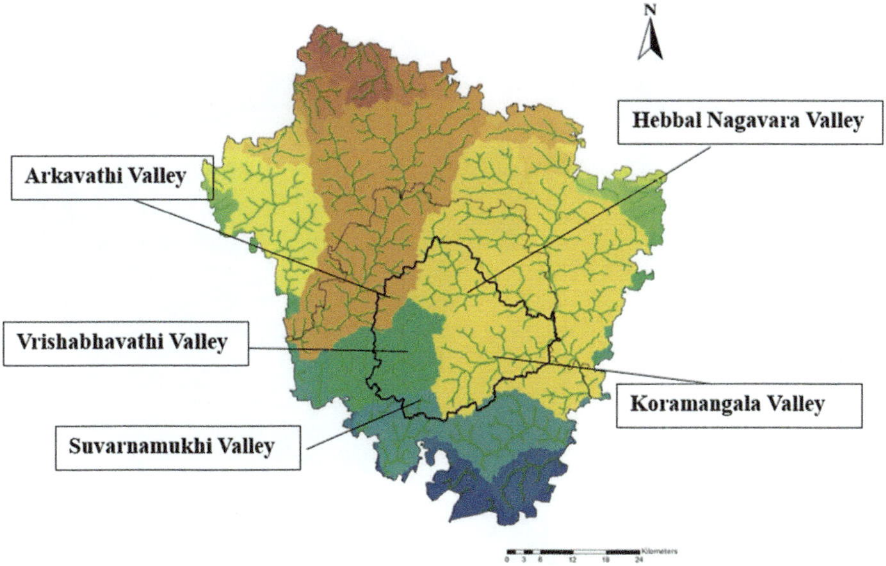

Fig. 11.14 Identified regional basins. *Source* Author

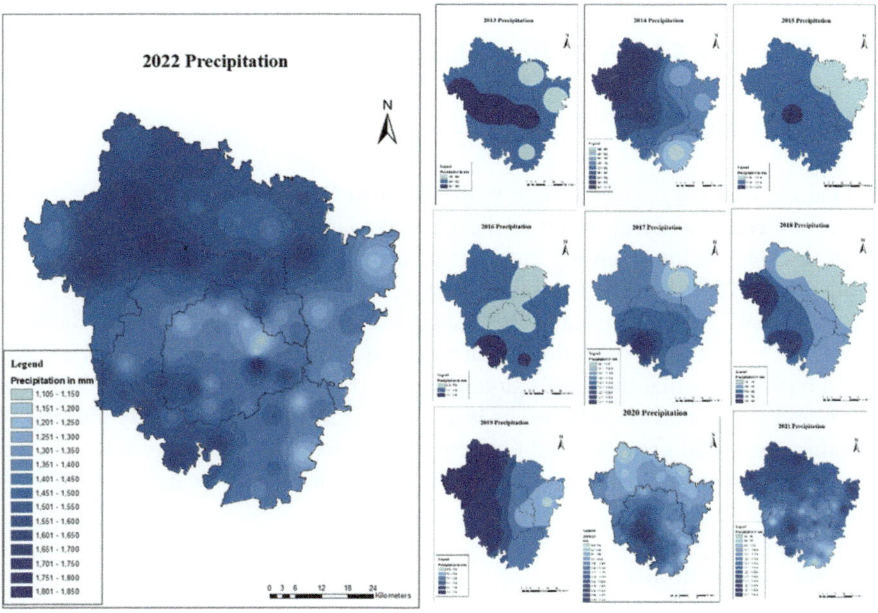

Fig. 11.15 Spatial distribution of Bengaluru district precipitation 2013–2022. *Source* Author

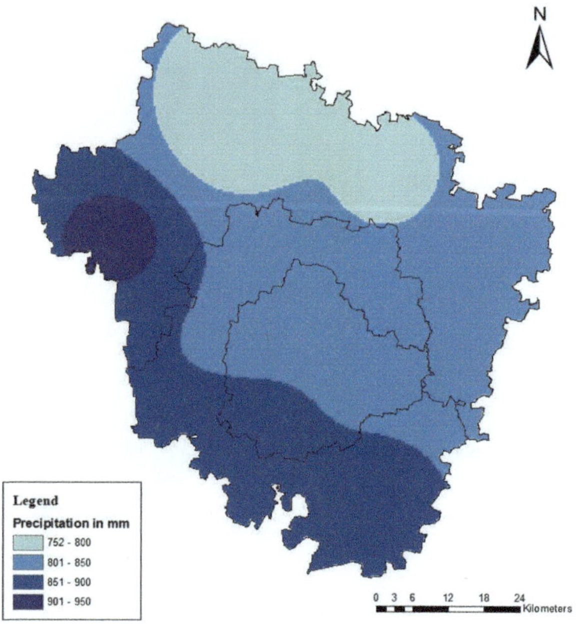

Fig. 11.16 Spatial distribution of Bengaluru's average annual rainfall. *Source* Author

11.4.5 Water Supply in Bengaluru

11.4.5.1 Availability

Bengaluru's primary water source is the Krishna Raja Sagara (KRS) Dam on the Cauvery River, supplying water to the Thoraikadana (T.K.) Halli Water Treatment Plant, located 90 km from the city, has a capacity of 1,310 million litres per day (MLD). The KRS Dam has a capacity of 49.452 thousand million cubic feet (TMC), of which 9.89 TMC (20% of capacity) is dead storage. The dam typically fills between June and October, with depletion occurring from November to May. Historical data Fig. 11.19 reveals that the dam's storage fell to 0 TMC for six months in 2012 and 2013, and in the past decade, water levels remained below 9.89 TMC for over 39 months. This highlights the need for Bengaluru to explore alternative water sources and adopt decentralised planning to enhance resilience against water scarcity.

11.4.5.2 Accessibility

In Bengaluru, only 70% of households have piped water connections, with the city core receiving water every other day for 6 to 18 h, while urban local bodies (ULBs) get water every three days. Newly incorporated villages and informal settlements, including slums, face severe water shortages, with some informal neighbourhoods

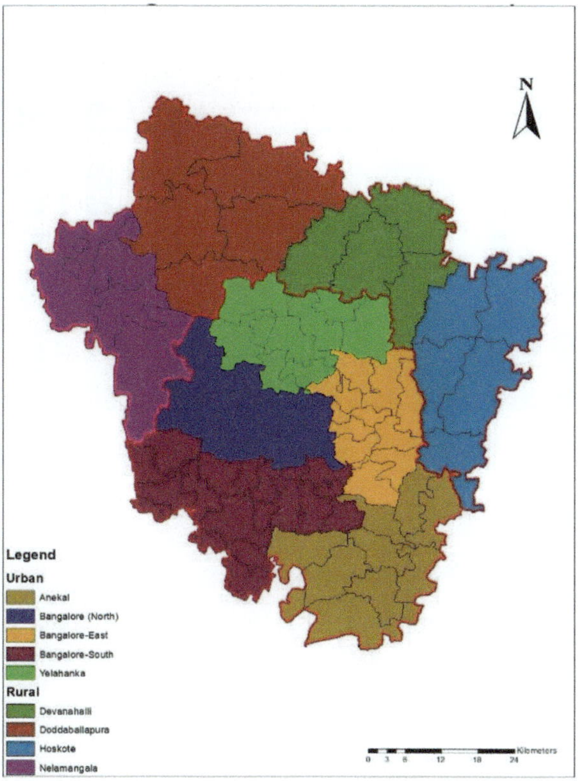

Fig. 11.17 Bengaluru district hobli distribution. *Source* Karnataka GIS

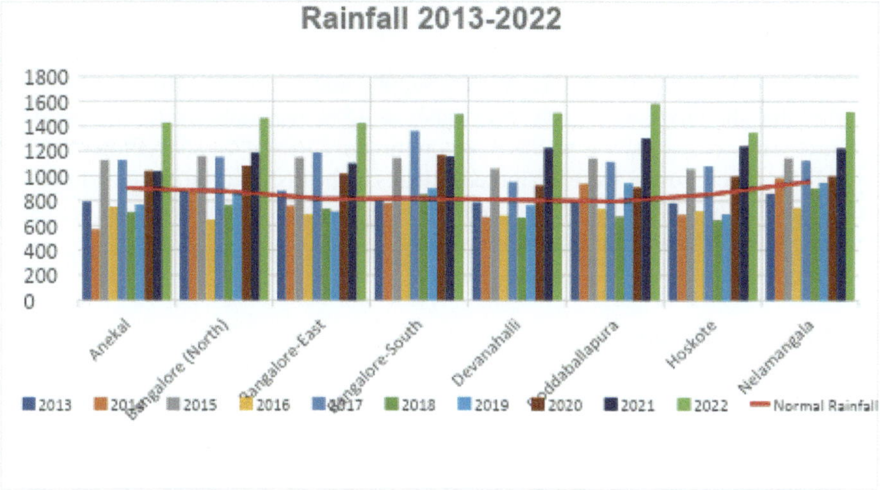

Fig. 11.18 Precipitation analysis for Bengaluru region from 2013–2022. *Source* KSNDMC

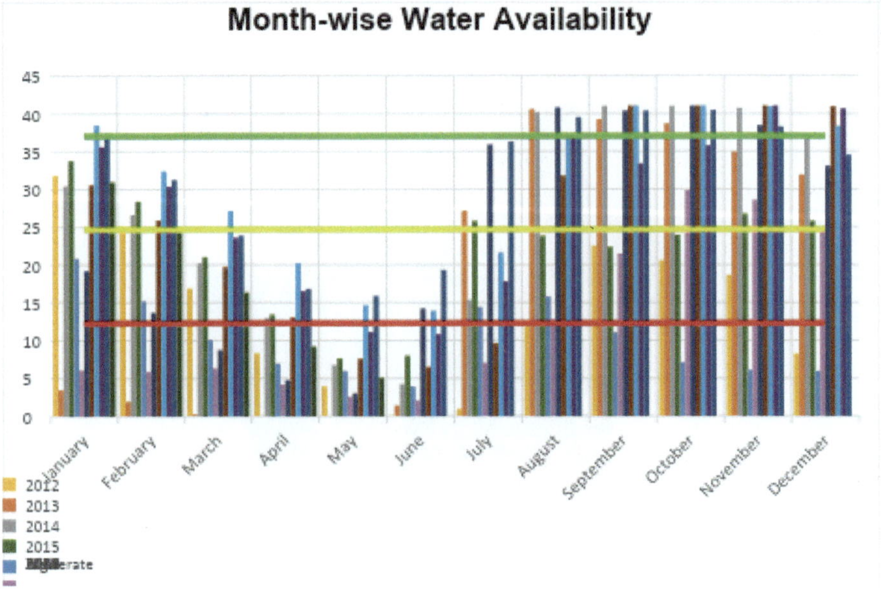

Fig. 11.19 Month-wise water availability in KRS Dam from 2012–2023. *Source* Karnataka Water Resources Department (WRD)

receiving less than two hours of water per day at low-pressure and poor quality (Du et al. 2019), increasing reliance on tanker trucks and borewells. A survey revealed that 58% of residents use piped water, 14% rely on borewells, and 30% use both.

11.4.5.3 Efficiency

The Chairman of the Bangalore Water Supply and Sewerage Board (BWSSB) reported that the city's water loss, previously at 45%, was reduced to 38% in 2021 and further decreased to 29% by 2023, amounting to a loss of 420.5 million litres per day (MLD). These losses were attributed to leaks in the distribution system, defective meters, public taps, water theft, and inefficiencies in supply to informal settlements, including slums (Bangalore Water Supply 2024).

11.4.5.4 Affordability

The cost of water for domestic use in Bengaluru varies based on consumption. Residents using less than 8,000 L per month are charged Rs. 7 per kilolitre, 8,001 to 25,000 L pay Rs. 11 per kilolitre, 25,001 to 50,000 L are charged Rs. 25 per kilolitre, and consumption exceeding 50,001 L is billed at Rs. 45 per kilolitre. Additionally, the city has installed drinking water ATMs offering 20 L for Rs. 5, with private

ATMs charging Rs. 8. These ATMs primarily serve informal settlements, slums, and residents. In areas lacking BWSSB piped water, such as ULBs and the newly incorporated 110 villages, residents rely on tanker water, which is significantly more expensive. The cost of tanker water is over 12 times that of piped water.

11.4.5.5 Safety

Safety is a major concern, as the city's centralised treatment plant, located 90 km away, can be vulnerable to disruptions, as seen during a 2022 flood that halted supply for 48 h. Additionally, over 6% of treated water is lost to leakage, leading to contamination. As per the Monitoring of Indian National Aquatic Resource System (MINARS), all the lakes inside the city are unfit for drinking even with conventional treatment.

11.4.5.6 Water Supply and Demand Calculation

Bengaluru's population, over 8.4 million in 2011, is projected to reach 10.5 million by 2021 and 20.5 million by 2051, doubling over 30 years Fig. 11.20. This rapid growth will significantly strain the city's infrastructure and increase resource demand. Based on Indian Standards (IS), per capita water consumption is estimated at 150 L per day for non-automated systems and 200 L for automated systems. With the Stage-V water supply scheme, the city will receive 2,070 MLD by 2031. However, water demand will rise sharply, from 1,646.4 MLD to approximately 4,000 MLD by 2051 for a consumption rate of 150 L per person and from 2,195.2 MLD to 5,350 MLD for 200 L per person. The shortfall, or gap in water supply, is expected to grow, peaking at 3,258.33 MLD by 2051 (Fig. 11.21). Given the history of droughts and rapid population growth, these figures will likely strain available water resources further, underscoring the need for diversified water sources and strategic planning.

11.4.6 Drainage Systems and Floods

11.4.6.1 Drainage Systems

Using data from 1972, the research examined Bengaluru's historical drainage system, as depicted in Fig. 11.22, which outline the city's natural drainage network. The analysis highlights an extensive system with stream widths ranging from 150 to 600 m. However, rapid, unplanned urbanisation has progressively encroached upon and narrowed this drainage network by 2021. Significant LULC changes have occurred in the frequently flooded southeast region, with vegetation, barren land, and dried waterbodies converted into built-up areas. This urban expansion has violated Bengaluru Development Authority (BDA) guidelines for maintaining buffer zones

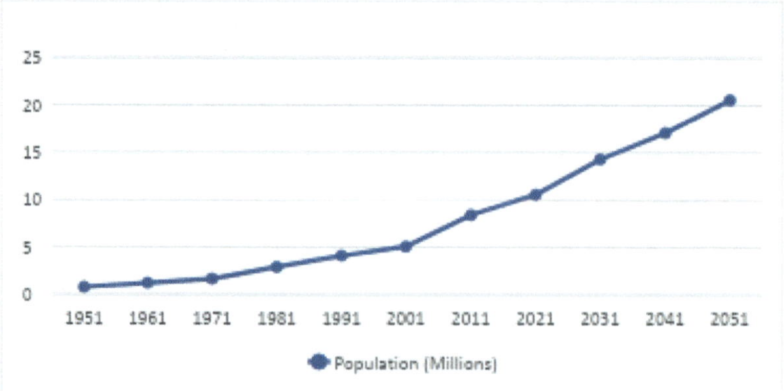

Fig. 11.20 Bengaluru population prediction. *Source* Census + BWSSB

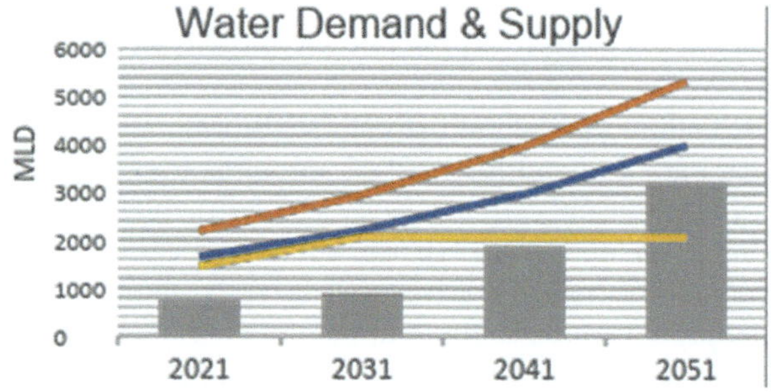

Fig. 11.21 Water demand and supply prediction. *Source* Author

around drainage networks and lakes. Figures 11.23, 11.24, 11.25 and 11.26 illustrate the spatial–temporal encroachment of built-up areas onto drainage networks, increasing impervious surfaces, raising runoff volumes, and heightening flood risks.

11.4.6.2 Encroachment on Lakes and Ponds

Over the past five decades, Bengaluru has experienced significant encroachment on its lakes. Historical data from a 1935 master plan (Fig. 11.27) identified approximately 69 lakes in the city. However, by 2021, LULC analysis and real-time imagery reveal that only five lakes remain intact, nine have been reduced in size due to encroachments, and 55 have been entirely converted into built-up areas, largely due to unplanned development (Fig. 11.28). Over 80% of the city's lakes have been encroached upon in the past 80 years. The analysis further indicates that these waterbodies were once interconnected through natural channels, which have disappeared or significantly narrowed.

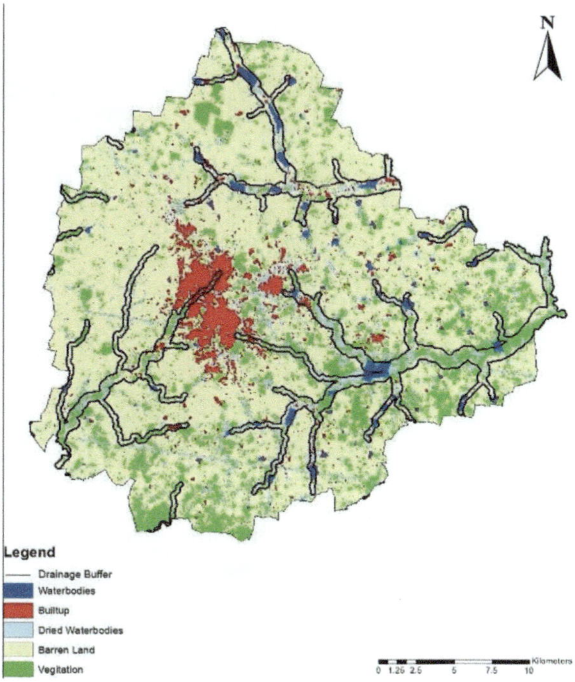

Fig. 11.22 Bengaluru 1972 drainage network. *Source* Author

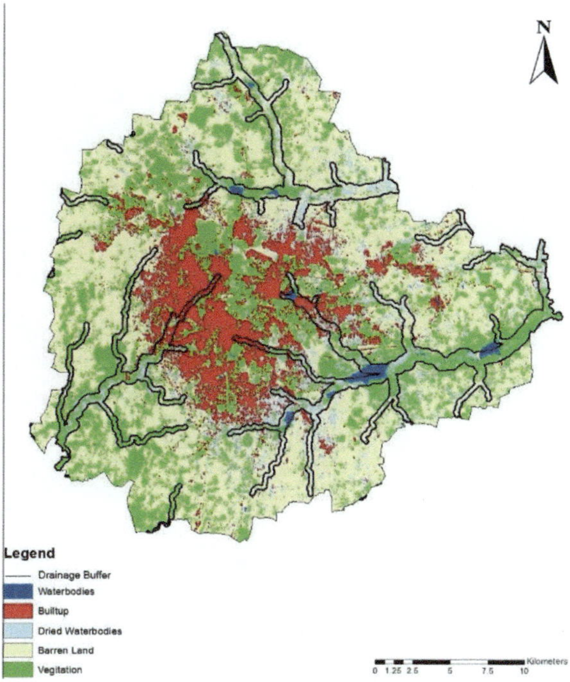

Fig. 11.23 Bengaluru 1991 drainage network. *Source* Author

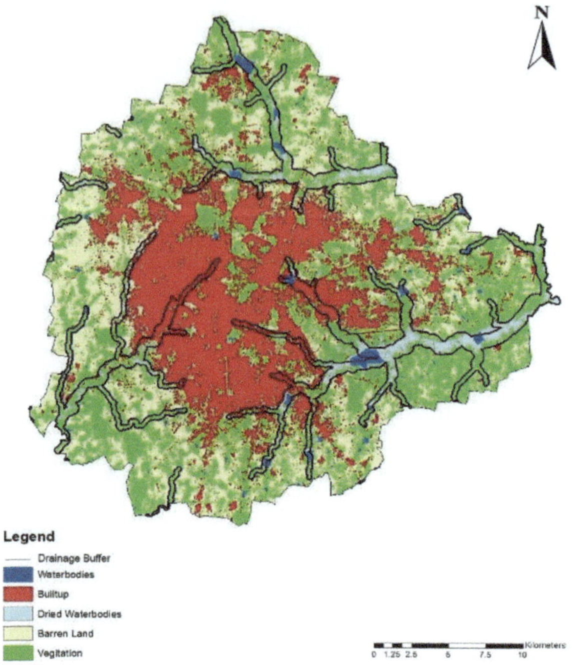

Fig. 11.24 Bengaluru 2001 drainage network. *Source* Author

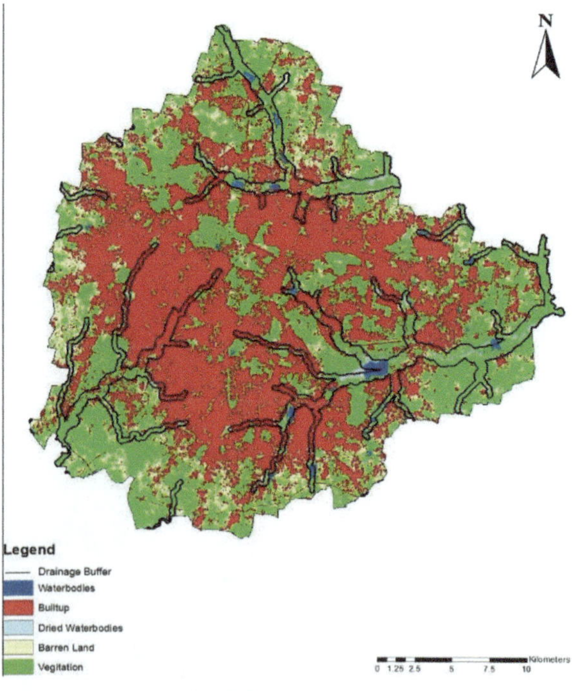

Fig. 11.25 Bengaluru 2011 drainage network. *Source* Author

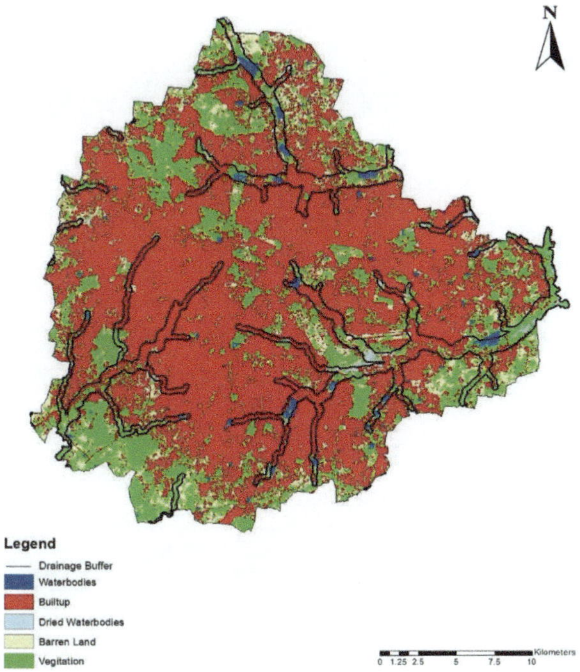

Fig. 11.26 Bengaluru 2021 drainage network. *Source* Author

11.4.6.3 Identification of Flood-Prone Areas

Flood-prone areas were identified and mapped using secondary data, then overlaid onto the city's stream order (Fig. 11.29), revealing that most flood-prone zones align with the drainage network. This suggests that encroachments are a primary cause of flooding in Bengaluru. Figure 11.30 presents flood mapping for 2017, 2018, 2020, 2021, and 2022, generated using Google Earth Engine, showing that most floods occurred in the southeastern region, particularly in the Koramangala Valley. The analysis indicates that implementing targeted policies and strategies can enhance the city's flood resilience.

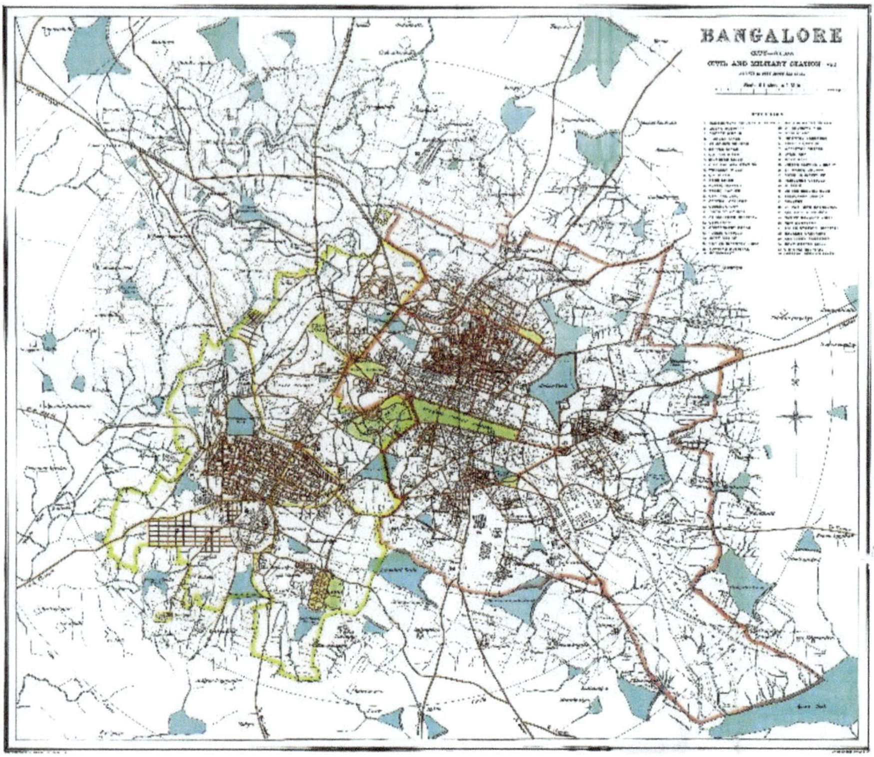

Fig. 11.27 Bengaluru 1935 map. *Source* British Library archives

Fig. 11.28 Lake and pond encroachment analysis. *Source* Author

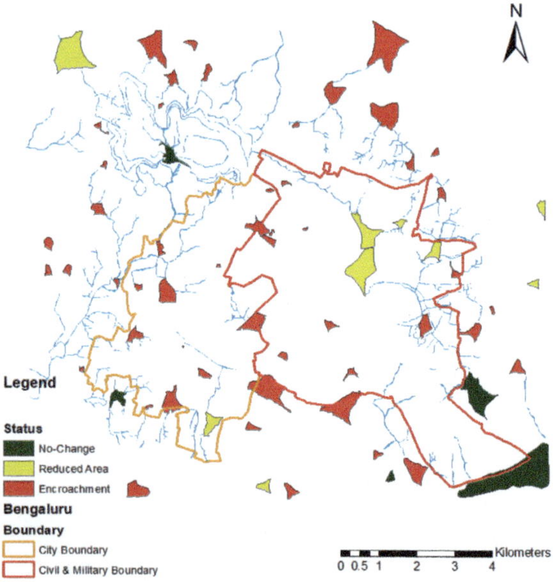

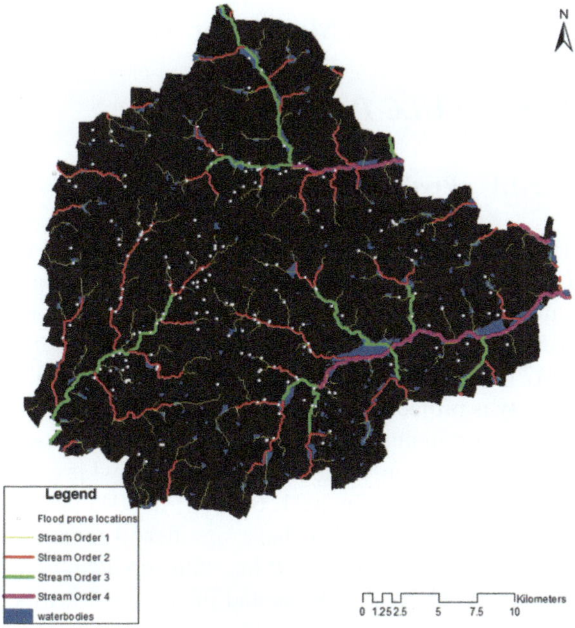

Fig. 11.29 Flood-prone points. *Source* Author and KSNDMC

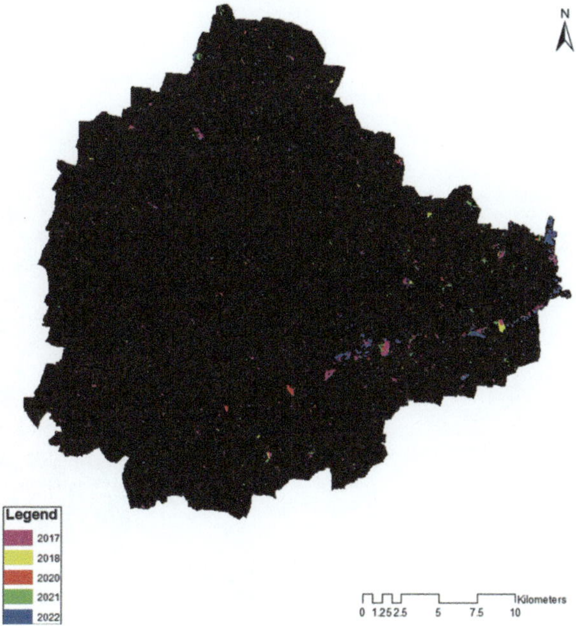

Fig. 11.30 Flood mapping. *Source* Author

11.5 Prediction Models for Multiple Scenarios

11.5.1 LULC Prediction

11.5.1.1 Prediction Model

The 2021 predicted LULC map (Fig. 11.32) was validated against the actual 2021 LULC map (Fig. 11.31) using the Kappa validation method. The K_{no} and $K_{location}$ values were 96.51% and 84.22%, respectively, while the $K_{standard}$ and $K_{locationStrata}$ were 82.94 and 84.22%. Future LULC maps were simulated for 2031, 2041, and 2047 based on current development trends (Figs. 11.33, 11.34 and 11.35). By 2047, the city is projected to be approximately 91.1% built-up (653 km^2), with only 2.6% (19 km^2) remaining as water bodies or dried waterbodies and 6.4% (46 km^2) as vegetation and barren land. The urban sprawl is expected to expand from 434.3 km^2 in 2021 to 543 km^2 in 2031, 616 km^2 in 2041, and 653 km^2 by 2047. In 2021, most natural drainage channels have significantly reduced in width, and by 2047, nearly all will have been converted into built-up areas. This increase in built-up land will expand impervious surfaces and increase stormwater runoff, raising the frequency of urban flooding.

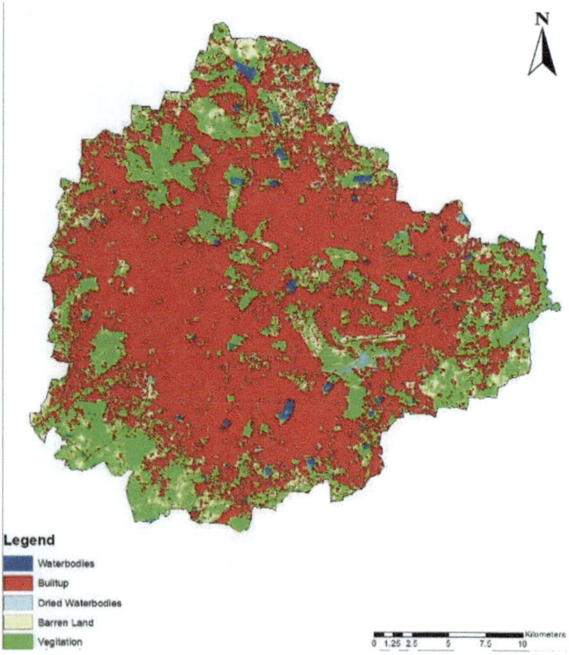

Fig. 11.31 Real/existing Bengaluru LULC map—2021. *Source* Author

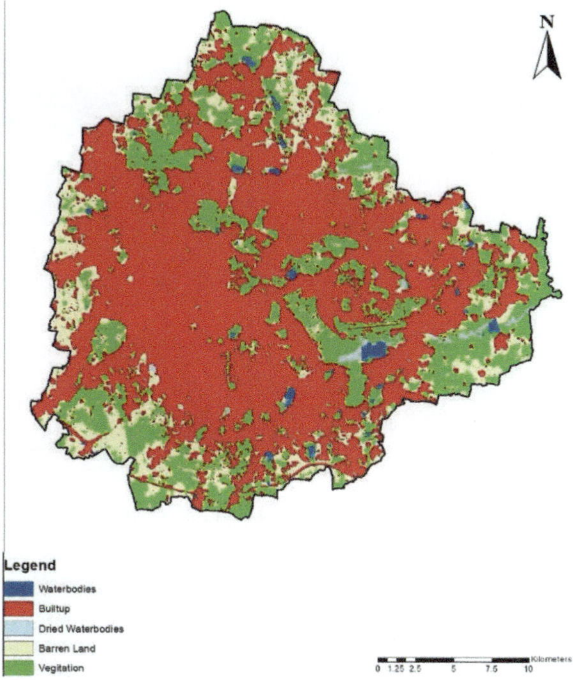

Fig. 11.32 Predicted Bengaluru LULC map—2021. *Source* Author

11.6 Strength, Weaknesses, Opportunities, and Threats (SWOT)

Bengaluru's rapid population growth and urbanisation pose significant water management challenges and create opportunities for large-scale implementation of water-sensitive practices. Effective planning and policies can integrate sustainable water infrastructure into new developments, promote water-efficient practices, and harness substantial rainfall through rainwater harvesting for groundwater recharge. The city's 20 wastewater treatment plants provide opportunities to treat and recycle wastewater for non-potable uses. Despite the degradation of many water bodies, restoration efforts could enhance sustainable water management. As a technology hub, Bengaluru can leverage its strengths in innovation, partnering with educational institutions and research centres to develop smart water management solutions.

However, the city faces key weaknesses. Its centralised water supply system is vulnerable to disruptions and inequitable distribution, especially in rapidly expanding fringe areas. Outdated infrastructure leads to significant water loss, while high-built density and extensive paving reduce rainwater infiltration, increasing runoff and decreasing groundwater recharge. Encroachments on water bodies disrupt stormwater flow, causing frequent flooding. Pollution from untreated sewage and industrial waste has degraded water quality, and inadequate investment in sustainable infrastructure exacerbates water scarcity. Climate change further strains the

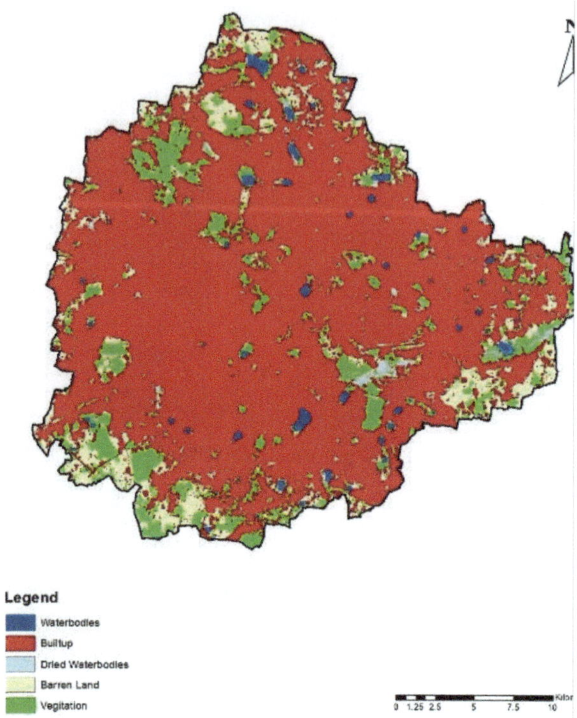

Fig. 11.33 Predicted Bengaluru LULC map—2031. *Source* Author

city's resources, while limited public engagement and weak policy enforcement hinder water management efforts.

Despite these challenges, there are opportunities for improvement. Upgrading Bengaluru's water infrastructure with smart, adaptive technologies can enhance efficiency and reduce reliance on centralised systems. Expanding rainwater harvesting and groundwater recharge efforts can bolster water supply, particularly during the monsoon. Public awareness campaigns and community involvement can foster water-conscious practices while preserving wetlands and incentivising water-sensitive developments, which can improve water management. Expanding wastewater recycling and adopting an integrated water management approach while leveraging technology will enhance sustainability and resilience, helping the city adapt to climate change impacts.

Bengaluru's reliance on centralised water systems makes it susceptible to water scarcity during system disruptions. Inadequate drainage networks exacerbate flooding risks, particularly in the Koramangala and Vrishabhavathi Valleys, while vegetation loss and urbanisation increase surface runoff and reduce natural infiltration. Groundwater overexploitation and pollution from untreated waste have led to resource depletion and water quality degradation. Encroachment on water bodies further reduces natural storage capacity and heightens flood risks.

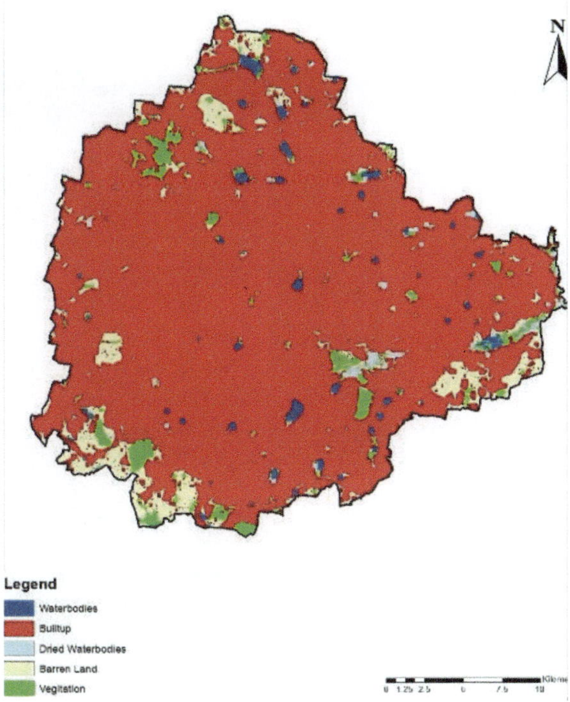

Fig. 11.34 Predicted Bengaluru LULC map—2041. *Source* Author

To address these issues, Bengaluru must adopt strategic planning, invest in sustainable water infrastructure, implement climate-resilient measures, and engage the community to mitigate the risks and challenges associated with its water-sensitive city aspirations. Collaborative efforts involving the government, residents, and other stakeholders are essential to devise and implement effective solutions for safeguarding the city's water resources and reducing vulnerabilities in the face of water-related adversities.

11.7 Recommendations

The key recommendations for Bengaluru's water management focus on ensuring clean and safe water supply for all, improving availability, accessibility, and reliability and exploring alternative sources such as rainwater harvesting and wastewater recycling. Efforts to develop a sustainable and efficient water system focus on diversifying water sources, implementing smart infrastructure, and constructing interconnected systems while considering social, environmental, and economic impacts. Enhancing resilience involves upgrading drainage infrastructure, reinforcing buffer zones, and implementing safety measures to ensure the sustainability of water and sanitation services amidst natural disasters and environmental changes. Additionally,

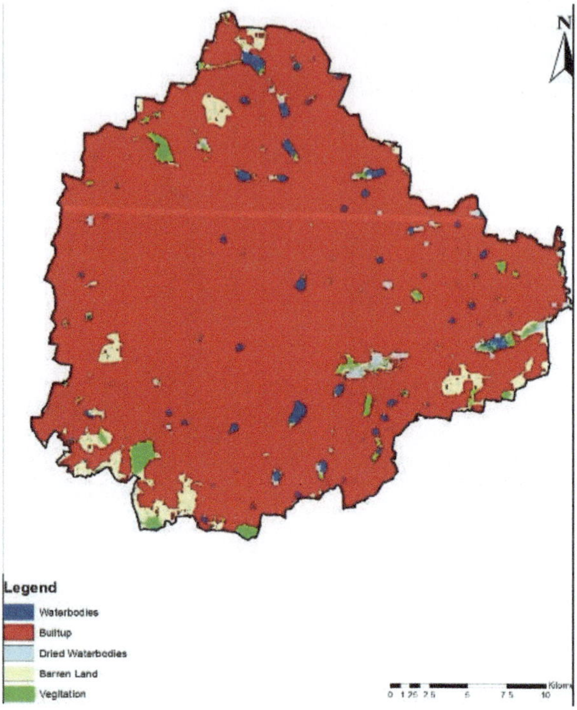

Fig. 11.35 Predicted Bengaluru LULC map—2047. *Source* Author

maintaining ecological health by preserving resilient catchments, promoting blue-green environments, and integrating water dynamics with urban planning is crucial for harmonising water systems with the urban landscape.

11.7.1 Storage Tunnel and Vault

Storage tunnels and vaults are key components for managing stormwater in urban areas. These structures are built-in phases and simultaneously convert the existing combined sewer system into a separate system (Fig. 11.36). They control flooding, reduce runoff, and improve water resource management by following natural drainage patterns. Phase 1 is projected to store 2,350 million litres (ML) of water to meet the Koramangala region's demand for 117–157 days (Table 11.5).

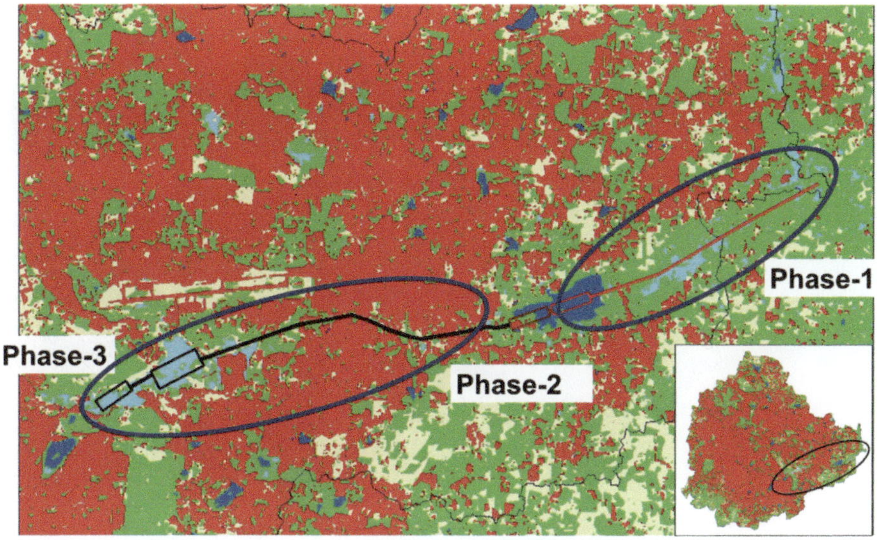

Fig. 11.36 Tunnel and vault location on spatial map. *Source* Author

Table 11.5 Calculation for tunnel and vault

Calculations for the phase-1 of water storage		South zone water demand calculation	
Length of one tunnel	5,150 m	Population (2011 census)	1,02,607
Volume of one tunnel	145,540 m^3	Water consumption for 150/Lt/ person/day	15.4 MLD
Total number of tunnels	3	Water consumption for 200/Lt/ person/day	20.52 MLD
Total volume from all 3 tunnels	436,620 m^3	Supply fulfilment for 150lt/ person	157 days
Total volume of the vault	1,920,000 m^3		
Total volume stored in Phase-1	2,356,620 m^3	Supply fulfilment for 200lt/ person	117 days
Conversion from m^3 to litres	2,356 ML		

Source Author

11.7.2 Rejuvenating Water Bodies and Buffers Along the Natural Drains

Restoring water bodies and establishing buffers along natural drains helps restore ecological balance, reduce water pollution, and mitigate urban flood risks. These efforts improve drainage efficiency and enhance urban livability, aesthetics, and sustainability. Figure 11.37 shows the encroachments into no-development buffer zones with widths ranging from 50 to 20 m along natural drains, which worsened from 2011 to 2021 despite buffer restriction guidelines introduced in 2014. Restoring

Fig. 11.37 Encroachment along the identified drainage network. *Source* Author

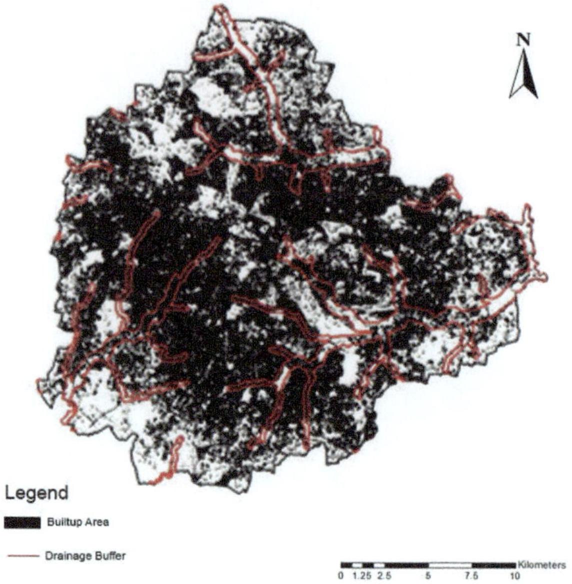

these areas will enhance flood prevention and improve the city's overall quality of life.

11.7.3 Water from Air

Atmospheric water generators (AWGs) provide a sustainable solution to water scarcity by extracting moisture from the air and converting it into potable water. Targeting urban fringes, informal settlements, and critical institutions like hospitals and schools, AWGs reduce groundwater exploitation and ensure affordable clean water, especially in resource-limited areas. Powered by renewable energy, they offer an eco-friendly approach to water supply, supported by government subsidies and public awareness initiatives.

11.7.4 Multi-Purpose Infrastructure

Multi-purpose infrastructure for water-sensitive urban design (WSUD) integrates water management with urban development, particularly in lakes like Varthur, Bellandur, and Madiwala. These infrastructures support rainwater harvesting and stormwater management, enhancing water storage capacity and quality of life. Recreational spaces within lake buffer zones serve as stormwater storage during heavy

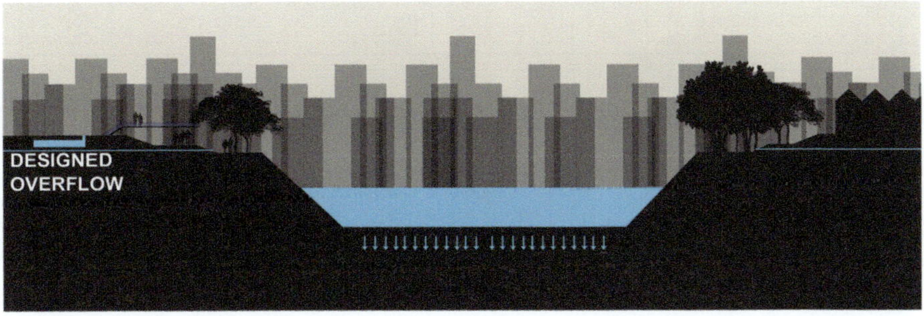

Fig. 11.38 Graphical representation of multi-purpose infrastructure—summer. *Source* Author

rains, preventing overflow and interconnecting with drainage networks (Figs. 11.38, 11.39 and 11.40). This approach promotes sustainable urban water management and serves as a model for WSUD strategies.

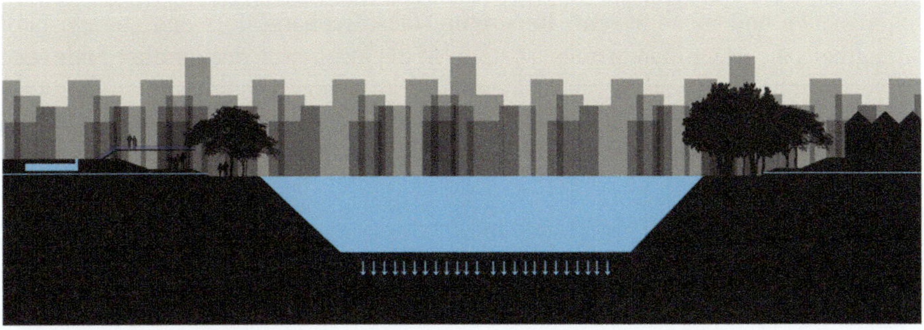

Fig. 11.39 Graphical representation of multi-purpose infrastructure—rainy period. *Source* Author

Fig. 11.40 Graphical representation of multi-purpose infrastructure—stormy period. *Source* Author

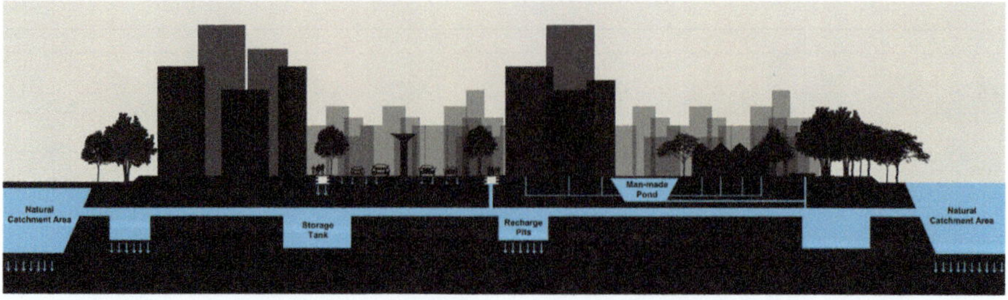

Fig. 11.41 Graphical representation of overall water-sensitive urban water design. *Source* Author

11.7.5 Water-Sensitive Urban Water Design

The water-sensitive urban design integrates water infrastructure with natural systems, linking water bodies to recharge pits, storage tanks, catchment areas to replenish groundwater, light impact development (LID) like permeable pavements, and green roofs. Strategically placed shafts along major roads facilitate more manageable maintenance and management. Integrating these recharge pits and storage tanks strengthens the water management system. At the local level, decentralised rainwater harvesting channels water from residential areas to ponds, increasing groundwater recharge and utilising stored water. Figure 11.41 illustrates this design, showcasing its effectiveness for water regulation and harvesting in Bengaluru.

11.8 Conclusion

The study's results and discussions underscore the urgent need for Bengaluru to transition towards a water-sensitive urban paradigm to effectively address the challenges posed by rapid urbanisation, unsustainable land use practices, and climate change impacts. The analysis revealed a significant reduction in water bodies, a high urban growth rate, and limited water availability, highlighting the city's pressing water-related issues. Over recent decades, rapid urbanisation and unplanned development have resulted in significant water management challenges, particularly water scarcity, pollution, and increased flooding. Between 1991 and 2011, the city experienced a dramatic transition, losing approximately 48% (340 km^2) of its open spaces to built-up areas. The city can proactively mitigate risks and enhance its water security and environmental sustainability by emphasising strategic planning, sustainable water infrastructure investment, community engagement, and climate-resilient measures.

The proposed vision, mission, and goals outlined in the recommendations serve as a roadmap for guiding Bengaluru towards a water-sensitive future. By integrating innovative solutions, stakeholder collaboration, and adaptive governance frameworks, the city can safeguard water resources, promote resilience, and ensure equitable access to clean and safe water for all residents. The SWOT analysis provides valuable insights into leveraging Bengaluru's assets, technological capabilities, and institutional capacities to foster a water-sensitive urban landscape that aligns with ecological integrity, social well-being, and economic prosperity.

Ultimately, the study's findings emphasise the importance of adopting a holistic approach guided by sustainability, efficiency, resilience, and environmental stewardship. By embracing these principles and implementing the recommended strategies, Bengaluru can embark on a transformative journey towards becoming a model water-sensitive city, setting a precedent for sustainable urban development worldwide.

Acknowledgements The authors extend their gratitude to NIT Calicut for providing the opportunity to undertake this study. Special thanks to the reviewers for their valuable comments and suggestions, which had a profound impact on enhancing the quality of the paper.

References

Amini Parsa V, Yavari A, Nejadi A (2016) Spatio-temporal analysis of land use/land cover pattern changes in Arasbaran biosphere reserve: Iran. Model Earth Syst Environ 2(4):1–13. https://doi.org/10.1007/S40808-016-0227-2/TABLES/8

Asulabha RTV, Sincy KS, Sudarshan V, Bhat P, Aithal BH (2016) Wetlands: treasure of Bangalore Vijnanapura lake (polygon with red represents encroachments). http://ces.iisc.ernet.in/energy

Behera MD, Borate SN, Panda SN, Behera PR, Roy PS (2012) Modelling and analyzing the watershed dynamics using cellular automata (CA)-Markov model—A geo-information based approach. J Earth Syst Sci 121(4):1011–1024. https://doi.org/10.1007/S12040-012-0207-5/FIGURES/7

Bangalore Water Supply: Bengaluru loses 29% water supply daily; vehicles main culprits|Bengaluru News—Times of India. Accessed 23 Sept 2024. https://timesofindia.indiatimes.com/city/bengaluru/bengaluru-loses-29-water-supply-daily-vehicles-main-culprits/articleshow/98319801.cms

Bangalore, India Metro Area Population 1950–2023|MacroTrends. Accessed 11 July 2023. https://www.macrotrends.net/cities/21176/bangalore/population

Bengaluru's IT Dominance: The Silicon valley of India - coding ninjas. Accessed 11 July 2023. https://www.codingninjas.com/studio/library/bengalurus-it-dominance-the-silicon-valley-of-india

City of Melbourne WSUD guidelines applying the model WSUD guidelines an initiative of the inner Melbourne action plan contents

Carter T, Fowler L (2008) Establishing green roof infrastructure through environmental policy instruments. Environ Manage 42(1):151–164. https://doi.org/10.1007/S00267-008-9095-5/FIGURES/2

Chester M et al (2021) Infrastructure resilience to navigate increasingly uncertain and complex conditions in the Anthropocene. npj Urban Sustain 1(1):1–6. https://doi.org/10.1038/s42949-021-00016-y

Cui M, Ferreira F, Fung TK, Matos JS (2021) Tale of two cities: How nature-based solutions help create adaptive and resilient urban water management practices in Singapore and Lisbon. Sustainability (Switzerland) 13(18). https://doi.org/10.3390/su131810427.

D'Andrea M (2020) Urban water management: city of Toronto a case study

Du J, Mitlin D, Beard VA, Satterthwaite D (2024) We're grossly underestimating the world's water access crisis, 2019. Accessed 23 Sept 2024. https://www.wri.org/insights/were-grossly-underestimating-worlds-water-access-crisis

Eastman JR (2006) IDRISI Andes tutorial

Field CB et al (2014a) Climate change 2014 impacts, adaptation, and vulnerability part a: global and sectoral aspects working group II contribution to the fifth assessment report of the intergovernmental panel on climate change. Accessed 17 July 2023. www.cambridge.org/9781107641655

Field CB et al (2014b) Climate change 2014—impacts, adaptation and vulnerability: part a: global and sectoral aspects: working group II contribution to the IPCC fifth assessment report. Climate change 2014 impacts, adaptation and vulnerability: part a: global and sectoral aspects, pp 1–1131. https://doi.org/10.1017/CBO9781107415379

Flörke M, Kynast E, Bärlund I, Eisner S, Wimmer F, Alcamo J (2013) Domestic and industrial water uses of the past 60 years as a mirror of socio-economic development: a global simulation study. Glob Environ Change 23(1):144–156. https://doi.org/10.1016/j.gloenvcha.2012.10.018

Flörke M, Schneider C, McDonald RI (2018) Water competition between cities and agriculture driven by climate change and urban growth. Nat Sustain 2017 1(1):51–58. https://doi.org/10.1038/s41893-017-0006-8

Hallegatte S, Green C, Nicholls RJ, Corfee-Morlot J (2013) Future flood losses in major coastal cities. Nat Clim Change 3(9):802–806. https://doi.org/10.1038/nclimate1979

Hua AK (2017) Application of CA-Markov model and land use/land cover changes in Malacca river watershed. Malaysia. https://doi.org/10.15666/aeer/1504_605622

Irvine K, Chua L, Eikass HS (2014) The four national taps of singapore: a holistic approach to water resources management from drainage to drinking water. J Water Manag Model. https://doi.org/10.14796/JWMM.C375

'India Population (2024) Worldometer. Accessed 23 Sept 2024. https://www.worldometers.info/world-population/india-population/

Kar SM (2016) View of locating Bengaluru as India's silicon valley. Accessed 23 Sept 2024. https://journals.christuniversity.in/index.php/artha/article/view/1663/1391

Kamusoko C, Aniya M, Adi B, Manjoro M (2009) Rural sustainability under threat in Zimbabwe—simulation of future land use/cover changes in the Bindura district based on the Markov-cellular automata model. Appl Geogr 29(3):435–447. https://doi.org/10.1016/j.apgeog.2008.10.002

Kumar S, Radhakrishnan N, Mathew S (2014) Land use change modelling using a Markov model and remote sensing 5(2):145–156. http://www.tandfonline.com/action/journalInformation?show=aimsScope&journalCode=tgnh20#.VsXodSCLRhE. https://doi.org/10.1080/19475705.2013.795502

Land, Water and Planning (2016) The state of Victoria department of environment. In: Managing extreme water shortage in Victoria. Accessed 23 Sept 2024. www.relayservice.com.au

Low KG et al (2015) Fighting drought with innovation: Melbourne's response to the millennium drought in southeast Australia. Wiley Interdiscip Rev Water 2(4):315–328. https://doi.org/10.1002/WAT2.1087

Melbourne Water System Strategy (2017)

McDonald RI et al (2011) Global urban growth and the geography of water availability, quality, and delivery. Ambio 40(5):437–446. https://doi.org/10.1007/s13280-011-0152-6

Mitlin D, Beard VA, Satterthwaite D, Du J (2019) Unaffordable and undrinkable: rethinking urban water access in the global south

Mondal A, Kundu S, Chandniha DK, Shukla R, Chandniha SK, Mishra PK (2012) Comparison of support vector machine and maximum likelihood classification technique using satellite imagery. Int J Remote Sens GIS 1(2):116–123. www.rpublishing.org

Nouri J, Gharagozlou A, Arjmandi R, Faryadi S, Adl M (2014) Predicting urban land use changes using a CA-markov model. Arab J Sci Eng 39(7):5565–5573. https://doi.org/10.1007/S13369-014-1119-2/METRICS

Pantelis Adamopoulou J, Apostolos Frantzana A, Pantelis Adamopoulos I (2023) Addressing water resource management challenges in the context of climate change and human influence. Eur J Sustain Dev Res 2023(3):2542–4742. https://doi.org/10.29333/ejosdr/13297

Pontius GR, Malanson J (2007) Comparison of the structure and accuracy of two land change models 19(2):243–265. https://doi.org/10.1080/13658810410001713434

Pontius RG, Schneider LC (2001) Land-cover change model validation by an ROC method for the Ipswich watershed, Massachusetts, USA. Agric Ecosyst Environ 85(1–3):239–248. https://doi.org/10.1016/S0167-8809(01)00187-6

Regmi RR , Saha SK , Subedi DS (n.d.) Geospatial analysis of land use land cover change modeling in Phewa lake watershed of Nepal by using GEOMOD model

Rogers BC et al (2020) Water sensitive cities index: a diagnostic tool to assess water sensitivity and guide management actions. Water Res 186:116411. https://doi.org/10.1016/J.WATRES.2020.116411

Rohr HE (2019) Evidence based water sensitive planning: securing water sustainability through innovative spatial planning tools

Realising the vision of a water sensitive city—the source. Accessed 23 Sept 2024. https://thesourcemagazine.org/realising-the-vision-of-a-water-sensitive-city/

Recommended Practice: Flood mapping and damage assessment using sentinel-1 SAR data in google earth engine|UN-Spider knowledge portal. Accessed 14 July 2023. https://www.un-spider.org/advisory-support/recommended-practices/recommended-practice-google-earth-engine-flood-mapping

Sang L, Zhang C, Yang J, Zhu D, Yun W (2011) Simulation of land use spatial pattern of towns and villages based on CA-Markov model. Math Comput Model 54(3–4):938–943. https://doi.org/10.1016/j.mcm.2010.11.019

Shivamurthy V (2017) Frequent floods in Bangalore: causes and remedial measures. https://doi.org/10.13140/RG.2.2.17517.90088.

Singh SK, Mustak S, Srivastava PK, Szabó S, Islam T (2015) Predicting spatial and decadal LULC changes through cellular automata Markov Chain models using earth observation datasets and geo-information. Environ Process 2(1):61–78. https://doi.org/10.1007/S40710-015-0062-X/TABLES/10

Subedi P, Subedi K, Thapa B (2013) Application of a hybrid cellular automaton–Markov (CA-Markov) model in land-use change prediction: a case study of saddle creek drainage Basin, Florida. Appl Ecol Environ Sci 1(6):126–132. https://doi.org/10.12691/AEES-1-6-5

Tarboton DG, Bras RL, Rodriguez-Iturbet Ralph IM (1991) On the extraction of channel networks from digital elevation data

Tortajada C, Joshi YK, Biswas AK (2013) The Singapore water story: sustainable development in an urban city-state. Routledge, Taylor & Francis Group. Accessed 23 Sept 2024. https://www.routledge.com/The-Singapore-Water-Story-Sustainable-Development-in-an-Urban-City-state/Tortajada-Joshi-Biswas/p/book/9780415657839

United Nations Department of Economic and Social Affairs and Pollution Division: World urbanization prospects the 2018 revision (2018)

Wada Y et al (2016) Modeling global water use for the 21st century: the water futures and solutions (WFaS) initiative and its approaches. Geosci Model Dev 9(1):175–222. https://doi.org/10.5194/gmd-9-175-2016

Wang J, Sandberg LA (2014) Beyond the pipes: planning for sustainable stormwater management in the city of Toronto

Wang Y, Zhang X (2001) A dynamic modeling approach to simulating socioeconomic effects on landscape changes. Ecol Modell 140(1–2):141–162. https://doi.org/10.1016/S0304-3800(01)00262-9

Weng Q (2002) Land use change analysis in the Zhujiang delta of China using satellite remote sensing, GIS and stochastic modelling. J Environ Manage 64(3):273–284. https://doi.org/10.1006/JEMA.2001.0509

Yang X, Zheng XQ, Chen R (2014) A land use change model: Integrating landscape pattern indexes and Markov-CA. Ecol Modell 283:1–7. https://doi.org/10.1016/J.ECOLMODEL.2014.03.011

Zaman S, Sayemuzzaman M (2014) Modeling of future land cover land use change in North Carolina using Markov chain and cellular automata model. Am J Eng Appl Sci 7(3):292–303. https://doi.org/10.3844/ajeassp.2014.292.303

Chapter 12
Transitioning Sri Lanka's Urban Transport: A System Dynamics Approach to Electrification and Decarbonization

M. Mavin De Silva, Navoda Fernando, Amal S. Kumarage, T. Sivakumar, A. G. T. Sugathapala, and Tadachika Nakayama

Abstract The transport sector, dominated by road transport, is a major contributor to greenhouse gas (GHG) emissions and climate change, both globally and locally. It accounts for 14%-24% of global GHG emissions. In Sri Lanka, the transport sector contributes approximately 50% of carbon dioxide (CO_2) equivalent emissions from fossil fuel use. Transitioning toward decarbonized mobility in Sri Lanka includes various challenges related to mobility, socioeconomic and environmental factors. A novel system dynamics (SD) approach that integrates cause-and-effect analysis and feedback loops is employed to address these complexities to develop effective policies targeting fuel economy and vehicular emissions. Three alternative vehicle growth trajectories up to 2050 are formulated based on the output from the SD model: minimal, moderate, and aggressive e-mobility strategies. This chapter outlines the pathways Sri Lanka should take to achieve its electric vehicle (EV) transition targets. By comparing these scenarios, the study identifies the most feasible and impactful strategies for reducing transport-related emissions tailored to Sri Lanka's unique context. The findings underscore the importance of tailored policy interventions and strategic planning to overcome existing barriers and leverage opportunities for sustainable transport. This chapter provides actionable insights for policymakers,

M. M. De Silva (✉) · T. Nakayama
Extreme Energy-Density Research Institute, Nagaoka University of Technology, Nagaoka, Niigata, Japan
e-mail: s225063@stn.nagaokaut.ac.jp

T. Nakayama
e-mail: nky15@vos.nagaokaut.ac.jp

M. M. De Silva · N. Fernando · A. S. Kumarage · T. Sivakumar
Department of Transport Management and Logistics Engineering, Faculty of Engineering, University of Moratuwa, Katubedda, Sri Lanka

A. G. T. Sugathapala
Department of Mechanical Engineering, Faculty of Engineering, University of Moratuwa, Katubedda, Sri Lanka

L. K. Dashora et al. (eds.), *Urban Sustainability*, Disaster Risk Reduction,
https://doi.org/10.1007/978-981-96-4890-0_12

stakeholders, and researchers, aiming to support Sri Lanka's transition toward a low-carbon transport future and contribute to broader global sustainability efforts.

Keywords System dynamics · E-Mobility policies · Decarbonization · Transition · Sri Lanka

12.1 Introduction

Urban transportation is a key driver of economic growth, enhancing access to jobs, reducing travel costs, and boosting productivity (Pierce et al. 2019; Pokharel et al. 2021). However, it is also a major source of greenhouse gas (GHG) emissions, particularly carbon dioxide (CO_2), contributing to global warming and climate change. To address these issues, many countries are adopting policies to promote compact city design, improve public transportation, and shift toward electric mobility as part of comprehensive decarbonization strategies (Lah et al. 2019). Despite these global trends, Sri Lanka's urban transport system remains heavily reliant on internal combustion engine (ICE) vehicles, contributing significantly to the country's GHG emissions (ADB 2019; Central Environment Authority Sri Lanka 2023). The sector's dependence on fossil fuels, particularly in road transport, accounts for the majority of the country's energy consumption and emissions (Our World in Data 2023). Rapid urbanization and growth in vehicle ownership, especially in cities like Colombo, have led to severe traffic congestion, air pollution, and increased carbon emissions (Jayasinghe et al. 2021).

While Sri Lanka's urban transport sector plays a crucial role in economic development, it also presents several significant challenges that urgently need to be addressed. The continued reliance on fossil fuels not only exacerbates environmental issues but also exposes the economy to fluctuations in the global energy markets (Pan et al. 2024). This reliance on ICE vehicles contributes to rising greenhouse gas emissions, traffic congestion, and public health concerns linked to air pollution. Compounding these challenges is the absence of a comprehensive policy framework to support the transition to more sustainable urban mobility. Therefore, there is a pressing need to explore integrated solutions that address these challenges and align Sri Lanka's transport policies with global decarbonization efforts (Kumarage and De Silva 2021).

In recent years, urban transport improvement efforts in Sri Lanka have focused on enhancing public bus services, addressing traffic congestion, and introducing electric buses to reduce carbon emissions. For instance, studies on the Colombo-Galle bus route highlight a decline in public transport usage due to inadequate service quality, resulting in increased traffic congestion as more people turn to private vehicles (Herath and Sivakumar 2022). Similarly, in Battaramulla, efforts to improve bus routes have been limited in scope, focusing solely on bus passengers without integrating other transport modes, leaving gaps in overall connectivity (Jayawardane 2021). Although the introduction of electric buses aligns with the United Nations' Sustainable Development Goals, financial and political challenges have slowed their progress, keeping the system largely dependent on fossil fuels (Thilakshan et al. 2023). While these initiatives mark progress, the absence of a comprehensive, well-coordinated policy framework has hindered the long-term sustainability of Sri Lanka's urban transport systems.

This study addresses these critical gaps by exploring strategic solutions for decarbonizing and electrifying Sri Lanka's urban transport sector. The primary objective of this study is to identify and analyze the key variables influencing Sri Lanka's urban transport sector, both at the microeconomic and macroeconomic levels. This study will map the complex interactions between these variables by employing system dynamics modeling techniques, particularly causal loop diagrams. Additionally, the research aims to develop and assess strategic solutions to determine the most effective pathways for decarbonizing and electrifying Sri Lanka's urban transport system.

The outcomes of this research have the potential to impact policymaking and urban planning in Sri Lanka significantly. By providing a detailed analysis of transport sector variables and developing strategies for sustainability, the study will contribute to reducing dependence on fossil fuels, lowering greenhouse gas emissions, and improving urban air quality. Furthermore, the findings will support decision-makers in creating policies that align with global sustainability goals, fostering a more resilient and efficient transport infrastructure.

12.2 Conceptual Framework

12.2.1 Transport Electrification and Decarbonization

Transport electrification and decarbonization are critical for mitigating climate change, reducing GHG emissions, and improving urban air quality. Numerous studies highlight the environmental benefits of transitioning from ICE vehicles to EVs. For instance, research in Dongguan City found that replacing ICE vehicles with EVs can reduce urban air pollution and significantly lower GHG emissions (Wu et al. 2023). This transition is particularly effective when combined with the use of renewable energy sources for electricity generation. In China, shifting to a more renewable electricity mix is projected to reduce well-to-wheel GHG emissions by 60–70% compared to current ICE vehicles by 2030 (Shen et al. 2019). While these findings are promising, the studies primarily focus on projections and lack real-world implementation data, which limits the immediate applicability of the results to policymaking in countries like Sri Lanka.

Several studies discuss the economic implications of transport electrification. A cost–benefit analysis of EV adoption in Illinois shows that the initial investment in EV infrastructure can lead to long-term savings in fuel and maintenance costs, as well as health benefits from reduced air pollution. The study projects significant fuel savings and GHG emission reductions by 2050. It also estimates that EV owners could save over $1,000 per year in fuel and maintenance costs compared to owning a gasoline vehicle (Bradley and Associates 2017). However, the study is region-specific and does not fully address challenges unique to developing nations, such as financial barriers to EV adoption and infrastructure development. Additionally, it may not fully account for variability in energy prices and policy shifts impacting the cost-effectiveness of EV adoption over time.

To smooth the transition to EV usage, a study in Jakarta emphasizes the importance of government support measures like EV subsidies, tax deductions, and investment in charging infrastructure had significant positive effects on EV adoption (Hasudungan et al. 2024). The study articulates how high upfront costs of EVs are a major barrier, and state interventions, such as subsidies, are necessary to make EVs more financially viable for consumers.

Moreover, addressing technical barriers is equally critical to encouraging widespread EV integration. Various researchers discuss the need for robust charging infrastructure, grid capacity enhancements, and advancements in battery technology (Platform for Electromobility 2023). These studies highlight that overcoming these technical challenges is crucial for the widespread adoption of EVs and achieving substantial decarbonization in the transport sector.

By better interlinking these aspects—environmental benefits, economic implications, the significance of government incentives, and the necessity to overcome technical challenges—the discourse on transport electrification becomes more cohesive, paving the way for a more seamless transition toward a cleaner, sustainable transportation future.

12.2.2 System Dynamics in Transport Modeling

SD is a methodology used to understand the behavior of complex systems over time. It employs causal loop diagrams (CLDs) and stock-flow diagrams (SFDs) to visualize and simulate dynamic system behavior, aiding decision-making and policy development (Abdelbari and Shafi 2019). SD models are widely used in urban transportation planning to evaluate policies aimed at reducing traffic congestion and improving public transportation. These models provide insights into factors affecting urban mobility and are instrumental in forecasting traffic patterns and evaluating policy scenarios (De Silva et al. 2024; Suryani et al. 2020).

A study titled "Electric Vehicle Diffusion in the Indonesian Automobile Market: A System Dynamics Modelling" analyzed the adoption of EVs in Indonesia. It found that EV adoption is expected to grow rapidly in the coming decades, with key factors being government policy support and consumer purchasing power. The study emphasized the importance of continued government incentives and strategic planning for successful EV integration into the Indonesian market (Lonan and Ardi 2020). Additionally, system dynamics modeling allows researchers to simulate different scenarios and assess the long-term impacts of various policy interventions, showing that higher EV adoption, supported by renewable energy integration, significantly improves air quality and reduces GHG emissions.

SD offers advantages over other methods like agent-based modeling (ABM) or discrete event simulation when dealing with system-wide interactions and long-term behaviors (Lempert 2002). Unlike ABM, which focuses on individual agent behaviors, SD captures feedback loops across the entire transport system, making it ideal for studying large-scale policy impacts over time. Additionally, methods like optimization models or multi-criteria decision analysis, although valuable for decision-making, cannot often model the dynamic feedback that SD excels at (Yu et al. 2017). Thus, SD provides a more comprehensive framework for analyzing policy interventions and emission reduction strategies, particularly in complex, interdependent sectors like urban transport and energy systems.

Additionally, SD provides a more comprehensive framework for analyzing policy interventions and emission reduction strategies in complex, interdependent sectors like urban transport and energy systems.

While specific studies applying SD models to Sri Lanka's transport sector are limited, the methodologies and insights from global studies can be adapted to address the unique challenges of Sri Lanka's urban transport system. SD models can help in understanding the complex dynamics of urban transport in Sri Lanka, capturing both microeconomic and macroeconomic aspects. These models provide comprehensive insights into the impacts of policy interventions and market dynamics, thereby supporting the development of sustainable and effective transportation strategies.

12.2.3 Global Policies and Strategies

Globally, numerous countries have implemented policies and strategies to promote transport electrification and decarbonization. The European Union, for instance, has set ambitious targets for reducing CO_2 emissions from new cars and vans, aiming for a 55% reduction by 2030 compared to 2021 levels (Directorate-General for Climate Action 2018). While these targets are ambitious, they are supported by strong financial backing and public infrastructure development—factors that are challenging for Sri Lanka to replicate without international financial support (Kumarage and De Silva 2021).

Sri Lanka can learn from Indonesia, Malaysia, Thailand, and India on how to promote EV adoption. For example, Sri Lanka could collaborate regionally to access EV battery technology like Indonesia (Huda et al. 2019; Veza et al. 2022), align renewable energy initiatives with EV demand like Malaysia (Veza et al. 2022), incentivize EV adoption through fiscal policies like Thailand (Chinda 2022; Kongklaew et al. 2021), and create targeted subsidies for urban and rural areas like India's FAME program (Singh et al. 2021).

In Sri Lanka, the government has articulated its commitment to reducing GHG emissions and promoting sustainable transportation through various policies and initiatives (SLoCaT 2018; UNDP Sri Lanka 2023). The National Policy on Sustainable Transport emphasizes the need to develop a multimodal transport system prioritizing public and non-motorized transport (Ministry of Transport 2008). However,

Sri Lanka still relies on the outdated 2008 National Transport Policy, presenting a significant barrier to modernizing the transport sector and adopting electric vehicles. Inconsistent policy implementation and challenges like the 2019 vehicle import ban have affected the availability of newer, more efficient models. Although the government has introduced incentives for electric and hybrid vehicles, progress has been slow.

The Sustainable Energy Authority of Sri Lanka has also been active in promoting the use of renewable energy sources for EV charging and collaborates with international organizations like the UNDP for developing low-carbon transport strategies (UNDP Sri Lanka 2023).

Despite the global advancements in transport electrification, Sri Lanka's transport sector faces unique challenges including infrastructure gaps, outdated policies, and financial constraints (Kumarage et al. 2023). Learning from regional approaches, a more tailored policy framework could benefit Sri Lanka, supporting the development of sustainable and effective transportation strategies through comprehensive insights and system dynamics modeling.

12.3 Methodology

12.3.1 Study Design

The study utilizes an SD approach to model and analyze the urban transport system of Sri Lanka. The approach involves the identification of key microeconomic and macroeconomic variables through a comprehensive literature review, followed by the construction of CLDs to understand the dynamic interactions between these variables.

12.3.2 Data Collection

The study utilized various data sources, including government reports, industry publications, and academic studies. The data collection methods included:

- Government Reports: Data from the Ministry of Transport and the Ministry of Environment.
- Industry Publications: Reports by vehicle manufacturers, transport associations, and energy providers.
- Academic Studies: Peer-reviewed articles and conference papers on EV adoption and transport decarbonization.

12.3.3 Key Variables

To understand and facilitate the adoption of EVs and the decarbonization of the transport sector, it is crucial to identify and analyze critical variables that influence these processes. These variables, derived from extensive literature and industry practices, are categorized into microeconomic and macroeconomic factors, each playing a significant role in shaping the dynamics of transport decarbonization.

The microeconomic variables (Table 12.1) directly impact individual consumer decisions and the operational aspects of transport. These factors are essential for policymakers, industry stakeholders, and practitioners to consider when developing and implementing strategies for sustainable transportation.

Table 12.1 Key microeconomic variables

Variable	Description	Literature
Purchase cost	Costs associated with purchasing ICE vehicles, public transport vehicles, and EVs	Kumar and Chakrabarty (2020), Manisha (2020), Nerurkar et al. (2023)
Operating cost	Ongoing costs of operating vehicles, including maintenance and fuel/electricity costs	Manisha (2020), Nerurkar et al. (2023), Weldon et al. (2018)
Fuel efficiency	Efficiency of fuel usage in ICE vehicles versus electricity usage in EVs	Jochem et al. (2016), Kumar and Chakrabarty (2020), Manisha (2020)
Consumer preference	Influences on consumer preferences for different types of vehicles	Manisha (2020), Nerurkar et al. (2023), Ottesen and Banna (2021)
Infrastructure investments	Investments in road infrastructure, public transport, and EV charging infrastructure	Chen (2019), Hardman et al. (2017)
Environmental impact	Environmental consequences of using different types of vehicles	Nerurkar et al. (2023), Siragusa et al. (2022)
Accessibility	Ease of access to transportation modes, including public transport and EV charging stations	Rudolph (2016; Van Dijk et al. (2021)
Service frequency	Frequency of public transport services, impacting user convenience and wait times	Lunke et al. (2021), Tyndall (2018)
Fare price	Cost of using public transport, influencing affordability and usage rates	Fowri and Seyedabrishami (2020), Matas et al. (2020)
Reliability	Consistency and dependability of transportation services affecting user satisfaction	Pulugurtha et al. (2022), Tyndall (2018)

Table 12.2 Key macroeconomic variables

Variable	Description	Literature
Consumer income levels	Impact on purchasing power and demand for different types of vehicles	Burguillo et al. (2017), Weldon et al. (2018)
Economic growth	Influence on overall transport demand and investment capacity	Paltsev et al. (2021), Rashid Khan et al. (2018), Zhao et al. (2019)
Government policies and incentives	Effect on the adoption of EVS and investments in transport infrastructure	Jenn et al. (2018); Peng and Lin (2022), Yan (2018)
Foreign exchange rate	Impact on the cost of vehicle imports and fuel	Ghoddusi et al. (2019), Modisaatsone and Motlaleng (2013)
Fuel prices	Influence on the operational costs of ICE vehicles and the competitiveness of EVs	He et al. (2017), Leung et al. (2019)
Import taxes	Effect on the cost of importing vehicles and components	Ma and Mayburov (2021), Nadirov et al. (2020)
Gross domestic product (GDP)	Reflecting the overall economic health and its influence on transport sector investments	De Silva et al. (2022), Feng et al. (2019), Kumarage and Weerawardana (2013), Rashid Khan et al. (2018)
CO_2 emissions	Considering the CO_2 emissions from different transport modes	Ivče et al. (2018), Solaymani (2019)

The macroeconomic variables (Table 12.2) have a broader impact on the transport sector and the economy as a whole. These factors are vital for developing long-term policies and investment strategies for sustainable transportation.

12.3.4 Multifactorial Assessment Criteria

The following Table 12.3 outlines the multifactorial criteria used to assess the different strategies developed for the transition to electrification and decarbonization of Sri Lanka's urban transport system.

12.3.5 System Dynamics Approach

The study employs an SD approach to understand the transport system's complex interactions and feedback loops. This method is instrumental in representing the transport system's current and potential future states through the development of

Table 12.3 Multifactorial assessment criteria

Assessment criterion	Description
Cost of fleet additions	Annual costs of adding new vehicles to the fleet
Vehicle spares	Costs associated with spare parts and maintenance
Fuel cost	Costs of fuel consumption at 80 USD per barrel of crude oil (BBL)
Electricity cost	Costs of electricity consumption at 0.2 USD per kilowatt hour (KWh)
Road infrastructure cost	Investments required for road infrastructure improvements
Public transport infrastructure cost	Investments required for public transport infrastructure improvements
CO_2 emissions	CO_2 emissions associated with different transport modes
Energy consumption	Total energy consumption by different types of vehicles
Environmental impact	Overall environmental impact of the transport system
Economic impact	Influence on the overall economy, including job creation and GDP
Social impact	Impact on social factors, such as accessibility and affordability of transport
Technological feasibility	Feasibility of implementing new technologies, such as EV infrastructure

CLDs. Four distinct CLDs were initially constructed using the identified key variables in Sect. 12.3.3. The first CLD focuses on the microeconomic aspects of the ICE Vehicle Market. The second diagram integrates the microeconomic aspects of the ICE Vehicle Market with the Public Transport Model Share. The third CLD expands further to include the potential EV Market alongside the ICE Vehicle Market and Public Transport Model Share. Finally, the fourth diagram incorporates macroeconomic aspects, providing a comprehensive view that includes the ICE Vehicle Market, Public Transport Model Share, and the potential EV Market.

Creating scenarios based on different strategies for electrification and decarbonization of Sri Lanka's urban transport system is a critical step in this chapter. Each scenario is assessed using a multifactorial approach, considering the key criteria identified in Sect. 12.3.4. These scenarios help visualize and evaluate the potential outcomes of various strategies, thereby aiding in formulating effective policies and interventions.

12.3.6 Data Analysis

The analysis of the CLDs plays a pivotal role in identifying feedback loops and complex relationships among the variables. This analysis helps in understanding how different factors influence the transport system and its decarbonization. By examining these interactions, the study can uncover the underlying dynamics that drive the system's behavior.

Excel is used for statistical analysis to compare different strategies and evaluate the impact of various variables on the adoption of EVs and the shift toward sustainable transport modes. This quantitative analysis provides a detailed evaluation of the proposed scenarios, highlighting the effectiveness and feasibility of each strategy.

By combining the system dynamics approach with a multifactorial assessment of scenarios, this study aims to provide a comprehensive understanding of the pathways toward decarbonizing the urban transport system in Sri Lanka. This integrated methodology ensures that all critical aspects are considered, offering a robust framework for future policy development and implementation.

12.4 Results

12.4.1 Results of Causal Loop Diagrams (CLDs)

The CLDs constructed for this study illustrate the complex interactions between various factors influencing the urban transport system in Sri Lanka. Four key diagrams were developed:

- Microeconomic aspects of the ICE Vehicle Market,
- Microeconomic aspects of the ICE Vehicle Market with Public Transport Model Share,
- Microeconomic aspects of the ICE Vehicle Market, Public Transport Model Share with potential EV Market,
- Macroeconomic aspects of the ICE Vehicle Market, Public Transport Model Share with potential EV Market.

12.4.1.1 Microeconomic Aspects of the ICE Vehicle Market

The analysis of the CLD for the microeconomic aspects (Fig. 12.1) of the ICE Vehicle Market reveals several critical balancing and reinforcing loops that highlight the complex interplay of variables influencing consumer behavior and market dynamics.

The analysis of these CLDs provides valuable insights into the microeconomic factors affecting the ICE Vehicle Market. The balancing loops illustrate how taxation and environmental concerns can temper market growth, while the reinforcing loop demonstrates the potential for infrastructure investments to drive increased vehicle

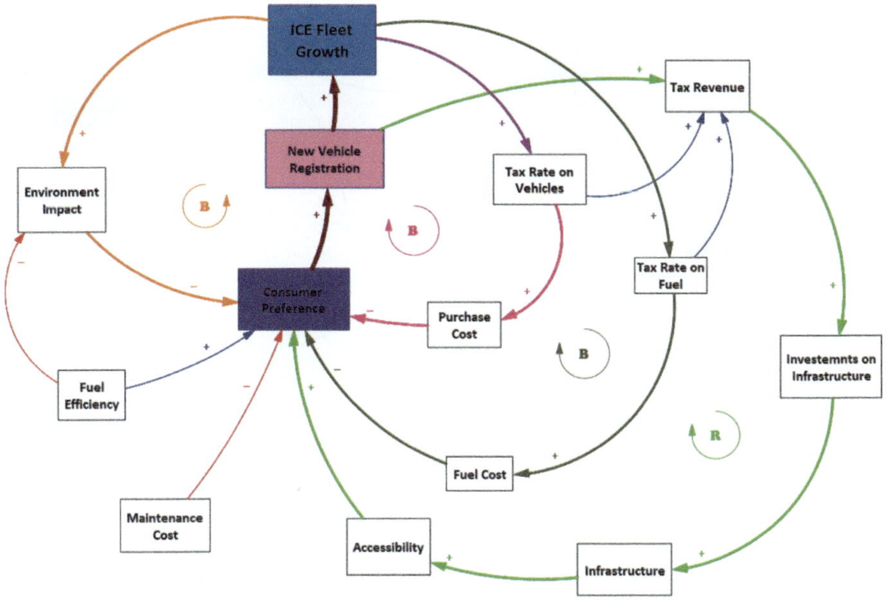

Fig. 12.1 CLD of microeconomic aspects of the ICE market. *Source* This figure was created by the author using variables sourced from studies cited in Table 12.1 of the main text

registrations. Understanding these dynamics is crucial for policymakers aiming to regulate the ICE Vehicle Market and promote sustainable transportation alternatives.

12.4.1.2 Microeconomic Aspects of the ICE Vehicle Market with Public Transport (PT) Model Share

The integration of public transport investments into the ICE Vehicle Market through the CLDs (Fig. 12.2) reveals several reinforcing loops that highlight the dynamic relationships between PT improvements and consumer preferences. This section delves into the key reinforcing loops and their implications for the transport system.

The analysis of these reinforcing loops illustrates how investments in PT can significantly improve various aspects of the system, including accessibility, service frequency, quality, and affordability. These enhancements collectively increase the public transport mode share, attracting more users and reducing the demand for new ICE vehicle registrations. This shift toward PT promotes a more sustainable urban transport system by decreasing reliance on private vehicles and supporting environmentally friendly alternatives. Understanding these dynamics is essential for policymakers aiming to enhance public transport infrastructure and encourage a transition to sustainable urban mobility solutions.

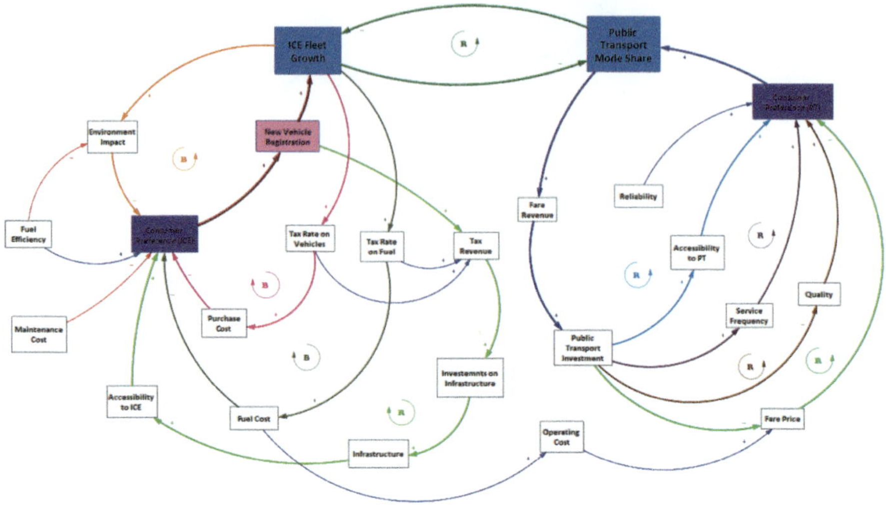

Fig. 12.2 CLD of microeconomic aspects of ICE market with public transport mode share. *Source* This figure was created by the author using variables sourced from studies cited in Table 12.1 of the main text

12.4.1.3 Microeconomic Aspects of the ICE Vehicle Market, Public Transport (PT) Model Share with Potential EV Market

The integration of EVs into the transport system alongside ICE vehicles and public transport is a pivotal step toward decarbonizing the urban transport sector. The analysis of the CLD (Fig. 12.3) for this combined market reveals a mixture of balancing and reinforcing loops that highlight the dynamic interactions between government policies, consumer preferences, and infrastructure investments.

The analysis of these loops reveals several key insights into the microeconomic aspects of integrating EVs with the ICE Vehicle Market and public transport. Lower taxes and government subsidies are crucial in reducing the purchase cost of EVs, thereby increasing adoption rates and reducing the dominance of ICE vehicles. Investments in EV infrastructure, particularly charging facilities, are vital in enhancing the accessibility and attractiveness of EVs, further boosting their market share. However, there is a competitive dynamic between EVs and public transport, where higher adoption of one can potentially reduce the use of the other. Policymakers need to balance these dynamics carefully to promote a holistic and sustainable urban transport system that includes EVs, public transport, and reduced reliance on ICE vehicles.

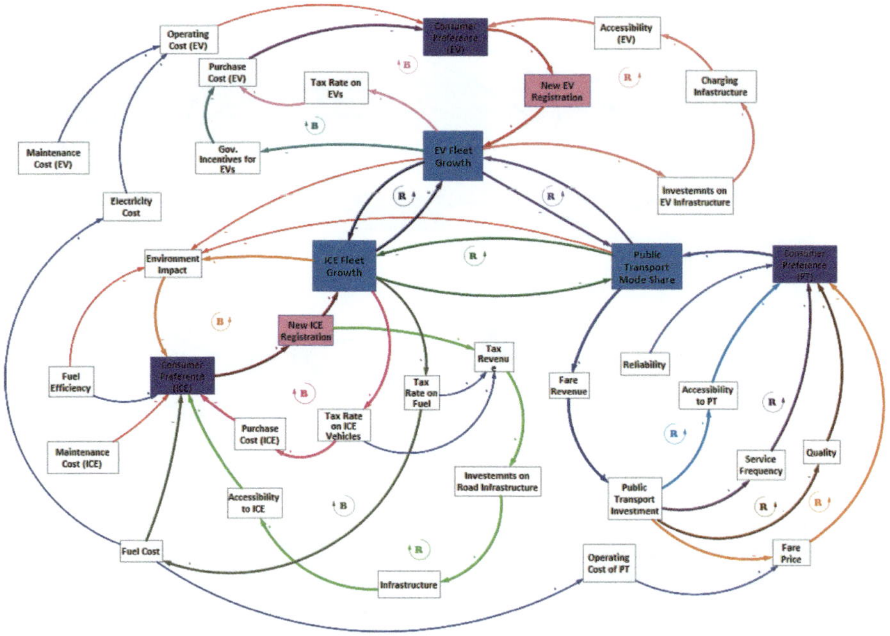

Fig. 12.3 CLD of microeconomic aspects of ICE vehicle market, public transport model share with potential EV market. *Source* This figure was created by the author using variables sourced from studies cited in Table 12.1 of the main text

12.4.1.4 Macroeconomic Aspects of the ICE Vehicle Market, Public Transport Model Share with Potential EV Market

The macroeconomic aspects of the transport system involve complex interactions between government policies, economic growth, and investments in EVs and public transport. The analysis of the CLD (Fig. 12.4) for this combined market highlights both reinforcing and balancing loops, which demonstrate the dynamic relationships that shape the overall economic and environmental outcomes.

Government policies and incentives are crucial in the initial stages of transitioning to sustainable transportation. These interventions attract investments in EV infrastructure and public transport, making them more accessible and appealing. As EV and public transport adoption increases, it reduces the growth of the ICE fleet, lowering vehicle and fuel imports. This reduction alleviates the economic burden, fostering economic growth and increasing GDP. Over time, the system becomes self-sustaining, reducing the need for continuous government interventions. Understanding these dynamics helps policymakers design effective strategies for a sustainable and balanced urban transport system.

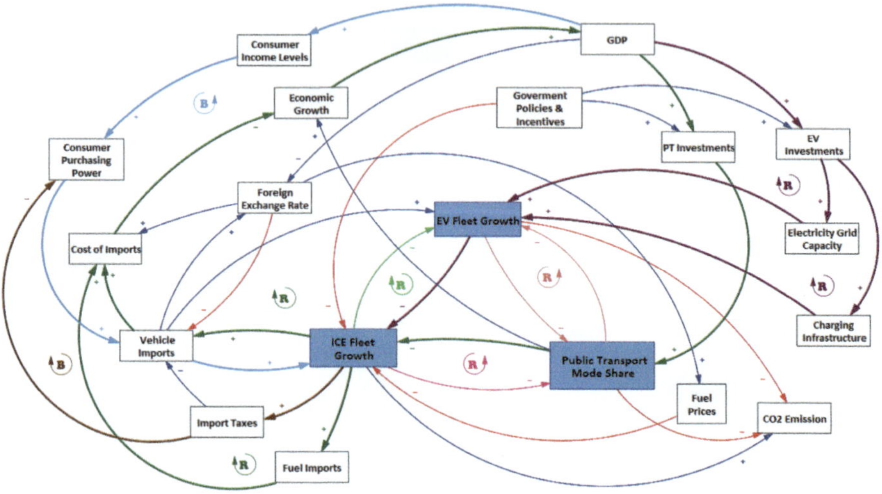

Fig. 12.4 CLD of macroeconomic aspects ICE vehicle market, public transport model share with potential EV market. *Source* This figure was created by the author using variables sourced from studies cited in Table 12.2 of the main text

By understanding these complex interactions and feedback loops, policymakers can design more effective strategies for transitioning to sustainable transportation, ensuring that the system evolves toward greater economic and environmental sustainability.

12.4.1.5 Summary of Key Insights of CLDs

The CLDs analysis offers key insights into the transport system's dynamics, focusing on consumer preferences, public transport investments, and the balance between ICE vehicles and EVs.

- Consumer preferences are strongly influenced by purchase and fuel costs driven by taxes. Increasing taxes on ICE vehicle purchases and fuel can reduce registrations and encourage sustainable behavior.
- Enhancing public transport infrastructure, service frequency, affordability, and accessibility can boost public transport usage, benefiting the microeconomy and environmental goals.
- Reducing new ICE vehicle registrations is more effective in the short term than investing heavily in EVs. Prioritizing public transportation improvements decreases reliance on ICE vehicles.
- Imposing higher purchase and fuel costs on ICE vehicles is more efficient than subsidizing EVs. This strategy encourages EV adoption, supports sustainable consumer behavior, and aligns with macroeconomic emission reduction goals.

- Heavy government investment in new EV infrastructure is unnecessary, as existing roads can accommodate EVs. The private sector will likely invest in charging stations as EV registrations increase, allowing the market to adapt naturally.

In quantifying the causal effects identified in the CLDs, three sub-models will be developed. The execution plan includes the evaluation of three strategies within these sub-models. The sub-models include e-mobility policy sub-model, fuel efficiency sub-model, and vehicle operational fleet sub-model.

12.4.2 Application of CLDs in Assessing Strategies

Arising from the CLDs, integrating different sub-models allows for a comprehensive assessment of the scenarios. The flow of information between the sub-models is as follows:

1. **E-Mobility Policy → Vehicle Operational Fleet:**
   - Policies such as subsidies or infrastructure investments increase EV adoption rates, changing the fleet composition.
2. **Vehicle Operational Fleet → Fuel Efficiency:**
   - A higher proportion of EVs in the fleet improves overall fuel efficiency and reduces emissions.
3. **Feedback Loop:**
   - Improved fuel efficiency and lower emissions provide data back to the e-mobility policy sub-model, allowing for adjustments in policy measures to achieve desired outcomes.

12.4.2.1 E-Mobility Policy Sub-model

The e-mobility policy sub-model discusses the scenarios formulated depending on the extent of policies in the policy interventions. These scenarios are as follows (Table 12.4).

12.4.2.2 Fuel Efficiency Sub-model

The fuel efficiency sub-model assesses the overall fuel efficiency of the vehicle fleet. It considers several key factors mentioned in Table 12.5 (Fig. 12.5).

Table 12.4 Scenarios under e-mobility policy sub-model

E Mobility policy	Description
1. Minimal e-mobility intervention	No specific e-mobility policy intervention allows electrification to be driven by market forces, continuing with the current fiscal and transport policies Public transport investment remains low, with gradual increases in the share of public transport
2. Moderate e-mobility intervention	Aiming for 100% EV registrations by 2050 with moderate public transport investment and higher vehicle/fuel taxes Public transport shares increased steadily, with significant electrification by the mid-century
3. Aggressive e-mobility policy	Targeting 100% EV registrations by 2040, aligned with a low-carbon development strategy Significant public transport investment and high vehicle/fuel taxes led to rapid electrification and reduced vehicle fleet growth

Table 12.5 Key component of vehicle fleet sub-model

Component	Description
Fuel consumption rates	Fuel consumption rates for ICE vehicles and electricity consumption rates for EVs
Fuel imports	Volume and fuel types imported based on fleet composition and evaluate the impact of reducing fuel imports on the economy and environment
Emission factors	Emission factors for different types of vehicles

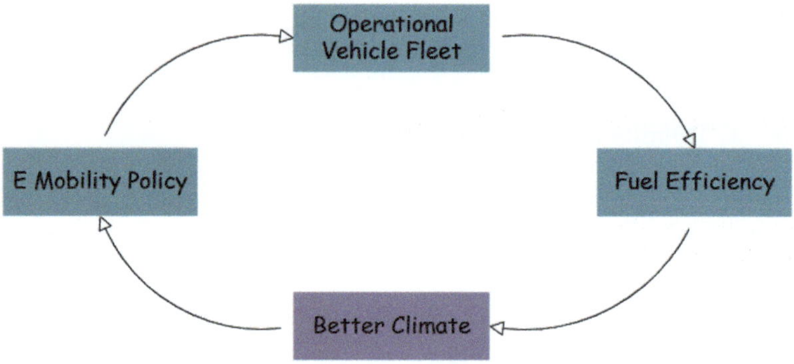

Fig. 12.5 Integration of sub-models

Table 12.6 Key components of vehicle operational sub-model

Component	Description
Growth rate	Historical and projected growth rates of the vehicle fleet and the impact of policies on these growth rates
Operational fleet	Total number of vehicles in operation, with a breakdown by vehicle type (ICE versus EV)
New vehicle registration	Annual registration data for new vehicles and the impact of policies on EV registrations
Scrappage rate	Rates at which old vehicles are scrapped and replaced and its impact on fleet renewal and composition
Fleet composition	Proportion of EVs versus ICE vehicles in the fleet and tracks changes in fleet composition over time due to policy interventions

12.4.2.3 Vehicle Operational Fleet Sub-model

The Vehicle Operational Fleet Sub-Model analyzes the composition and dynamics of the vehicle fleet under different policy scenarios (Table 12.6).

12.4.3 Strategies Assessment

The strategies assessment evaluates the impact of different developed e-mobility strategies on Sri Lanka's urban transport system. The assessment is based on key assumptions (Table 12.7).

In assessing the potential strategies for e-mobility, Tables 12.8 and 12.9 provide a comprehensive overview of both environmental and economic impacts. Table 12.8 presents the outcomes of various e-mobility strategies, ranging from non-intervention to aggressive intervention, by examining key metrics such as fleet composition, annual fuel use, electric vehicle EV penetration, and GHG emissions.

Building on these environmental metrics, Table 12.9 translates the strategies into their associated economic costs. This table details the financial burden of implementing each strategy, breaking down the expenses related to fuel, electricity, vehicle maintenance, and infrastructure investments.

Figure 12.6 depicts the projected cost per passenger kilometer across three e-mobility strategies—non-intervention, moderate intervention, and aggressive intervention—over the periods of 2030, 2040, and 2050. The chart shows that the cost per passenger km decreases across all strategies over time, but the aggressive intervention strategy consistently yields the lowest cost at every time point. As time progresses, the cost of aggressive intervention drops significantly, reaching 0.0408 USD/km by 2050. The moderate intervention strategy follows a similar trend but remains more expensive than aggressive intervention, while the non-intervention strategy remains the costliest, with minimal reductions over the 2050 horizon.

Table 12.7 Metrics used to evaluate the impact of e-mobility strategies on Sri Lanka's urban transport system

Metric	Description
Bus and railway share	The market share is carried by all buses and railways
Motor vehicle fleet	Number of units based on the shift to public transport
Annual fuel use	Estimated vehicle fuel consumption
EV penetration of fleet	Percentage of EVs in each vehicle category
Annual electricity consumption	Energy consumed by EVs in Terawatt hours (TWh)
Annual CO_2 emissions	Tail-pipe emissions in million tons for ICE vehicles
Annual passenger kilometers (km)	The total distance traveled by all passengers using the transport system annually is measured in billions of km
Annual economic cost of transport inputs	Includes the impact of emission cost, fuel prices, electricity prices, vehicles/spare/battery cost, and investments in road and public transport infrastructure costs
Emission cost	The monetary cost is assigned to the carbon emissions vehicles produce. For this assessment, the carbon price is set at \$20 per ton of CO_2, based on the findings of Herath and Jung (2021)
Cost per passenger km	The total annual economic cost of the transport system is divided by the total passenger km, providing a per-km transport cost for each passenger

Figure 12.7 compares the cost per passenger kilometer for the moderate and aggressive intervention strategies against the non-intervention strategy for 2030, 2040, and 2050. The moderate intervention strategy shows a cost reduction to 94.2% of the non-intervention cost in 2030, further decreasing to 93.5% by 2040, and reaching 92.8% by 2050. In contrast, the aggressive intervention strategy offers even greater savings, with costs reduced to 80.4% of the non-intervention level in 2030, dropping further to 77.6% in 2040, before slightly increasing to 79.0% in 2050. These results highlight the long-term economic benefits of adopting more proactive e-mobility strategies, with the aggressive intervention consistently providing the greatest cost savings over time.

These charts collectively emphasize that while all e-mobility strategies reduce costs over time, adopting a more aggressive approach leads to the most substantial long-term financial savings. This underscores the importance of proactive intervention in reducing the economic burden of transportation on a per-passenger basis.

Table 12.8 E-mobility strategies and their impact on fleet composition, energy use, and emissions

E-Mobility strategy	Year	Bus and railway share	Motor vehicle fleet (mn)	Annual fuel use (mn liters)	EV penetration of fleet				Annual electricity consumption GWh	Annual GHG (CO_2 Gg)
					Cars and dual purpose %	2/3 wheel %	Buses %	Goods vehicles %		
Baseline										
Last year of normal operations recorded	2019	49%	5.6	4,027	0	0	0	0	0	9,943
Non-intervention										
Business-as-usual transport policy: no specific EV targets	2030	41%	8.0	5,397	2%	9%	2%	0%	223	13,109
	2040	46%	11.1	7,452	10%	39%	8%	2%	1,689	17,897
50% Public transport by 2050	2050	52%	13.6	8,158	22%	80%	17%	8%	4,839	19,554
Moderate intervention										
Reach 100% EV registrations by 2050	2030	44%	7.2	4,984	3%	12%	4%	2%	404	12,149
	2040	50%	9.4	5,663	23%	52%	19%	10%	3,223	13,723
50% Public transport by 2040	2050	53%	11.8	4,208	58%	96%	46%	37%	10,407	10,215
Aggressive intervention										
Reach 100% EV Registrations by 2040	2030	50%	6.5	4,423	7%	12%	5%	2%	564	10,842
	2040	56%	8.1	3,915	43%	52%	27%	26%	4,751	9,619
50% Public transport by 2030	2050	61%	10.2	1,915	80%	95%	58%	64%	12,485	4,753

Mn: million, bn: billion, GWh: Giga watt-hours, Gg: Giga gram
Source Based on authors analysis

Table 12.9 Economic cost of transport inputs under different e-mobility strategies

Strategy	Year	Annual passenger km (bn)	Annual GHG (CO_2 Gg)	Annual economic cost of transport inputs (USD bn)								Cost per passenger km (USD/ km)
				Emission cost	Petroleum fuel @ current 80 USD/bbl	Electricity @ current 0.2 USD / kWh	Vehicles	Vehicle spare/ battery etc	Investment in road infrastructure	Investment in Public Transport	Total cost	
Baseline												
Last year of normal operations recorded	2019	148	9,943	0.20	5.6	0.0	1.7	0.7	1.3	0.4	9.8	0.0662
Non-intervention												
Business-as-usual transport policy: no specific EV targets: 50% Public transport by 2050	2030	202	13,109	0.26	5.3	0.0	3.8	1.3	0.9	0.4	12.1	0.0599
	2040	287	17,897	0.36	7.5	0.3	5.1	2.2	1.0	0.4	16.9	0.0590
	2050	363	19,554	0.39	8.3	1.0	5.0	3.0	0.7	0.5	18.8	0.0517
Moderate intervention												
Reach 100% EV registrations by 2050 50% Public transport by 2040	2030	192	12,149	0.24	4.9	0.1	3.1	1.2	0.8	0.4	10.8	0.0564
	2040	260	13,723	0.27	5.7	0.6	4.5	1.9	0.8	0.5	14.3	0.0551
	2050	329	10,215	0.20	4.3	2.1	5.4	2.7	0.5	0.5	15.8	0.0479
Aggressive intervention												
Reach 100% EV registrations by 2040 50% Public Transport by 2030	2030	190	10,842	0.22	4.3	0.1	2.1	1.1	0.8	0.5	9.1	0.0481
	2040	239	9619	0.19	3.9	1.0	3.2	1.5	0.6	0.5	10.9	0.0458
	2050	299	4753	0.10	2.0	2.5	4.3	2.2	0.5	0.6	12.2	0.0408

Source Based on authors analysis

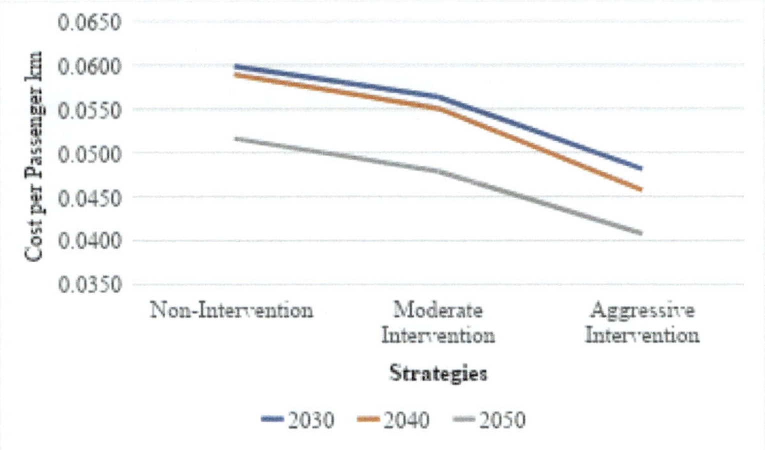

Fig. 12.6 Projected cost per passenger km by strategies proposed *Source* based on authors analysis

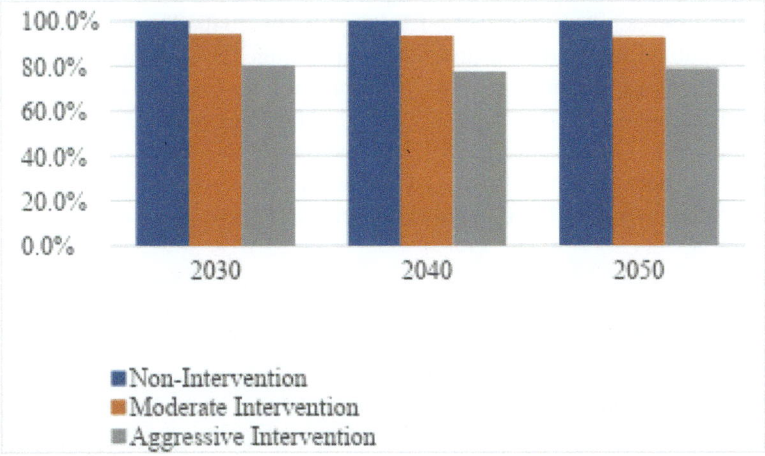

Fig. 12.7 Cost per passenger km compared to non-intervention strategy. *Source* Based on authors analysis

12.5 Discussion

The results from the CLDs and scenario assessments highlight the complex interactions within Sri Lanka's urban transport system, emphasizing the need for integrated policy measures.

Firstly, taxing ICE vehicles and fuel, along with government subsidies, are crucial for boosting EV adoption and supporting sustainable transitions in the transport sector.

Secondly, investing in PT infrastructure significantly increases public transport use and reduces private vehicle demand, promoting sustainable urban mobility. Enhancing the quality, frequency, and accessibility of PT is key.

Thirdly, the competitive dynamic between EV adoption and PT use underscores the need for a balanced approach. Investments in both EV infrastructure and PT are necessary for achieving sustainable urban mobility. The Aggressive E-Mobility Policy emerges as the most effective strategy for cutting emissions and supporting Sri Lanka's low-carbon strategy. This finding aligns with global trends, where aggressive decarbonization efforts yield substantial benefits.

However, this study acknowledges limitations, data constraints, model assumptions, and urban focus and suggests refining models with more comprehensive data and exploring long-term policy impacts in future studies.

12.6 Recommendations

Based on the analysis of e-mobility strategies in Sri Lanka, it is recommended to adopt a proactive approach targeting 100% EV adoption by 2040. The focus should be on enhancing public transport infrastructure to encourage the shift to more sustainable transport modes, rather than extensive government investment in EV infrastructure. Implementing financial incentives such as higher fuel and vehicle taxes for ICE vehicles, along with moderate subsidies for EV adoption, will drive behavioral change toward sustainable transport options. Continuous reassessment of fuel prices and carbon pricing is essential to maximize the long-term benefits of e-mobility.

12.7 Conclusion

This study has explored the transition toward sustainable urban transport in Sri Lanka through the lens of electrification and decarbonization using an SD approach. The chapter has demonstrated that transitioning to EVs and enhancing public transport systems can significantly reduce GHG emissions, improve urban air quality, and reduce fuel costs. The study evaluated three alternative vehicle growth trajectories (minimal, moderate, and aggressive e-mobility strategies), with the aggressive e-mobility strategy yielding the lowest economic cost per passenger. This strategy aligns with Sri Lanka's current policy of a climate-smart-green/low-carbon development and targets a 50% public transport share by 2030, achieving substantial CO_2 reduction and fostering early economic recovery.

This chapter contributes to the existing body of knowledge by providing a comprehensive analysis of the socioeconomic and environmental impacts of transitioning to e-mobility in an urban context. The application of the SD model in this study offers a robust framework for understanding the complex interactions between various factors influencing transport dynamics. The insights from this chapter can inform policy-makers on effective strategies for promoting EV adoption, enhancing public transport, and achieving decarbonization goals. The emphasis on government incentives, infrastructure development, and renewable energy integration provides actionable recommendations for sustainable urban transport planning.

The transition to sustainable urban transport is essential for mitigating climate change and improving the quality of life in urban areas. This study underscores the importance of adopting aggressive e-mobility strategies to achieve significant environmental and economic benefits. By implementing comprehensive policies and investing in public transport modernization, Sri Lanka can pave the way for a resilient and sustainable e-mobility future. Continued studies and policy support are crucial for overcoming technical and financial barriers and ensuring a smooth transition to a decarbonized transport system.

References

Abdelbari H, Shafi K (2019) A system dynamics modeling support system based on computational intelligence. Systems 7(4). https://doi.org/10.3390/SYSTEMS7040047

ADB (2019) Energy sector assessment, strategy, and road map: Sri Lanka. Asian Development Bank (December). https://doi.org/10.22617/TCS190557-2

Bradley MJ, Associates (2017) Electric vehicle cost-benefit analysis plug-in electric vehicle cost-benefit analysis. Illinois

Burguillo M, del Río P, Romero Jordán D (2017) Car use behaviour of Spanish households: differences for quartile income groups and transport policy implications. Case Stud Transp Policy 5(1):150–158. https://doi.org/10.1016/J.CSTP.2016.09.005

Central Environment Authority Sri Lanka (2023) Air quality index for Sri Lanka (AQI-SL), Calculation and Guideline (v1.0)—CEA.LK. https://www.cea.lk/web/images/pdf/airqulity/AQI-SL_Calculation_Guideline_CEA.LK_V1.0.pdf

Chen C (2019) Essays on the effectiveness of environmental and urban transportation policies. University of British Columbia. https://doi.org/10.14288/1.0380426

Chinda T (2022) Long-term trend of electric vehicle sales in Thailand. Eng Manag Prod Serv 14(1):13–25. https://doi.org/10.2478/EMJ-2022-0002

De Silva MM, Kumarage AS, Dilini KP, Amalan T (2022) Modelling gamma coefficient in the Gompertz curve to determine the variability of vehicle ownership saturation. J South Asian Logist Transp 2(2):1–17. https://doi.org/10.4038/JSALT.V2I2.47

De Silva MM, Herath OK, Nakayama T, Kumarage AS (2024) A system dynamics model for vehicle fleet transformation towards energy efficiency and low-carbon development: a case study of Sri Lanka and its strategies. Transp Policy 147: 244–258. https://doi.org/10.1016/j.tranpol.2024.01.004

Directorate-General for Climate Action (2018) CO_2 emission performance standards for cars and vans. European Commission. https://climate.ec.europa.eu/eu-action/transport/road-transport-reducing-co2-emissions-vehicles/co2-emission-performance-standards-cars-and-vans_en

Feng Z, Cai T, Xiang K, Xiang C, Hou L (2019) Evaluating the impact of fossil fuel vehicle exit on the oil demand in China. Energies 12(14). https://doi.org/10.3390/EN12142771

Fowri HR, Seyedabrishami S (2020) Assessment of urban transportation pricing policies with incorporation of unobserved heterogeneity. Transp Policy 99:12–19. https://doi.org/10.1016/J.TRANPOL.2020.08.008

Ghoddusi H, Morovati M, Rafizadeh N (2019) Foreign exchange shocks and gasoline consumption. Energy Econ 84:104472. https://doi.org/10.1016/J.ENECO.2019.08.005

Hardman S, Chandan A, Tal G, Turrentine T (2017) The effectiveness of financial purchase incentives for battery electric vehicles—a review of the evidence. Renew Sustain Energy Rev 80:1100–1111. https://doi.org/10.1016/J.RSER.2017.05.255

Hasudungan A, Tandean B, Aurelius E, Widarsyah R, Artha IKDS (2024) The impact of government incentives on electric vehicle adoption in the metropolitan Jakarta area. Jurnal Ekonomi Pembangunan 21(2):191–199. https://doi.org/10.29259/JEP.V21I2.23050

He LY, Yang S, Chang D (2017) Oil price uncertainty, transport fuel demand and public health. Int J Environ Res Public Health 14(3). https://doi.org/10.3390/IJERPH14030245

Herath E, Sivakumar T (2022) Modelling the level of service of public transport : as a case study along Colombo–Galle buses. Bolgoda Plains 2(1):22–23. https://doi.org/10.31705/BPRM.V2(1).2022.6

Herath HMSA, Jung TY (2021) Carbon pricing and supporting policy tools for deep decarbonization; case of electricity generation of Sri Lanka. Carbon Manag 12(5):465–484. https://doi.org/10.1080/17583004.2021.1966514

Huda M, Aziz M, Tokimatsu K (2019) The future of electric vehicles to grid integration in Indonesia. Energy Procedia 158:4592–4597. https://doi.org/10.1016/J.EGYPRO.2019.01.749

Ivče R, Zekić A, Mohović D (2018) Comparison of CO_2 emissions for road and sea transport on the specific route. In: International symposium ELMAR, pp 19–22. https://doi.org/10.23919/ELMAR.2018.8534689

Jayasinghe P, Raghavan V, Yonezawa G (2021) Exploration of expansion patterns and prediction of urban growth for Colombo city, Sri Lanka. Spat Inf Res 29(4):465–478. https://doi.org/10.1007/S41324-020-00364-4

Jayawardane W (2021) Bus service improvement plan for Battaramulla along with location analysis for a transport hub. J South Asian Logist Transp 1(2):57–72. https://doi.org/10.4038/JSALT.V1I2.34

Jenn A, Springel K, Gopal AR (2018) Effectiveness of electric vehicle incentives in the United States. Energy Policy 119:349–356. https://doi.org/10.1016/J.ENPOL.2018.04.065

Jochem P, Doll C, Fichtner W (2016) External costs of electric vehicles. Transp Res Part D Transp Environ 42:60–76. https://doi.org/10.1016/J.TRD.2015.09.022

Khan HUR, Siddique M, Zaman K, Yousaf SU, Shoukry AM, Gani S, Sasmoko KA, Hishan SS, Saleem H (2018) The impact of air transportation, railways transportation, and port container traffic on energy demand, customs duty, and economic growth: evidence from a panel of low-, middle-, and high-income countries. J Air Transp Manag 70:18–35. https://doi.org/10.1016/J.JAIRTRAMAN.2018.04.013

Kongklaew C, Phoungthong K, Prabpayak C, Chowdhury MS, Khan I, Yuangyai N, Yuangyai C, Techato K (2021) Barriers to electric vehicle adoption in Thailand. Sustainability 13(22). https://doi.org/10.3390/SU132212839

Kumar P, Chakrabarty S (2020) Total cost of ownership analysis of the impact of vehicle usage on the economic viability of electric vehicles in India. Transp Res Rec 2674(11):563–572. https://doi.org/10.1177/0361198120947089

Kumarage AS, De Silva M (2021) A gap analysis of EV policies for promoting e-mobility in Sri Lanka. https://www.researchgate.net/publication/358200899

Kumarage A, Sugathapala T, De Silva S (2023) Development of policy framework for EV transition for Sri Lanka and implementation plan volume II- technical notes (Draft Final) II(November)

Kumarage AS, Weerawardana J (2013) System cost-based multi-criteria analysis for urban transport solutions. Int J Urban Sci 17(2):212–225. https://doi.org/10.1080/12265934.2013.776285

Lah O, Fulton L, Arioli M (2019) Decarbonization scenarios for transport and the role of urban mobility. In: Sustainable urban mobility pathways: policies, institutions, and coalitions for low carbon transportation in emerging countries, pp 65–80. https://doi.org/10.1016/B978-0-12-814897-6.00003-X

Lempert R (2002) Agent-based modeling as organizational and public policy simulators. Proc Natl Acad Sci USA 99(Suppl 3):7195–7196. https://doi.org/10.1073/PNAS.072079399

Leung A, Burke M, Cui J, Perl A (2019) Fuel price changes and their impacts on urban transport—a literature review using bibliometric and content analysis techniques, 1972–2017. Transp Rev 39(4):463–484. https://doi.org/10.1080/01441647.2018.1523252

Lonan ES, Ardi R (2020) Electric vehicle diffusion in the indonesian automobile market: a system dynamics modelling. In: 2020 IEEE international conference on industrial engineering and engineering management (IEEM), pp 43–47. https://doi.org/10.1109/IEEM45057.2020.9309988

Lunke EB, Fearnley N, Aarhaug J (2021) Public transport competitiveness vs. the car: impact of relative journey time and service attributes. Res Transp Econ 90. https://doi.org/10.1016/J.RETREC.2021.101098

Ma J, Mayburov I (2021) Using did model to analyse the impact of vehicle purchase tax on the sales of alternative fuel vehicle. IOP Conf Ser Earth Environ Sci 849(1). https://doi.org/10.1088/1755-1315/849/1/012006

Manisha (2020) Consumer preference towards various brands of cars : a case study of Dehradun. Alochana Chakra J IX:8806. https://www.researchgate.net/publication/344172371_CONSUMER_PREFERENCE_TOWARDS_VARIOUS_BRANDS_OF_CARS_A_CASE_STUDY_OF_DEHRADUN

Matas A, Raymond JL, Ruiz A (2020) Economic and distributional effects of different fare schemes: evidence from the Metropolitan region of Barcelona. Transp Res Part A Policy Pract 138:1–14. https://doi.org/10.1016/J.TRA.2020.05.014

Ministry of Transport (2008) Draft national policy on transport in Sri Lanka, pp 1–37

Modisaatsone NC, Motlaleng GR (2013) Impact of exchange rate volatility on Botswana's imports. E3 J Bus Manag Econ 4(5). https://www.econbiz.de/Record/impact-of-exchange-rate-volatility-on-botswana-s-imports-modisaatsone-naledi/10010878101

Nadirov O, Vychytilová J, Dehning B (2020) Carbon taxes and the composition of new passenger car sales in Europe. Energies 13(18). https://doi.org/10.3390/EN13184631

Nerurkar V, Barge P, Bhatt V, Rastogi S, Agarwal B (2023) Factors influencing consumer decision to purchase a car. Mark Manag Innov 14(1):24–40. https://doi.org/10.21272/MMI.2023.1-03

Ottesen A, Banna S (2021) Why so few EVs are in Kuwait and how to amend it. Int J Eng Technol 10(2):181. https://doi.org/10.14419/IJET.V10I2.31754

Our World in Data (2023) CO_2 emissions from transport. https://ourworldindata.org/grapher/co2-emissions-transport

Paltsev S, Ghandi A, Morris J, Chen H (2021) Global electrification of light-duty vehicles: impacts of economics and climate policy. Econ Energy Environ Policy 11(1). https://doi.org/10.5547/2160-5890.11.1.SPAL

Pan A, Xu S, Zaidi SAH (2024) Environmental impact of energy imports: natural resources income and natural gas production profitability in the Asia-Pacific economic cooperation countries. Geosci Front 15(2):101756. https://doi.org/10.1016/J.GSF.2023.101756

Peng J, Lin Z (2022) Eco-innovation in new energy vehicle supply chains under government subsidies. Sustainability 14(22). https://doi.org/10.3390/SU142215216

Pierce D, Shepherd S, Johnson D (2019) Modelling the impacts of inter-city connectivity on city specialisation. Int J Syst Dyn Appl 8(4):47–70. https://doi.org/10.4018/IJSDA.2019100104

Platform for Electromobility (2023) Solutions for a smooth integration of e-mobility into the grid. https://www.platformelectromobility.eu/

Pokharel R, Bertolini L, te Brömmelstroet M, Acharya SR (2021) Corrigendum to spatio-temporal evolution of cities and regional economic development in Nepal: does transport infrastructure matter? J Transp Geogr 90. https://doi.org/10.1016/J.JTRANGEO.2020.102904

Pulugurtha SS, Mishra R, Jayanthi SL (2022) Does transit service reliability influence ridership? Mineta Transportation Institute, 73 p. https://transweb.sjsu.edu/sites/default/files/2118-Pulugurtha-Transit-Reliability-Ridership-Road-Network.pdf

Rudolph C (2016) How may incentives for electric cars affect purchase decisions. Transp Policy 52:113–120. https://doi.org/10.1016/J.TRANPOL.2016.07.014

Shen W, Han W, Wallington TJ, Winkler SL (2019) China electricity generation greenhouse gas emission intensity in 2030: implications for electric vehicles. Environ Sci Technol 53 10(10), 6063–6072. https://doi.org/10.1021/ACS.EST.8B05264

Singh V, Singh V, Vaibhav S (2021) Analysis of electric vehicle trends, development and policies in India. Case Stud Transp Policy 9(3):1180–1197. https://doi.org/10.1016/J.CSTP.2021.06.006

Siragusa C, Tumino A, Mangiaracina R, Perego A (2022) Electric vehicles performing last-mile delivery in B2C e-commerce: an economic and environmental assessment. Int J Sustain Transp 16(1):22–33. https://doi.org/10.1080/15568318.2020.1847367

SLoCaT (2018) Transport and climate change global status report 2018 partnership, sustainable, low carbon transport (SLoCaT). SLoCaT (2018). Transport and climate change global status report 2018 7(2):107–115. http://slocat.net/tcc-gs

Solaymani S (2019) CO_2 emissions patterns in 7 top carbon emitter economies: the case of transport sector. Energy 168:989–1001. https://doi.org/10.1016/J.ENERGY.2018.11.145

Suryani E, Hendrawan RA, Adipraja PFE, Indraswari R, Stmik A, Malang I (2020) System dynamics simulation model for urban transportation planning: a case study. Int J Simul Model 19:5–16. https://doi.org/10.2507/IJSIMM19-1-493

Thilakshan T, Sugathapala AGT, Bandara JMSJ (2023) Implementation of electric buses into an existing public transit network: a Sri Lankan perspective. In: Moratuwa engineering research conference (MERCon), pp 794–799. https://doi.org/10.1109/MERCON60487.2023.10355524

Tyndall J (2018) Bus quality improvements and local commuter mode share. Transp Res Part a: Policy Pract 113:173–183. https://doi.org/10.1016/J.TRA.2018.04.011

UNDP Sri Lanka (2023) Sri Lankalclimate promise. https://climatepromise.undp.org/what-we-do/where-we-work/sri-lanka

Van Dijk J, Farsi M, Weber S (2021) Travel mode choices in a greening market: the impact of electric vehicles and prior investments. Transp Res Rec 2675(11):1205–1218. https://doi.org/10.1177/03611981211025279

Veza I, Abas MA, Djamari DW, Tamaldin N, Endrasari F, Budiman BA, Idris M, Opia AC, Juangsa FB, Aziz M (2022) Electric vehicles in Malaysia and Indonesia: opportunities and challenges. Energies 15(7). https://doi.org/10.3390/EN15072564

Wei F, Walls WD, Zheng X, Li G (2023) Evaluating environmental benefits from driving electric vehicles: the case of Shanghai, China. Transp Res Part d: Transp Environ 119(2020):103749. https://doi.org/10.1016/j.trd.2023.103749

Weldon P, Morrissey P, O'Mahony M (2018) Long-term cost of ownership comparative analysis between electric vehicles and internal combustion engine vehicles. Sustain Cities Soc 39:578–591. https://doi.org/10.1016/J.SCS.2018.02.024

Wu LM, Chen BY, Ou LC, Bai YJ, Liu KX, Wang WW, Peng B, Wang XM (2023) Impact of accelerated electrification under the low carbon path in Dongguan city on the coordinated emission reduction of CO_2 and pollutants. Huanjing Kexue/Environ Sci 44(12):6653–6663. https://doi.org/10.13227/J.HJKX.202211028

Yan S (2018) The economic and environmental impacts of tax incentives for battery electric vehicles in Europe. Energy Policy 123:53–63. https://doi.org/10.1016/J.ENPOL.2018.08.032

Yu X, Xu Z, Hu J, Liu S (2017) Systematic decision making: a extended multi-criteria decision making model. Technol Econ Dev Econ 23(1):157–177. https://doi.org/10.3846/20294913.2016.1212121

Zhao F, Liu F, Liu Z, Hao H (2019) The correlated impacts of fuel consumption improvements and vehicle electrification on vehicle greenhouse gas emissions in China. J Clean Prod 207:702–716. https://doi.org/10.1016/J.JCLEPRO.2018.10.046

Chapter 13
Can Urban Environment Studies Help in Designing Environmentally Conducive Neighbourhoods? Case Study from Three Indian Cities

Swati Dutta, Suchandra Bardhan, and Sanjukkta Bhaduri

Abstract One of the most enduring themes behind the search for more sustainable urban forms is that of the pattern or density of development. The chapter identifies eight neighbourhoods each representing a particular residential pattern from the three case study cities of Amritsar, Chandigarh and Gurugram and presents a detailed study of the relationship between the different physical density variables (denoting the pattern of development) and indicators of environmental quality. It shows that all the measures of density (people density, building density and spatial density) have significant impact on environmental quality. Crowding and congestion increases with increase in people density. Activity intensity and activity diversity also show an increasing trend with increasing residential density thus suggesting that more people engage in different activities and greater social interaction happens in high density neighbourhoods. Other aspects like presence and condition of sidewalks, streetlights, outdoor furniture, level of cleanliness, etc. determine peoples' desire to reside in, walk through and utilize spaces. The arrangement of buildings and open spaces in a neighbourhood have considerable effect on the micro-climate. Both shade and naturalness—measures of nature-based solutions increase the outdoor comfort conditions for the residents especially during the summer months. Also, proximity to daily needs, good condition of sidewalks and open spaces along with optimum level of services helps in creating walkable neighbourhoods. The study thus gives objectivity to an otherwise abstract and subjective term 'environmental quality'. It validates the conceptual framework and helps formulate guidelines for environmentally conducive development of residential neighbourhoods for future sustainable cities.

S. Dutta (✉)
Amity University Punjab, Mohali, India
e-mail: sdutta@pb.amity.edu

S. Bardhan
Jadavpur University, Kolkata, India
e-mail: suchandra.bardhan@jadavpuruniversity.in

S. Bhaduri
School of Planning and Architecture, New Delhi, India
e-mail: sanjukktabhaduri@gmail.com

L. K. Dashora et al. (eds.), *Urban Sustainability*, Disaster Risk Reduction,
https://doi.org/10.1007/978-981-96-4890-0_13

Keywords Residential patterns · Physical density · Neighbourhood environmental quality · Environmentally conducive patterns · Future sustainable cities

13.1 Introduction

India is well endowed with cities spread evenly across the country. Historically, Indian cities have been situated in regions with ample water resources and along prominent trade routes. As in many other countries, these cities have concentrated economic wealth and functioned as 24-hour marketplaces. The high concentration of people living has brought variety of resources, products and services to these cities (Bhandari 2006). Over the years, the megacities have occupied significant mind, media and policy space in terms of urbanization in India. Post-independence also urbanization in India has been dominated by the largest cities. However, with time, these cities have grown in size but quality of life has been severely compromised (Tripathi 2021). It has also been observed that most of the country's urban population has been residing in smaller towns and cities. According to the 2011 census, out of approximately 8,000 urban centres, only 53 have populations exceeding 1 million. Furthermore, only 20% of the urban population lives in cities with populations of 10 million or more, while the remaining 80% resides in towns and cities with populations of 1 million or less. Hence, the most meaningful approach is to anticipate the rapid growth of these cities and put the Class I (0.1–1 million) cities centrally on the development agenda. A study by the Indian Institute of Human Settlements (IIHS) for the India Urban Conference 2011: Evidence and Experience (IUC 2011)[1] shows this fact that in the twentieth century the rapid growth of the largest metropolitan cities has slowed down, whereas the smaller cities are expanding.

Considering the contribution to the gross domestic product (GDP) or output of the country, it is seen that while the top 10 and 53 million plus cities account for 15% and 31% of the GDP; and hold 8% and 13% of the population, respectively, the top 100 cities with 16% of the population and having approximately 47 cities in the 0.1–1 million population range contribute 41–43% to the GDP thus additionally justifying their potential as key cities for future urbanization in India (IIHS 2011).

Most studies pertaining to causes and effects of urban expansion, nationally and internationally examine issues at the citywide scale. They do not consider the issues at a smaller scale and understand the differences in patterns of urbanization and its effect on the environment in different neighbourhoods of a city (Dutta et al. 2013).

The literature review effectively outlines various methodologies for assessing urban environmental quality across different contexts. It also helps in understanding

[1] Urban India 2011: Evidence—A book produced by Indian Institute of Human Settlements (IIHS) for the India Urban Conference 2011: Evidence and Experience (IUC 2011) comprising of a series of events designed to raise the salience of urban challenges and opportunities in the on-going debate on India's development.

the various indicators that are important to be considered for assessment of environmental quality—physical, social, economic, geographical, psychological, etc. The study by Delsante (2016) evaluates medium-density neighbourhoods in Lodi and Genoa, Italy, by establishing adaptable indicators for assessing urban environmental quality based on different population densities. It underscores the need for flexible evaluation tools that can address specific urban challenges, such as high-density areas or urban sprawl. Similarly, the research by Discoli et al. (2014) develops a model to assess urban life quality (ULQ) by integrating basic services, infrastructure and environmental factors in La Plata, Argentina. This model combines qualitative and quantitative measures to gauge resident satisfaction, offering a thorough understanding of urban living conditions. Likewise, Shieh et al. (2011) investigate residential satisfaction in two neighbourhoods in Tehran, focusing on several factors. The study finds that dwelling size is the most critical factor influencing residential satisfaction. It highlights that, alongside psychosocial, economic, and physical factors, attributes of the built environment are crucial components of environmental quality. In the Asian context, Majumdar et al. (2007) study that residents' perceptions of environmental quality vary by income and are influenced by infrastructure quality and management in Chittagong Metropolitan City, Bangladesh. This highlights the role of socioeconomic factors in shaping residents' experiences and satisfaction. Kimhi's (2005) work addresses the challenge of balancing increased residential densities with maintaining quality of life in Israel. It emphasizes the importance of physical parameters in urban planning to mitigate the effects of suburbanization and maintaining high living standards. These studies among others throw light on the fact that physical aspects of the built environment and their impact on urban environmental quality are not often accounted for as compared to social, economic, geographical and psychological factors. Also, studies with physical aspects are sparsely reported and quite scanty in number thus justifying the scope of the present work.

Physical density plays a key role in minimizing environmental impact by reducing land usage, encouraging apartment living, clustering public services and transportation, and promoting walkability. However, much of the urban sustainability discourse in India focuses on 'green codes' for buildings. The issue with these codes is that they primarily aim to optimize individual buildings rather than addressing broader urban design factors that lead to substantial sustainability gains. The crucial elements are: Is the city compact or sprawling? Do people live in apartments or individual houses? Is the city designed to support public transport? For example, energy consumption can be reduced by over 30% simply by shifting from houses to apartments, even without following green building codes (World Wide Fund for Nature 2010). Similarly, public transportation systems fail to operate efficiently in sprawling cities where commuters cannot easily access bus or metro stations. Moreover, while TERI's Green Guidelines for Large Developments (2012) and GRIHA (Green Rating for Integrated Habitat Assessment, 2015) provide various criteria for sustainable neighbourhood development, they do not offer clear guidelines on the ideal physical density parameters for urban areas (Zia 2012; Vij 2015).

Hence, a research on environmental quality with an objective analysis of the physical aspects of a neighbourhood can really be useful. With this intent, the present study tries to understand the impact of different intra-city residential patterns on the neighbourhood environmental quality and search for an alternative approach for future sustainable growth of our cities. Accordingly, the key research questions are:

1. What are the different residential patterns identifiable in rapidly urbanizing Indian cities?
2. How does Population/Residential Density impact environmental quality in different residential patterns?
3. How does FAR and Plot Coverage impact environmental quality in different residential patterns?
4. How does Spatial Density impact environmental quality in different residential patterns?

13.1.1 Aim and Objectives of the Study

The present study aims to understand the impact of density on environmental quality with change in the residential pattern of the neighbourhood. Accordingly, the objectives are:

1. To identify different residential patterns and find representative neighbourhoods in the case study cities.
2. To establish the association between density variables and indicators of environmental quality.
3. To arrive at guidelines to achieve environmentally conducive patterns of residential development for future sustainable cities.

The scope is limited to the objective analysis of the physical aspects of the different residential neighbourhoods. Eight neighbourhoods with different residential patterns are identified from the case study cities and coefficient correlation analysis is undertaken to assess the relationship between density and environmental quality. The research is limited to cities with 0.75 million $\leq$ population $\leq$ 1.5 million since it is known from the literature and background study that there is dearth of comprehensive database of cities with population under 750,000 in India (WWF 2010; IIHS 2011). Slum and squatter settlements that pose a constant threat to the sustainable development of our cities are also kept out of the scope of the present study. Lastly, factors like the role of urban governance in the taxation and maintenance of municipal services and the resultant impact on environment quality and variations in socioeconomic status of people and its effect on the type of facilities and amenities in a residential neighbourhood are also kept for future study and analyses.

Thus, the present research is focused towards understanding and suggesting ways to achieve environmentally conducive neighbourhoods—one of the key considerations for achieving long term sustainability. The approach can be used by residents to highlight and communicate concerns, wishes and positive aspects of their area to fellow residents or decision makers (e.g. local authorities, planners, policy makers and organizations). The method can also be used for the planning and design of new residential areas of the city.

13.2 Identifying Case Study Cities

The main criteria for selecting the cities are:

1. The cities are emerging centres of economic activity.
2. They are located between other urban centres and well connected.
3. In near future, these cities will create better opportunities for citizens living in and around them.
4. Some out of them are budding and have the potential to turn into much larger centres.
5. Finally, they are steadily gaining the necessary scales in terms of population and market size.

Three cities namely Amritsar, Chandigarh and Gurugram are selected for the study purpose. It is observed that all the three cities have been growing constantly (see Table 13.1). The average decadal growth rate has been 2.3%, 11.4% and 10.24%, respectively. The steep rise as seen in Fig. 13.1 establishes the fact that these cities are expanding rapidly and hold the key to future urbanization in India.

The city of **Amritsar** situated in north-western part of the state of Punjab is recognized as one of the budding cities of India. It is a historic city that has attained the status of a global city due to it being a major pilgrimage/tourist destination. The present population of the city is 1.13 million (as per census 2011). Amritsar has imprints of all types of settlement patterns and associated densities. There are important differences between the older habitations in the core of the city—walled city of Amritsar (1665–1849) with post-colonial settlements outside the walled city.

Chandigarh, located in northern India, is a union territory, serving as the capital of the states of Punjab and Haryana. Established as the first planned city in India after independence in 1947, Chandigarh is renowned worldwide for its innovative architecture and urban design. The city's master plan was crafted by Swiss-French architect Le Corbusier. The majority of government buildings and housing in Chandigarh were designed by the team of Pierre Jeanneret, Jane Drew and Maxwell Fry. The

Table 13.1 Population growth in case study cities

Name of city	1951	1961	1971	1981	1991	2001	2011
Amritsar	336,114	390,055	454,805	594,844	708,835	1,011,327	1,133,000
Chandigarh	24,261	119,881	257,251	451,610	642,015	900,635	1,054,686
Gurugram	18,613	37,868	57,151	100,877	121,486	239,684	876,824

Source District Gazetteers and Census of India 1971, 1981, 1991, 2001 and 2011

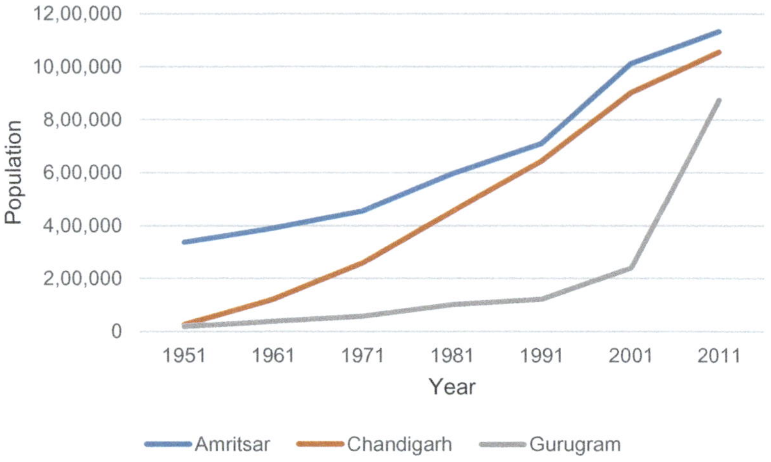

Fig. 13.1 Growth of case study cities by population. *Source* District Gazetteers and Census of India 1971, 1981, 1991, 2001 and 2011

city was reported to be the cleanest in India in 2010, based on a national government study[2] and the territory also headed the list of Indian states and territories according to Human Development Index (Meghalaya Human Development Report 2008; Chhabra 2010). The present population of the city is 1.05 million (as per census 2011). The metropolitan of Chandigarh-Mohali-Panchkula collectively forms a Tri-city, with a combined population of over 2 million.[3] This is the first smoke-free city in India (WHO 2011). The roads in Chandigarh are lined by trees and it has the third highest forest cover in India at 8.51% after Lakshadweep and Goa (Walk through India 2016 and Thakur, 2014).

Gurugram, a prominent industrial hub in India, is located in the National Capital Region (NCR) near New Delhi, in the state of Haryana. Positioned 32 kms south-west of New Delhi, Gurugram had a population of 0.88 million according to the 2011 Census. Witnessing rapid urbanization, Gurugram has become the city with the third highest per capita income in India (Julka 2011). Historically known as Guru Gram, the city's economic growth story started when the leading Indian automobile manufacturer Maruti Suzuki India Limited established a manufacturing plant in the 1970s (Kumar and Mishra 2012). Today, Gurugram is centre for more than 250 Fortune 500 companies.[4] In Gurugram, there is a growing trend towards multi-dwelling units, including apartments, condominiums and high-rise residential towers. Despite this, some residents live in slums where basic amenities such as proper sanitation, safe water supply, electricity and hygienic living conditions are inadequate. Gurugram

[2] India's cleanest: Where does your city stand? Rediff.com News, 13 May 2010. [online] Available at: http://www.rediff.com/news/slide-show/slide-show-1-chandigarh-cleanest-of-all/20100511.htm.

[3] [Online] Available at: https://en.wikipedia.org/wiki/Panchkula.

[4] [Online] Available at: https://en.wikipedia.org/wiki/Gurugram.

Table 13.2 Identification of case study cities

Criteria	Amritsar	Chandigarh	Gurugram
1. Category	Budding city	Upcoming city	Sibling city
2. Population	1.13 million	1.05 million	0.88 million
	All have population in the range 0.75 million–1.5 million (as per census 2011)		
3. Typology	Historic city	First planned city	Millennium city
	Being from different time periods, several residential patterns and their associated environmental quality are evaluated and compared to arrive at generalized and meaningful conclusions		
4. Climate Zone	Composite	Composite	Composite
	Same climate zone ensures consistency and comparability of the physical aspects of the built environment for environmental quality assessment		

Source Primary and secondary data, 2015

has a complex park system managed by Haryana Urban Development Authority (HUDA) but most parks are small and ill-maintained.

Table 13.2 sums up the important criteria considered for selecting the case study cities. In addition, ease of accessibility to the three cities (within 1–5 hours as author is based in Panchkula) has been an important decisive factor for selecting the cities.

13.3 Identifying Case Study Neighbourhoods

For detailed investigation, a matrix based on two critical factors namely building height and gross population density is generated and eight neighbourhoods are identified from the three case study cities as indicated in Table 13.3. Each neighbourhood represents a particular type of residential pattern commonly found in Indian cities with 0.75 million $\leq$ population $\leq$ 1.5 million. The common patterns found are low-rise low-density (LRLD), low-rise medium-density (LRMD), low-rise high-density (LRHD), medium-rise medium-density (MRMD), medium-rise high-density (MRHD), high-rise low-density (HRLD) and high-rise high-density (HRHD).

Generally, G+3 storeyed structures that do not require elevators are considered low rise. The International Building Code (IBC 2006) and the Paragraph 3.3.36.7 of the NFPA101®, Life Safety Code, 2012 edition (Hall 2013), define high-rise buildings as buildings 75 feet (23 m approximately) or greater in height measured from the lowest level of fire department vehicle access to the floor of the highest occupiable storey. Also, buildings with more than 8 storeys or 24 m in height with mandatory provision of diesel generators in case of an electricity failure are considered high rise as per the National Building Code 2016 Fire Safety and Fire Protection Norms. Therefore, from

Table 13.3 Case study neighbourhoods

Density Height	Low	Medium	High
Low	Sector 8, Chandigarh	Ranjit Avenue, Amritsar	Sector 38W, Chandigarh (Planned)
			Katra Dullo, Amritsar (Organic)
Medium	No identifiable neighbourhood	Marble Arch (Manimajra), Chandigarh	Sector 20 (Panchkula), Chandigarh
High	World Spa, Sector 30, Gurugram	No identifiable neighbourhood	Uniworld City, Sector 30, Gurugram

Source Primary and secondary data, 2015

the above two definitions, buildings having 5–8 storeys are considered as medium-rise structures for the study as seen in Table 13.4 and Plate 13.1.

The population density is categorized as low density with <200 pph, medium density with 200 to <400 pph, high density with 400 to <600 pph and very high density with $\geq$ 600 pph as listed in Table 13.5. This is based on the gross population densities observed in different neighbourhoods.

Further, the study considers gross residential densities because of the paucity of net density values and comparability issues with gross values generally given in different government documents, reports etc. in our scenario. Also, in case of old/core areas of the city, it becomes a challenge to calculate net densities as the boundaries are not clear. Since the streets are irregular and other spaces (excluding incidental spaces) are not well-defined, it becomes still more difficult to calculate net density values that are conventionally biased to formal settings with well-defined streets and spaces (Lupala 2001; Dovey 2016). Here, the residential density is categorized as low density with <50 DUs/ha, medium density with 50 to <100 DUs/ha, high density

Table 13.4 Neighbourhoods grouped according to height of residential buildings

Neighbourhood	No. of storeys	Range	Rise
Sector 8, Chandigarh	G, G+1, G+2	$\leq$G+3 storeys	Low
Ranjit Avenue, Amritsar	G, G+1, G+2		
Sector 38W, Chandigarh	G+2		
Katra Dullo, Amritsar	G, G+1, G+2, G+3		
Marble Arch (Manimajra), Chandigarh	G+4	5–8 storeys	Medium
Sector 20 (Panchkula), Chandigarh	G+5, G+6, G+7		
World Spa, Gurugram	G+12, G+16	>8 storeys	High
Uniworld City, Gurugram	G+12, G+13		

Source Primary survey, Author, 2015

Plate 13.1 Residential neighbourhoods in case study cities. *Source* Primary and secondary data, 2015

Table 13.5 Neighbourhoods grouped according to gross population density

Neighbourhood	Gross population density (pph)	Range	Level
Sector 8, Chandigarh	95	<200 pph	Low
World Spa, Gurugram	178		
Ranjit Avenue, Amritsar	256	200 to <400 pph	Medium
Marble Arch (Manimajra), Chandigarh	345		
Sector 38W, Chandigarh	441	400 to <600 pph	High
Sector 20 (Panchkula), Chandigarh	540		
Katra Dullo, Amritsar	641	≥600 pph	Very high
Uniworld City, Gurugram	735		

Source Primary and secondary data, 2015

with >100 DUs/ha as indicated in Table 13.6. This is based on the gross residential densities calculated for the different neighbourhoods.

Table 13.6 Neighbourhoods grouped according to gross residential density

Neighbourhood	Gross residential density (DUs/hectare)	Range	Level
Sector 8, Chandigarh	17	<50DUs/ha	Low
Ranjit Avenue, Amritsar	36		
World Spa, Gurugram	44		
Marble Arch (Manimajra), Chandigarh	65	50–100 DUs/ha	Medium
Sector 38W, Chandigarh	76		
Sector 20 (Panchkula), Chandigarh	98		
Uniworld City, Gurugram	147	>100DUs/ha	High
Katra Dullo, Amritsar	149		

Source Primary and secondary data, 2015

13.4 The Analytical Model for Assessing Environmental Quality

Physical density is disaggregated as per Cheng's (2010) study and a conceptual framework is constructed showing possible correlations between the density variables and indicators of environmental quality (Dutta et al. 2020). A bottom-up model is proposed to understand the impact of density on neighbourhood environmental quality. This model assesses the top-level attribute of environmental quality based on the following factors: (i) Crowding and Congestion, (ii) Nature and Use of Open Spaces, (iii) Shade and Ventilation, (iv) Temperature Variations, (v) Noise Levels, (vi) Quality requirements for Walking, (vii) Level of Cleanliness and (viii) Air Quality. Each of these is further disaggregated into a series of variables that can be objectively measured. The grouped variables provide a clear overview of the specific types of factors influencing each indicator of environmental quality as shown in Fig. 13.2.

Additionally, comparing the environmental quality across neighbourhoods with varying densities helps identify optimal density ranges that support favourable environmental conditions. This analysis aids in developing guidelines for creating environmentally conducive patterns of residential development in our cities.

13.4.1 Data Collection and Tools for Study

Data is gathered from both primary and secondary sources. Primary sources included detailed field studies, visual surveys and photographic documentation of the case study neighbourhoods with the help of base maps prepared from Google Earth images. Observation is employed as a primary method to study house forms, plot

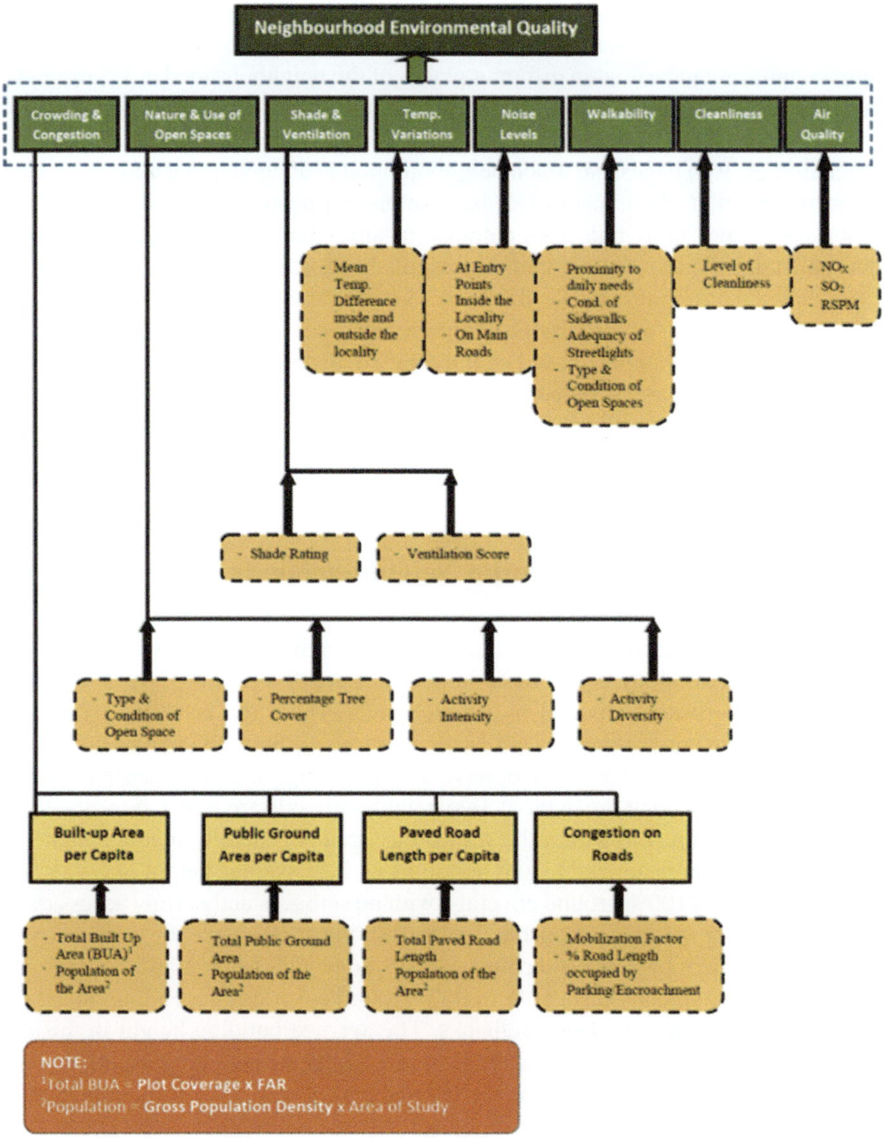

Fig. 13.2 Analytical model. *Source* Author's elaboration from literature review, 2015

sizes, configuration and plot characteristics, types of roads, sidewalks and open spaces and their uses, etc. Additionally, unstructured interviewing through focus group interviews, narratives and oral histories is also resorted to, to extract information. Information about certain aspects of the neighbourhood is also gathered through an interview schedule consisting of close-ended and open-ended questions. Certain mechanical instruments, such as temperature and humidity data loggers, as well as smartphone applications like deciBel, are used to record ambient temperatures and noise levels in various locations within and around the neighbourhood. The ambient temperatures are measured in different places of the neighbourhood at 1.5 m (5'–0″) street level and terrace level approximately. The secondary data is collected from various government and non-governmental documentary sources, newspaper and other articles, web pages, etc. Each neighbourhood is divided into a number of study units to determine the impact of density on environmental quality. On an average 5 households per study unit from amongst 64 study units are surveyed for primary data collection. Thus, the total sample size equals $64 \times 5 = 320$ households which is more than the minimum sample size ($n = 93$ or 93 samples) determined statistically and hence adequate for carrying out different analyses.

After the data collection, based on the detailed visual survey land-use maps of all the neighbourhoods are prepared. Also, the study units as mentioned are drawn on the maps of each neighbourhood. Simultaneously, the raw data is sorted and tabulated in Excel sheets for further study and analyses.

13.5 Study of the Residential Neighbourhoods

The assessment of environmental quality at the neighbourhood level with respect to the density variables is the most important part of this study. To begin with, the city of Amritsar is divided into three zones. The three zones are (i) Walled city forming the core of the city (ii) New and more cosmopolitan areas beyond the walled city and up to the northern bypass, and (iii) Fringe areas comprising of newly developed areas beyond the bypass. Two neighbourhoods are selected for the study. In Amritsar, the housing within the walled city area is generally in a dilapidated or poor condition, characterized by old buildings with an average height of G+3 floors. This area features 100% ground coverage with no setbacks and narrow access roads, leading to a poor spatial relationship. In contrast, the area outside the walled city but within the northern bypass is a medium-density zone with better housing conditions. Here, buildings typically have 60% to 80% ground coverage, depending on the plot size, and include front and side setbacks. The average building height in this area is G+1. No neighbourhood is selected from the fringe areas as the housing colonies are still under construction and hence uninhabited. The similar residential density patterns are considered later while studying the city of Gurugram. Katra Dullo, Ward No. 5 is selected within the walled city while Ranjit Avenue, Ward No. 43 near the northern bypass is selected from the new city.

In case of Chandigarh, the area of the chosen neighbourhoods varies from 35 to 105 hectares. They generally have low-rise structures varying from single storey to G+3 buildings with metaled roads, neighbourhood parks, open spaces, greenery and overall services. The newly developing peripheral areas like Manimajra and satellite cities like Panchkula and Mohali constituting the Chandigarh Urban Complex (CUC) in addition to the low-rise structures have medium- and high-rise structures coming up in recent years. The gated communities and group housing societies range from 2 to

130 hectares—the latter forming whole sectors in some places and are characterized by apartment blocks, community open spaces, shared facilities and amenities. The Chandigarh/ Panchkula Municipal Corporation looks after the sectors whereas inside the gated communities residents' welfare associations (RWAs) are responsible for upkeep and maintenance.

Gurugram—more popularly known as 'Gurgaon' was a village that transformed into a financial and industrial city post 1970s. The urban form generally consists of glass and concrete high-rise structures. The city features a scattering of office blocks, shopping malls and gated communities, but it lacks sufficient number of public spaces. The limited green areas available are primarily private golf courses, which do little to foster a vibrant urban community. There are few venues for general social interaction, such as public parks, sports complexes, temples or cultural hubs, which diminishes opportunities for communal engagement and vibrant urban life. The neighbourhoods considered for the study are primarily built and managed by private developers and examples of the very common type of residential pattern being adopted in all the cities so as to accommodate relatively more people on a piece of land and provide them with all amenities and facilities like parks, open spaces to interact and socialize, play areas, gym, pool, retail outlets catering to daily requirements, etc. within the boundary of the gated community. However, even though these communities have high quality physical infrastructure, there is a need to integrate the neighbourhoods by providing 'urban software' at the city level—a main thrust area of Atal Mission for Rejuvenation and Urban Transformation (AMRUT), 2015 that can have a positive impact on the urban environmental quality. The following pages show the plans (Figs. 13.3, 13.4, 13.5, 13.6, 13.7 and 13.8) and tabulate key features of all the neighbourhoods (Tables 13.7 and 13.8) under consideration along with some glimpses (Plate 13.2, 13.3, 13.4, 13.5, 13.6, 13.7, 13.8, 13.9, 13.10, 13.11, 13.12, 13.13, 13.14, 13.15, Source: Author, 2015–16) to understand the visual character of the study areas.

13.6 Results and Discussions

The urban form of the city is characterized by different configuration and types of built form and related population distribution across the geographical extent of the city. Alternatively, it can be stated that distribution of buildings, their typology and geometry, layout patterns, distribution/availability of open spaces, quality of infrastructure, etc. vary from one locality to the other within the city. Based on these findings, the study helps to identify eight different residential patterns from the cities of Amritsar, Chandigarh and Gurugram based on building height and gross population density and provides a detailed analysis of each. The neighbourhoods range in size from approximately 2 to 105 hectares. The data analysis includes comparison with threshold values from various standards and guidelines, including documents and reports from the URDPFI (2014), the Handbook of Service Level Benchmarking by

Fig. 13.3 Land-use maps of Katra Dullo and Ranjit Avenue, Amritsar. *Source* Author, 2016

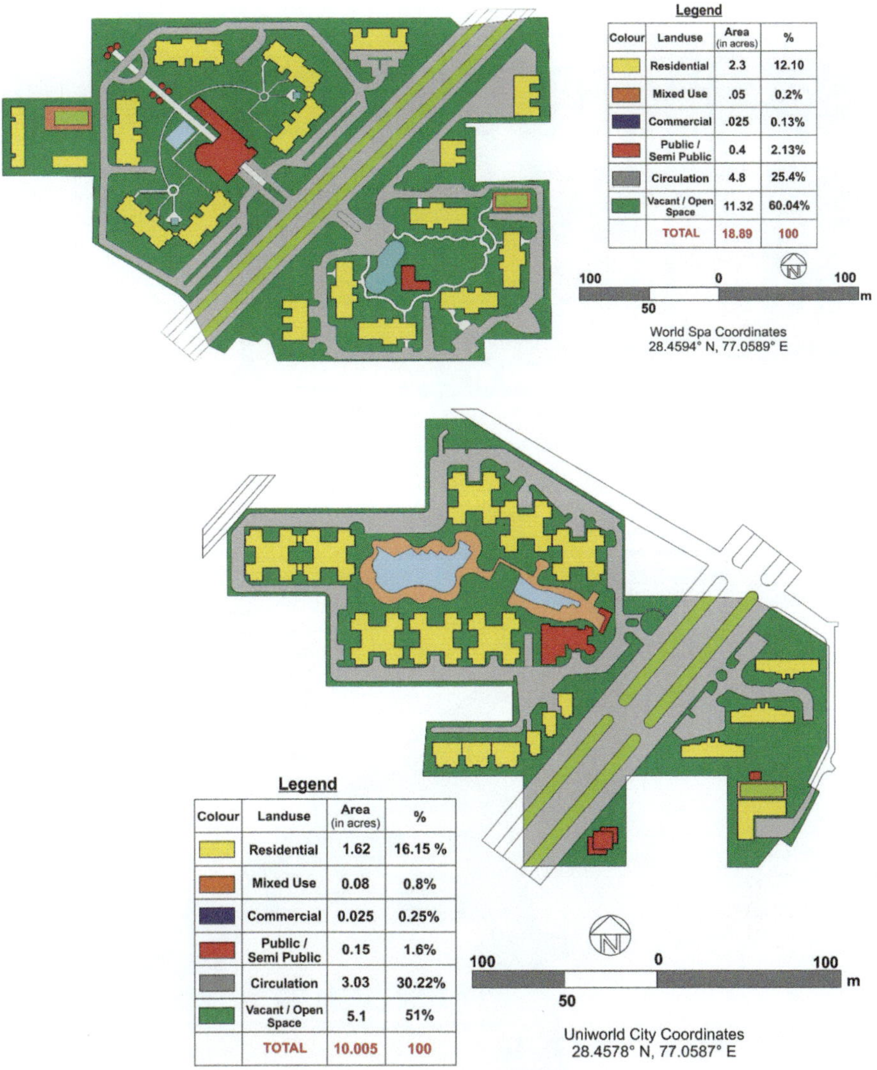

Fig. 13.4 Land-use maps of World Spa and Uniworld City, Gurugram. *Source* Author, 2016

the Ministry of Urban Development (MoUD) 2011, and norms suggested by government bodies such as the Central Pollution Control Board (CPCB). The comprehensive visual survey, along with data organization, study and analysis based on the conceptual framework (Dutta et al. 2020) and correlation coefficient analyses, yield the following results:

Fig. 13.5 Land-use maps of Sector 8 and Marble Arch (Manimajra), Chandigarh. *Source* Author, 2016

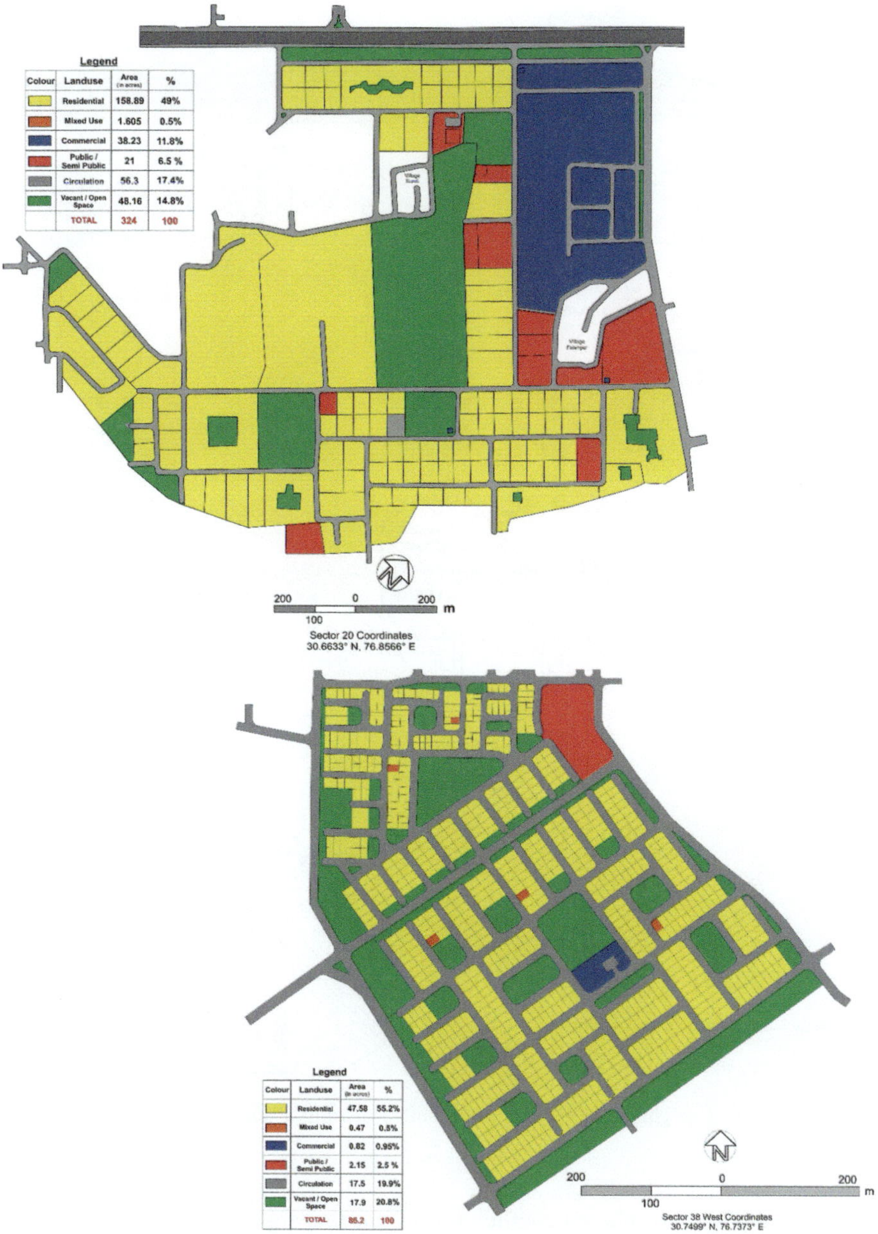

Fig. 13.6 Land-use maps of Sector 20, Panchkula and Sector 38W, Chandigarh. *Source* Author, 2016

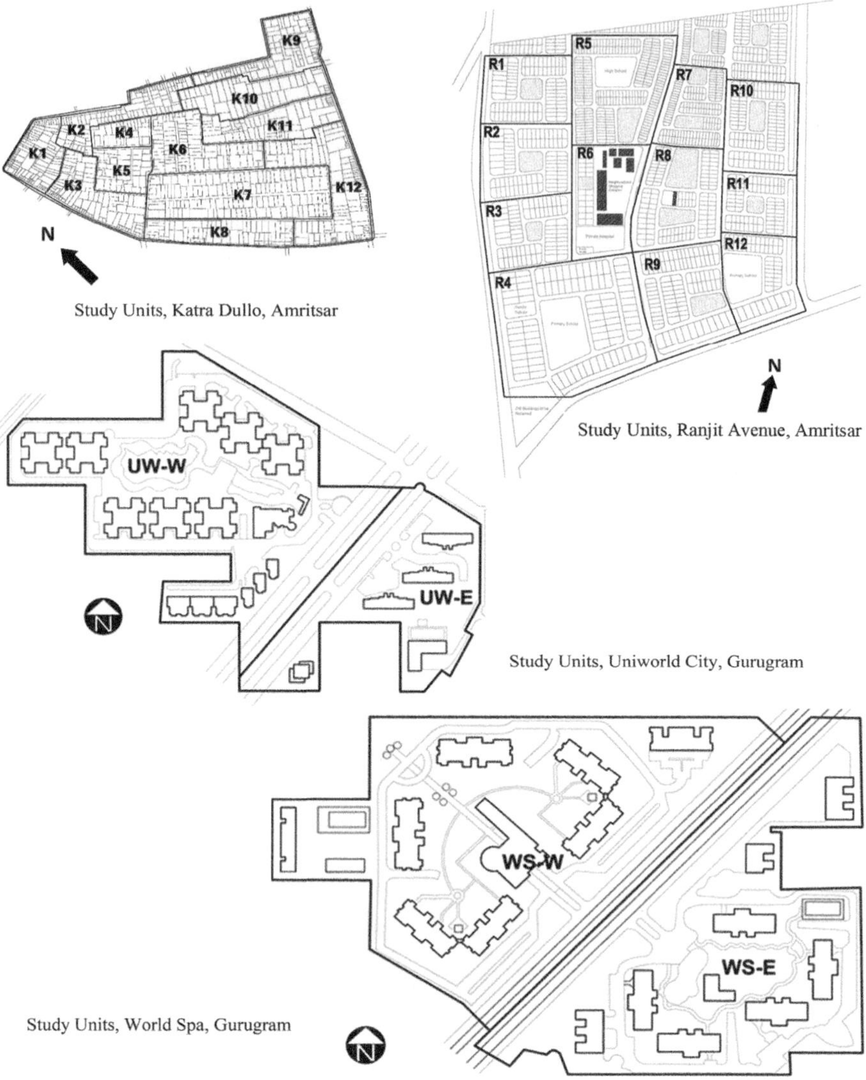

Fig. 13.7 Plans of neighbourhoods with study units in Amritsar and Gurugram. *Source* Author, 2016

The analysis of correlation coefficients reveals that significant positive and negative relationships exist between density variables and indicators of neighbourhood environmental quality (NEQ).

a. **Crowding and Congestion:** Crowding and congestion increase with increasing people density. Higher Floor Area Ratios (FARs) and increased plot coverage lead to a reduction in public ground area per capita (PGA/Capita) and paved

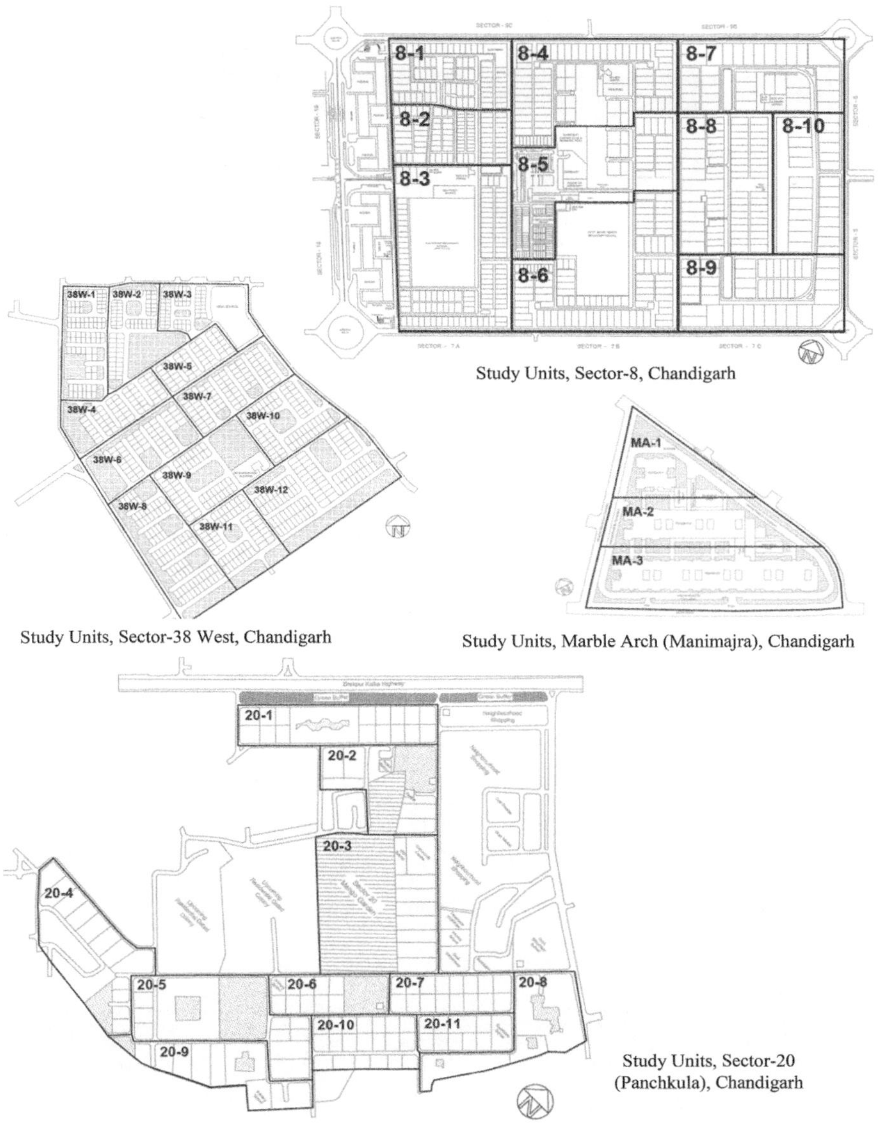

Fig. 13.8 Plans of neighbourhoods with study units in neighbourhoods of Chandigarh. *Source* Author, 2016

road length per capita (PRL/Capita), contributing to greater road congestion. It is seen that as the residential density increases from 20 DUs/hectare to 150 DUs/hectare approximately, FAR from 1 to 3 and plot coverage from 15% to 80–85%, the PGA/Capita falls from 35 sqm/person to less than 10 sqm/person. Similarly, the PRL/Capita reduces from over 2 m/person to less than 0.55 m/person thus

Plate 13.2 Poor condition of buildings, Katra Dullo, Amritsar. *Source* City development plan Amritsar (2025)

Plate 13.3 Garbage dumps, open drains, congested roads and scanty tree cover, Katra Dullo, Amritsar. *Source* Author, 2015–2016

Plate 13.4 Types of plots and residential buildings, Ranjit Avenue, Amritsar. *Source* Author, 2015–2016

Plate 13.5 Condition of roads and sidewalks, Ranjit Avenue, Amritsar. *Source* Author, 2015–16

Plate 13.6 Types of commercial establishments, Ranjit Avenue, Amritsar. *Source* Author, 2015–16

Plate 13.7 Open spaces and community facilities, World Spa, Sector – 30, Gurugram. *Source* Author, 2015–16

Plate 13.8 Open spaces and community facilities, Uniworld City, Sector–30, Gurugram. *Source* www.google.com and Author, 2015–16

Plate 13.9 Key features of Sector 8, Chandigarh. *Source* Author, 2015–16

Plate 13.10 Maintained parks, roads and greenery, Sector 38W, Chandigarh. *Source* Author, 2015–16

Plate 13.11 Key features of Sector 38W, Chandigarh. *Source* Author, 2015–16

indicating increasing crowding and congestion and deteriorating environmental quality (see Table 13.9).

b. **Nature and Use of Open Space:** Community interactions are influenced by the type and condition of open spaces. As both people density and building density increase, activity intensity rises significantly, ranging from 11 persons/hectare to over 500 persons/hectare. Additionally, activity diversity tends to increase, with a range from 0.03 to 0.15, as residential density grows from 50 DUs/hectare to 150 DUs/hectare, although there are some exceptions. This indicates that higher-density neighbourhoods generally foster greater social interaction and a wider variety of activities (see Table 13.10).

   Factors such as the presence and condition of sidewalks, streetlights, outdoor furniture, and the overall level of cleanliness play a crucial role in influencing people's willingness to reside in, walk through, and make use of various spaces (Zhang et al. 2022).

c. **Shade, Ventilation and Temperature Variations:** The arrangement of buildings and open spaces in a neighbourhood affect the radiation falling on the building surfaces, streets, etc. These in turn heat or cool the air temperature and have considerable effect on the micro-climate. The compact and organically woven urban form of the old city in Amritsar as compared to the grid iron plan of the new areas is more comfortable for the residents during the summer months because of the higher shade rating (average value = 4 on a scale of 1–5) and

Plate 13.12 Aerial view, open spaces and facilities of Marble Arch (Manimajra), Chandigarh. *Source* www.google.com and Author, 2015–16

Plate 13.13 Built form and hierarchy of parks, Sector 20 (Panchkula), Chandigarh. *Source* Author, 2015–16

Plate 13.14 Types of commercial establishments, Sector 20 (Panchkula), Chandigarh. *Source* Author, 2015–16

Plate 13.15 Roads and sidewalks, Sector 20 (Panchkula), Chandigarh. *Source* Author, 2015–16

lower temperatures (almost 2 °C) within the neighbourhood as compared to the surroundings. The study also shows that there is significant association between building height to road width ratio and shade rating. This means that as the height to width ratio increases, the shade increases in the neighbourhood. This is particularly observed in case of Katra Dullo, Amritsar, World Spa and Uniworld City, Gurugram with building height to road width ratio ≥ 3 and shade rating $=$ 3 and in some cases$\approx$4 (see Table 13.11).

The presence of trees helps to filter sunlight, lower air temperatures through evaporation, protect ground-level plants, and reduce glare from bright, overcast skies. In most of the neighbourhoods, because of the presence of parks and playgrounds (55–90%); and trees (varying from 30 to 50%), the temperatures are about 1 °C lower than the surrounding areas. The negative correlation between the distribution of open spaces and temperature variations validate the point that presence of naturalness and greenery regulates the temperatures, reduce fluctuations and helps in creating a comfortable microclimate. Thus, both shade and naturalness (nature-based measures) increase the outdoor comfort conditions for the residents especially during the summer months.

d. **Noise Levels:** Noise levels tend to be in the prescribed range (50–55 dB) for residential areas with higher plot coverage (30–85%) and distribution of roads (180–750 m/hectare), both situations contrary to the assumptions. This is because in compact neighbourhoods with high plot coverage, people tend to walk for their daily needs and short trips. Even the presence of more road length per hectare, does not prompt them to use their motorized vehicles thus lowering the noise

Table 13.7 Key features of case study neighbourhoods (Amritsar and Gurugram)

Brief description of the neighbourhoods			
Katra Dullo, Amritsar	*Ranjit Avenue, Amritsar*	*World Spa, Gurugram*	*Uniworld City, Gurugram*
Area—6 hectares. Organic form, poor housing condition, poor built to open space relationship and high intensity of development characterize Katra Dullo	Area—50 hectares. A planned neighbourhood located in the new part of the city outside the walled area. It is characterized by straight avenues and plotted development of row houses	Area—7.64 hectares. A high-rise low-density gated community located in Sector 30 along the National Highway NH-8	Area—4 hectares. A high-rise high-density gated community located in Sector 30 along the National Highway NH-8
Population density, residential density and residential pattern			
641pph, 149DUs/ha, LRHD	*256pph, 36DUs/ha, LRMD*	*178pph, 44DUs/ha, HRLD*	*735pph, 147DUs/ha, HRHD*
Most buildings are G+2 storeyed and have 100% plot coverage. Many of them are in a dilapidated condition	Most of the buildings are G+1 storeyed on varying plot sizes except for the flatted developments restricted to two pockets that are G+2 and G+3 storeyed	The project comprises of apartments primarily spread over 11 towers. Most of the towers are 17 storeyed except for two towers in the east that are 13 storeyed. The town houses are G+2 storeyed and constitute just 7.5% of the total dwelling units	The project comprises of 11 towers. The eastern part is more congested than the western part. All the towers are 13–14 storeyed while the town houses are G+2 storeyed and constitute just 3.7% of the total dwelling units
Plot sizes/apartment sizes			
The plots are irregular in shape and size varying from 20 to 67sqm approximately. All the houses are arranged in back to back manner having problem of light and ventilation	It is a planned neighbourhood that consists of different sizes of plots (1Kanal = 500sqm, 10Marla = 250sqm, 8Marla = 200sqm and 6Marla = 150sqm) arranged in a pin-wheel arrangement around a central open space	The project comprises of townhouses (208sqm), 4/5—bedroom apartments (381–462sqm) and penthouses (618–731sqm)	The community consists of apartment blocks with 3/4—bedroom apartments and penthouses. Size of DUs varies from 236 to 375sqm

(continued)

Table 13.7 (continued)

Open spaces			
There are no open spaces for the community or playgrounds for the children except for the small and unmaintained squares. Greenery is almost absent	Parks and playgrounds are maintained. Trees like neem, jamun, devil tree are found in the open spaces of the neighbourhood	Lush green lawns and abundantly planted landscaped areas with shared amenities and facilities covering 60% of the area are used for social interaction	Lush green lawns and abundantly planted landscaped areas covering 51% of the area are used as community interaction spaces
Roads and sidewalks			
Roads are narrow (1.2–1.5 m in the inner parts) and congested. Additionally, garbage dumps, absence of sidewalks, on-street parking of two-wheelers make walking an unpleasant experience	Roads are wide, well maintained and shaded. However, sidewalks are discontinuous or absent in some places. On-street parking causes congestion	Well-maintained roads and sidewalks covering almost 25% area with provision of open parking lots are laid out along the periphery	Well-maintained roads and sidewalks covering almost 30% area are laid out along the periphery
Level of services			
The level of services is also poor. Domestic waste is disposed on the roads. Choked open drains create insanitary conditions. Only 30% DUs are connected to sewerage system	The level of infrastructure and services is fairly well. Door to door collection of waste and road sweeping is done on daily basis	The level of services is very good with chutes for domestic waste collection, 100% stormwater and sewerage system in operation	
Miscellaneous			
Shops catering to daily requirements are in close proximity for the residents of the locality. Overall condition of infrastructure and services is quite poor	All daily need items are available in near proximity within a radius of 0.5 km. Ranjit avenue has overall good facilities	Both the neighbourhoods are examples of the very common type of residential pattern being adopted in all the cities so as to accommodate relatively more people on a piece of land and provide them with all amenities and facilities	

Source Primary and secondary data, 2016

levels. The presence of vegetated open spaces also helps in ameliorating noise pollution in the neighbourhoods (see Table 13.12).

e. **Level of Cleanliness:** A high standard of services, along with regular maintenance and upkeep, ensures a greater level of cleanliness, as observed in most of the case study neighbourhoods.

Table 13.8 Key features of case study neighbourhoods (Chandigarh)

Brief description of the neighbourhoods			
Sector 8	*Sector 38 West*	*Marble Arch, Manimajra*	*Sector 20, Panchkula*
Area—104 hectares It is laid out in the grid iron system. V2 and V3 roads define the sector whereas V5 (neighbourhood streets) and V6 (access roads) form the road network within. Two V4 (shopping streets) divide the sector into three parts	Area—32.2 hectares It is wedge shaped in plan. It is an extension of the main 38 sector and lies to its west, hence the name Sector 38 West	Area—2 hectares (approx.) The gated community is triangular in shape and located in Manimajra lying 200 m north of the Chandigarh-Kalka Highway. It is a pedestrian friendly neighbourhood with vehicular roads restricted to the periphery	Area—130 hectares It is irregular in shape. It lies along the Zirakpur-Kalka Highway and is one of the densest residential neighbourhoods
Population density, residential density and residential pattern			
95pph, 17DUs/ha, LRLD	*441pph, 76DUs/ha, LRHD*	*345pph, 65DUs/ha, MRMD*	*540pph, 98DUs/ha, MRHD*
G, G+1 and occasionally G+2 plotted development characterizes the area. One of the earliest planned sectors of Chandigarh by Albert Mayer. It abounds in large plots with bungalow type houses in addition to plotted row house development	Compactly placed G+2 flatted apartment style residential blocks in a pin-wheel and row house arrangement signify the area	G+4 storeyed apartment blocks interspersed with community open space and shared facilities and amenities identify this locality	Group Housing Societies developed as gated communities characterize the area. The plots are arranged in a systematic manner with medium-rise (5–8 storeyed) buildings
Plot sizes			
The sizes of the plots are 5 marlas, 8 marlas, 10 marlas, 16 marlas, 1 Kanal—5 Kanals[5]	The sizes of the plots are 8marlas and 10marlas	Consists of three rows of apartment blocks. Size of DUs varies from 205 to 265sqm	Plot sizes vary from 0.5 to 26 acres. Most common sizes are between 1 and 5 acres

(continued)

[5] 1 Marla = 25.293 sqm, 1 Kanal = 505.857 sqm and 1 Kanal = 20 Marlas—All these are traditional Indian units of land area especially used in northern India. Its use in India is in decline with urban land measurements standardizing the use of square meters.

Table 13.8 (continued)

Open spaces			
There are abundant open spaces and trees in the neighbourhood. However, there is lack of maintenance. Some spaces are used for playing and social interaction. People are more inclined towards using private gardens	Parks and open spaces are maintained and used by the children and residents. Level of infrastructure is good	Well-maintained parks and open spaces with paved pathways and good infrastructure. Used for gatherings and social interaction. A badminton and basketball court form part of the open spaces	There are sector level parks as well as parks/open spaces in all of the gated communities. A large mangrove acts as lungs for the high density neighbourhood
Roads and sidewalks			
The road pattern is in a grid form. All roads are metaled. However, sidewalks are discontinuous or absent in some places. This discourages people to walk	The road pattern is regular with dead end roads. On-street parking causes congestion in the neighbourhood	The vehicular road is restricted to the periphery. Sidewalks and pedestrian pathways through the open spaces connect the blocks	There are ample roads in the neighbourhood. However, due to heavy traffic because of high density, there is congestion on the roads
Level of services			
Overall level of services is good in all neighbourhoods in Chandigarh with 100% houses connected to the sewerage and drainage system and regular door-to-door collection of domestic waste. The problem observed in some neighbourhoods is the provision of adequate nos. of community bins or littered garbage near the bins causing filthy conditions, growth of insects and pathogens (flies and mosquitos), clogging of stormwater drains and foul smell thus affecting the environmental quality			
Miscellaneous			
The sector shopping, temple, Gurudwara, etc. serve the daily needs of the people	The on-street parking and congestion renders roads unsafe for children and elderly	People walk freely inside the gated community. However, they have to use vehicles for daily needs	Some people come to the neighbourhood parks, shopping etc. on their vehicles while some walk on the sidewalks

Source Primary and secondary data, 2016

f. **Walkability:** Proximity to daily necessities, along with well-maintained sidewalks and open spaces, contributes to the development of walkable neighbourhoods. It is seen that there is significant positive correlation between level of services and neighbourhood walkability. Neighbourhoods with score ≥ 3 for the level of services and other indicators on a 5-point scale are more walkable and sustainable. Therefore, as the cleanliness increases in the neighbourhood as a result of better services like efficient solid waste management, good stormwater drains and proper sewerage system, people are encouraged to walk for their daily

Table 13.9 People density; building density and spatial density; and indicators of crowding and congestion

Neighbourhood	Population Density (persons/ha)	Residential Density (DUs/ha)	FAR	Plot Coverage (%)	Dist. of Open Spaces (%)	Dist. of Roads (m/ha)	Average BUA/Capita (sqm/person)	Average PGA/Capita (sqm/person)	Average PRL/Capita (m/person)	Average Mobilization Factor (M.F.)
Sector 8, Chandigarh	95	17	1.00	49%	37%	189	94.84	48.99	1.78	1.26
Ranjit Avenue, Amritsar	256	36	1.00	30%	55%	354.38	23.14	20.30	0.99	0.41
World Spa, Gurugram	178	44	2.00	11%	89%	422	88.53	50.56	2.01	0.50
Marble Arch, Manimajra	345	65	1.57	31%	68%	232.79	44.93	23.65	0.67	1.00
Sector 38W, Chandigarh	441	76	2.00	70%	30%	403	28.45	10.36	0.68	1.36
Sector 20, Panchkula	540	98	2.00	35%	65%	179	22.23	19.36	0.33	0.85
Uniworld City, Gurugram	735	147	2.03	15%	85%	192	27.97	7.56	0.57	0.75
Katra Dullo, Amritsar	641	149	3.00	85%	15%	737	47.62	2.43	0.56	1.96
						Threshold Values	**20-30**	**12**	**0.91**	**≤1**
						Source of Information	Analyzing Urban Layouts **(Patel, 2011)**	URDPFI 2014	Amritsar Master Plan 2010-2031	Environment Quality Indicators **(Sarmento et al., 2000)**

Legend:

Less than threshold value	Within range or equal to threshold value	More than threshold value
Explanation of Terms:		
BUA/Capita – Built-up Area per Capita or Built floor area per person in the unit of study.		
PGA/Capita – Public Ground Area per Capita or all the open-to-sky area that is accessible to a person. It includes streets, parks, school play grounds etc.		
PRL/Capita – Paved Road Length per Capita or all the paved road area in the unit of study that is accessible to a person.		
Mobilization Factor (M.F) > 1 will imply more no. of exits from (entry into) the area compared to via routes thus indicating more traffic and greater chances of congestion.		

Source: Primary and Secondary Data, 2016

Table 13.10 People density and indicators of nature and use of open space

Neighbourhood	Population density (persons/ha)	Residential density (DUs/hectare)	Type and condition of open space (score)	Tree cover (%)	Activity intensity (persons/ha)	Activity diversity (score)
Sector 8, Chandigarh	95	17	0.24	28	11	0.04
Ranjit Avenue, Amritsar	256	36	0.30	49	28	0.05
World Spa, Gurugram	178	44	0.52	41	20	0.10
Marble Arch, Manimajra	345	65	0.44	35	48	0.14
Sector 38W, Chandigarh	441	76	0.36	23	45	0.03
Sector 20, Panchkula	540	98	0.31	25	19	0.03
Uniworld City, Gurugram	735	147	0.52	41	201	0.15
Katra Dullo, Amritsar	641	149	0.20	3	552	0.12

needs and short trips within and around the neighbourhood (see Table 13.13 and Fig. 13.9).

g. **Air Quality:** More number of vehicles and smoke generating appliances imply more pollution and deterioration of air quality. In compact neighbourhoods with less dependence on vehicular traffic; air quality improves as shown by the decreasing values of $RSPM_{10}$ in Katra Dullo, Amritsar. The increase in road distribution (m/ha) suggests more vehicles on the roads per unit area which in turn increases NOx values as found in Ranjit Avenue, Amritsar. Increasing NOx deteriorates the air quality and the overall environmental quality of the place (see Table 13.14). In this context, efforts are required to further develop a tool and monitoring methods for residential neighbourhoods similar to the Comprehensive Environmental Pollution Index (CEPI) for industrial clusters (CPCB 2009) to gauge the impact on people and the neighbourhood environment in addition to assessing the environmental pollution levels (air, surface water and land).

   Level of services also have an impact on air quality. Certain smells like foul smell from littered garbage, household and street dust, dust from development activities, etc. also affect the quality of air and deteriorate the environmental quality.

The important findings from the density and neighbourhood environmental quality analyses are: As people density, building density and spatial density go on

Table 13.11 Density variables and indicators of shade, ventilation and temperature variation

Neighbourhood	Residential density (DUs/ha)	FAR	Plot coverage (%)	Building height to road width (Ratio)	Dist. of open spaces (%)	Tree cover (%)	Shade rating	Ventilation score	Temp. variation (°C)
Sector 8, Chandigarh	17	1.00	49	0.40	37	28	3.00	1.00	0.9
Ranjit Avenue, Amritsar	36	1.00	30	0.93	55	49	3.00	1.00	1.1
World Spa, Gurugram	44	2.00	11	3.17	89	41	3.00	4.00	1.2
Marble Arch, Manimajra	65	1.57	31	0.50	68	35	3.00	1.00	1.1
Sector 38W, Chandigarh	76	2.00	70	0.83	30	23	3.00	2.00	0.8
Sector 20, Panchkula	98	2.00	35	1.40	65	25	1.00	1.00	0.3
Uniworld City, Gurugram	147	2.03	15	3.17	85	41	3.00	3.00	1.2
Katra Dullo, Amritsar	149	3.00	85	5.30	15	3	4.00	3.00	2.0

Table 13.12 Density variables and average noise level in case study neighbourhoods

Neighbourhood	Population density (persons/ha)	Residential density (DUs/hectare)	FAR	Plot coverage (%)	Dist. of open spaces (%)	Dist. of roads (m/ha)	Average noise level (dB)
Sector 8, Chandigarh	95	17	1.00	49	37	189	54
Ranjit Avenue, Amritsar	256	36	1.00	30	55	354.38	65
World Spa, Gurugram	178	44	2.00	11	89	422	67
Marble Arch, Manimajra	345	65	1.57	31	68	232.79	53
Sector 38W, Chandigarh	441	76	2.00	70	30	403	56
Sector 20, Panchkula	540	98	2.00	35	65	179	61
Uniworld City, Gurugram	735	147	2.03	15	85	192	68
Katra Dullo, Amritsar	641	149	3.00	85	15	737	50

Source (Tables 13.10, 13.11 and 13.12): Primary and secondary data, 2016

Table 13.13 Average scores of indicators of walkability

Indicators of walkability					
Neighbourhood	Average score				
	Proximity to daily needs	Condition of sidewalks	Adequacy of streetlights	Type and condition of open space	Aggregate walkability score
Sector 8, Chandigarh	4	2	5	2	3
Ranjit Avenue, Amritsar	5	3	4	3	4
World Spa, Gurugram	5	5	5	5	5
Marble Arch, Manimajra	1	3	5	4	3
Sector 38W, Chandigarh	5	2	5	4	4
Sector 20, Panchkula	3	2	5	3	3
Uniworld City, Gurugram	5	5	5	5	5
Katra Dullo, Amritsar	5	1	4	2	3

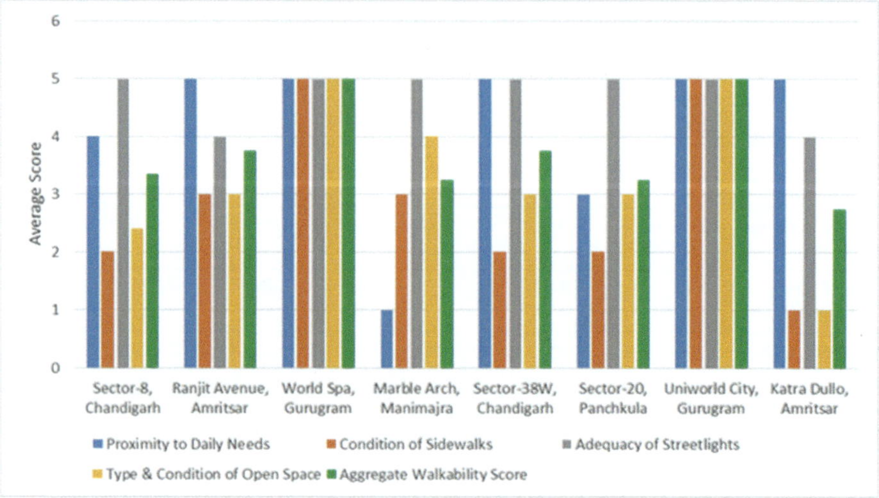

Fig. 13.9 Walkability indicators in different case study neighbourhoods

Table 13.14 Density variables and indicators of ambient air quality

Neighbourhood	Residential density (DUs/ha)	Dist. of open spaces (%)	Dist. of roads (m/ha)	Level of services (Score)	SO_2 (μg/m^3) 50	NOx (μg/m^3) 40	$RSPM_{10}$ (μg/m^3) 60	Source
Sector 8, Chandigarh	17	37	189	2.7	2	20	101	NAAQM report
Ranjit Avenue, Amritsar	36	55	354.38	2.33	15	40	210	NAAQM report
World Spa, Gurugram	44	89	422	3	6.73	15.67	112.71	HSPCB 2015–17
Marble Arch, Manimajra	65	68	232.79	3	2	19	110	NAAQM report
Sector 38W, Chandigarh	76	30	403	2.67	2	18	103	NAAQM report
Sector 20, Panchkula	98	65	179	2.55	6.42	23.68	53.93	HSPCB 2015–17
Uniworld City, Gurugram	147	85	192	3	6.73	15.67	112.71	HSPCB 2015–17
Katra Dullo, Amritsar	149	15	737	1.42	8.24	32.59	81.22	CPCB 2017

Source (this table and Table 13.13). Primary and secondary data, 2016

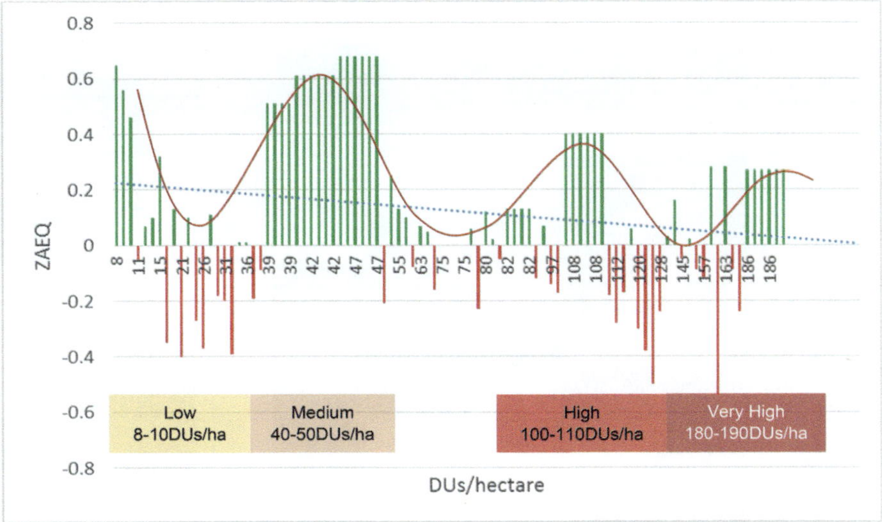

Fig. 13.10 Variation in environmental quality with residential density. *Source* Author, 2016

increasing across all the different residential patterns, the environmental quality fluctuates becoming positive or improving (favourable to very favourable) for particular combinations of the density variables and again deteriorating (unfavourable to very unfavourable) for other combinations of the variables (see Fig. 13.10).

a. **People Density**—Neighbourhoods with 40–50 pph and 8–10 DUs/ha; 160–210 pph and 40–50 DUs/ha; 500–540 pph and 100–110 DUs/ha; and 900–950 pph and 180–190 DUs/ha have favourable environmental quality. These can be suitably classified as low-, medium-, high- and very high-density patterns.
b. **Building Density**—The favourable values of FAR and plot coverage are found to be 1–1.5 with 35–40% coverage; and 2–3 with 11–20% coverage, respectively. The first group as observed in the case studies can be the low- and medium-rise neighbourhoods while the second group corresponds to the high-rise residential developments.
c. **Spatial Density**—Height to width ratio of $\approx$3 (w.r.t high-rise developments) and distribution of open spaces greater than equal to 35% is required for favourable environmental quality. Similarly, 60–70% of the neighbourhood should have continuous and good quality sidewalks to promote walkability. Streetlights should be 30 m c/c and domestic waste collection, stormwater drainage and sewerage system should be as per the norms laid down (100% coverage of waste collection, drainage and sewerage systems) for good environmental quality.

Therefore, it can be concluded that most neighbourhood environmental quality indicators show significant correlations with density variables, as detailed in

Table 13.15 Summary of correlation analysis

Density Variables	Environmental Quality Indicators							
	Crowding & Congestion	Nature & Use of Open Spaces	Shade & Ventilation	Temperature Variations	Noise Levels	Level of Cleanliness	Walkability	Air Quality
Population Density	Negative (65%)	Positive (44.5%)				Negative (27.8%)		Positive (39.9%)
Residential Density	Negative (54.66%)	Positive (57.7%)	Positive (44.6%)	Positive (41%)		Negative (38.5%)		Positive (49%)
FAR, Plot Coverage	Negative (20.56%)	Positive (53.2%)	Positive (27.5%)	Positive (54.25%)	Negative (52.6%)			
Height to Width Ratio			Positive (60.6%)	Positive (78.1%)			Positive (14.5%)	
Distribution of Open Spaces	Positive (33.6%)	Negative (50.6%)		Negative (61.7%)	Positive (64.2%)		Positive (44.7%)	
Distribution of Roads and Sidewalks	Negative (47.4%)				Negative (45.8%)		Negative (13.9%) Positive (75.1%)	Negative (35.5%)
Distribution of Services	Positive (41.33%)	Negative (60.6%)				Positive (66.9%)	Positive (30.92%)	Positive (36.2%)
Strong Correlation (>40%)		Weak Correlation (≤40%)			Empty Cells – No Apparent Correlation			

Source Author, 2016

Table 13.15. This confirms the validity of the study framework and assists in developing guidelines for environmentally friendly neighbourhoods, paving the way for more sustainable cities in the future.

13.7 Conclusion

It can be concluded that the neighbourhood serves as a valid level of aggregation for assessing environmental quality. Several policies, guidelines and development control regulations should be formulated at this level to promote environmentally sustainable growth and development in rapidly urbanizing Indian cities. Effective strategies may include:

a. A need to revise the density regulations depending on the residential pattern observed for future sustainable growth and development of the city. Moderate-to-high density neighbourhoods (40–50 DUs/ha or 170–180DUs/ha) should be promoted as they would use land efficiently to accommodate more people, have better access to services and facilities, feel more secure and develop as vibrant and active social units in the overall city structure.
b. Horizontal and vertical randomization of buildings coupled with high FAR and low plot coverage (FAR 2–3 and plot coverage 13–22%) should be adopted for environmentally conducive sustainable development. Equally distributed low

FAR is detrimental for the city's growth. The FAR should be regulated in accordance with the residential pattern envisaged for future sustainable development of the city.

c. Streets and sidewalks well connected to services and facilities like public transportation stops at the neighbourhood edges and open spaces supporting pedestrian access (taking lighting, paving, safety, etc. into account) should be designed. Relatively small design and planning decisions, such as street width and block size, can significantly impact urban sustainability. These elements should be carefully designed and integrated with emerging concepts such as mixed land use and smart neighbourhoods to enhance overall sustainability.
d. Management of shared greens or open spaces in high- and very low-density neighbourhoods is problematic as observed in Sector 8, Sector 38W and to some extent in Sector 20 (Panchkula), Chandigarh. Design of such neighbourhoods should frame certain guidelines for maintenance and supervision of open spaces as also shown by Dutta et al. (2023).
e. Creating a range of attractive open spaces centrally located and visually more integrated with other neighbourhood spaces may be taken up as an integral part of neighbourhood planning. This in turn may influence the number of social contacts, quality of social network and frequency of social interaction in the neighbourhood as also shown by Raman's (2010) study.
f. Additionally, children play areas, etc. may be provided near medium- and high-rise buildings to provide comfortable micro-climate by ensuring natural sun protection from mutual shading that may further enhance activity intensity and diversity (Shekhar and Tripathi, 2015).
g. For old areas of historical cities as seen in case of Walled City—Amritsar, the built forms reflect the history, culture, traditions, climate responsiveness and socio-economic patterns of those periods. In such cases, links to the past can be maintained by retaining the integrity of the urban form and benefitting from centuries of experience while adopting changes through appropriate retrofitting and revitalization strategies.
h. Programs, meetings and gatherings at the neighbourhood level should be organized to ensure the participation of people and local communities along with the government agencies to create inclusive neighbourhoods that continue to thrive and grow.

It is apparent from the study that neighbourhood environmental quality assessed on the basis of the physical attributes of the built environment gives different outcomes for different residential patterns. This shows that environmental policies and guidelines should be contextual and made at the neighbourhood level to cater to the issues at hand promptly and make communities safe, resilient and sustainable as envisioned in the United Nations Sustainable Development Goals: Goal 11—Sustainable Cities and Communities. In summary, adopting action policies and strategies that focus on effective land use and infrastructure planning, urban design, investment, and efficient management at the neighbourhood level can promote sustainable development in rapidly urbanizing Indian cities.

Acknowledgements We appreciate the valuable comments provided by the reviewers. The findings, interpretations and conclusions presented in this manuscript are solely those of the authors and do not reflect the views of their respective organizations or any entities that have provided institutional or organizational support for the preparation of this manuscript.

References

Bhandari L (2006) The diversity of the top 100 cities of India. In: Indicus analytics—an economic research firm. http://www.indicus.net

Central Pollution Control Board (CPCB) (2009) Comprehensive environmental assessment of industrial clusters. Ministry of Environment and Forests (MoEF), New Delhi. http://cpcb.nic.in/divisionsofheadoffice/ess/NewItem_152_Final-Book_2.pd

Chandigarh Master Plan (CMP) (2031) http://chandigarh.gov.in/cmp_2031.htm

Cheng V (2010) Understanding density and high density. In: Ng E (ed) Designing high density cities—for social and environmental sustainability. Earthscan, New York, pp 3–17

Chhabra A (2010) Chandigarh tops in HDI, GDI. India Today. https://www.indiatoday.in/latest-headlines/story/chandigarh-tops-in-hdi-gdi-70639-2010-03-31

City Development Plan Amritsar (2025) http://jnnurm.nic.in/wpcontent/uploads/2013/08/CDP-AMRITSAR.pdf

Delsante I (2016) Urban environment quality assessment using a methodology and set of indicators for medium-density neighbourhoods: a comparative case study of Lodi and Genoa. Ambiente Construído Porto Alegre 16(3):7–22

Discoli C, Martini I, San Juan G, Barbero D, Dicroce L, Ferreyro C, Esparza J (2014) Methodology aimed at evaluating urban life quality levels. Sustain Cities Soc 10:140–148

Dovey K (2016) Urban density matters—but what does it mean? The Conversation Blog posted on 20 May 2016. http://theconversation.com/urban-density-matters-but-what-does-it-mean-58977

Draft Master Plan Amritsar 2010–2031. http://www.indiaenvironmentportal.org.in/files/plan%20report%20new.pdf

Dutta S, Bardhan S, Bhaduri S (2013) Patterns of urbanisation and environmental quality in the context of Indian cities. Environ Urban Asia 4(2):287–299. https://doi.org/10.1177/0975425313510768

Dutta S, Bardhan S, Bhaduri S, Koduru S (2020) Understanding the relationship between density and neighbourhood environmental quality—a framework for assessing Indian cities. Int J Sustain Dev Plan 15(7):1067–1079. https://doi.org/10.18280/ijsdp.150711

Dutta S, Koduru S, Juganaru M (2023) Impact of physical density on nature and use of open spaces: a pilot study of two residential areas from Jaipur, India. J Inst Eng India Ser A 104(1):95–110. https://doi.org/10.1007/s40030-022-00700-x

Government of Meghalaya, Shillong (2008) Meghalaya human development report 2008, p 23. http://www.in.undp.org/content/dam/india/docs/human_development_report_meghalaya_2008_full_report.pdf

Hall JR (2013) High rise building fires. NFPA (National Fire Protection Association) Fire Analysis & Research, Quincy, MA. https://technokontrol.com/pdf/elpes/oshighrise.pdf

Handbook of Service Level Benchmarking (2011) Ministry of urban development, Government of India. http://www.wsp.org/sites/wsp.org/files/publications/service_benchmarking_india.pdf

Indian Institute for Human Settlements (IIHS) (2011) Urban India 2011: evidence. In: Indian urban conference 2011: evidence and experience. Autumn Worldwide (https://www.autumnworldwide.com); https://www.iihs.co.in/wp-content/uploads/2013/12/IUC-Book.pdf

International Code Council, Inc (2006) International building code (IBC), USA. https://www.bechtel.colorado.edu/~willam/4830%202006%20IBC.pdf

Julka H (2011) IT firms looking beyond Gurugram, Noida, and Greater Noida to other cities in North India. The Economic Times, ET Bureau. https://www.economictimes.indiatimes.com/realty-trends/it-firms-looking-beyond-Gurugram-noida-to-other-cities-in-northindia/articleshow/10182117.cms?intenttarget=no
Kimhi I (ed) (2005) Urban environmental quality. The Centre for Environmental Policy, Jerusalem Institute for Israel Studies, Jerusalem
Kumar KPN, Mishra U (2012) Gurugram: how not to build a city. Forbes India. http://www.forbesindia.com/article/real-issue/Gurugram-how-not-to-build-a-city/33444/1
Lupala JM (2001) Peri-urban land management in the rapidly growing cities—the case of Dar es salaam. PhD. Dissertation, University of Dortmund
Majumdar AK, Hossain ME, Islam MN, Sarwar MI (2007) Urban environmental quality mapping: a perception study on Chittagong metropolitan city. Kathmandu Univ J Sci Eng Technol 1(4):1–14
National Building Code (NBC) (2016) Bureau of Indian Standards (BIS), Government of India. https://www.archive.org/details/nationalbuilding01/in.gov.nbc.2016.vol1.digital/
Patel SB (2011) Analysing urban layouts—can high density be achieved with good living conditions? Environ Urban 23(2):583–595
Raman S (2010) Designing a liveable compact city: physical forms of city and social life in urban neighbourhoods. Built Environ (1978)36 (1):63–80
Sarmento R, Zorzal FMB, Serafim AJ, Allmenroedr LB (2000) Urban environmental quality indicators. In: Brebbia CA, Ferrante A, Rodiguez M, Terra B (eds) The sustainable city. WIT Press, United Kingdom
Shekhar S, Tripathi A (2015) Smart neighbourhood: a way to sustainable development. In: XVI international conference on governance: changing paradigms, 3–5 Jan 2015, Delhi, pp 773–789. http://www.internationalconference.in/XVI_AIC/TS5B-PDF/20Shashank_Shekhar%20(1).pdf
Shieh E, Sharifi A, Rafieian M (2011) Identification of factors that assure quality of residential environments, using environmental assessment indices: a comparative study of two of Tehran's neighbourhoods (Zafaranieh & Khaniabad). Int J Archit Eng Urban Plann 21(2):119–132
Thakur BS (2014) Chandigarh has third-highest tree cover in India. Hindustan Times. http://www.hindustantimes.com/chandigarh/chandigarh-has-third-highest-tree-cover-in-india/story-WS9QsTBjCuRTJhqo4AFDBN.html
Town and Country Planning Organisation (TCPO) (2014) Urban and regional development plans formulation and implementation (URDPFI) guidelines, Ministry of Urban Development, Government of India. http://www.moud.gov.in/upload/uploadfiles/files/URDPFI%20Guidelines%20Vol%20I.pdf
Tripathi S (2021) Towards sustainable urban system through the development of small towns in India. Reg Sci Policy Pract 13(3):777–798. https://doi.org/10.1111/rsp3.12424
Vij A (2015) GRIHA for large development. http://www.grihaindia.org/files/GRIHA-LD-V2015.pdf
Walk Through India (2016) Top 10 green cities of India. http://www.walkthroughindia.com/lifestyle/top-10-green-cities-of-india/
World Health Organisation (WHO) (2011) WHO smoke-free city case study. Experience of Chandigarh as a smoke-free city. http://www.who.int/kobe_centre/interventions/smoke_free/chandigarh_web_final.pdf?ua=1
World Wide Fund for Nature (WWF) (2010) Urbanisation and sustainability in India: an interdependent agenda. The Alternative Urban Futures Report 2010. http://www.assets.wwfindia.org/downloads/urbanisation_report.pdf
Zhang R, Zhang C, Lai PC, Kwan M (2022) Park and neighbourhood environmental characteristics associated with park-based physical activity among children in a high-density city. Urban For Urban Green 68. https://doi.org/10.1016/j.ufug.2022.127479
Zia H (2012) Looking beyond buildings: green guidelines for large developments. http://www.grihaindia.org/events/ncgd/2012/pdf/HINA_ZIA.pdf

Chapter 14
Landfill Site Selection Directives: A Landscape-Centric Perspective in the Context of Kerala

Aswajith K. Babu and M. Banu Chitra

Abstract As urbanization accelerates, the effective management of solid waste and the strategic placement of landfills are considered as crucial environmental concerns. This chapter focuses on the landfill site selection directives, in the context of Kerala, India, with a landscape-centric approach. The study, centered on the Brahmapuram landfill site in Ernakulam district, employs Geographic Information Systems (GIS) technology and the Analytic Hierarchy Process (AHP) to highlight the potential of a landscape-centric approach to transform landfills into assets that benefit both the environment and society. By evaluating the impacts of landfills on adjacent land and water bodies, we emphasize the importance of holistic environmental consideration in site selection processes. Based on case study comparisons and existing literature, the research proposes site selection directives derived from the AHP methodology. These guidelines consider factors like geology, hydrology, and vegetation and thus these directives aid in choosing landfill locations with minimal environmental disruption. This chapter contributes to the development of a systematic framework for landfill site selection, offering valuable insights for policymakers and planners. The proposed directives provide a practical tool for assessing potential sites and guiding decision-making processes, ultimately promoting sustainable waste management practices and environmental preservation in Kerala and beyond.

Keywords Analytic hierarchy process · GIS · Landfill · Landscape · Solid waste management

14.1 Introduction

Due to the escalating waste management issue in Kochi, India, a fundamental change in landfill selection strategies is essential. Current practices, often focuses on the technical and engineering considerations that have led to adverse environmental

A. K. Babu (✉) · M. B. Chitra
School of Planning and Architecture Vijayawada, Vijayawada, Andhra Pradesh, India
e-mail: aswajith4@gmail.com

L. K. Dashora et al. (eds.), *Urban Sustainability*, Disaster Risk Reduction,
https://doi.org/10.1007/978-981-96-4890-0_14

and social consequences, exemplified by the recent fire at the Brahmapuram landfill in 2023. This research advocates for a landscape-centric approach to landfill site selection, recognizing that landfill siting is not merely a technical exercise but one deeply connected with the surrounding environment and communities. By adopting a landscape-centric lens, this study aims to identify and prioritize site selection parameters that minimize negative impacts on natural ecosystems, water resources, and human well-being. Kochi, as a rapidly growing metropolitan area with unique geographic characteristics and a diverse cultural landscape, serves as a pertinent case study for developing a versatile framework that can be adapted to varying contexts.

Through a comprehensive analysis of existing waste management practices and the utilization of the Analytic Hierarchy Process (AHP), this research seeks to formulate a robust decision-making tool for landfill site selection. This framework will integrate landscape-based parameters, providing a more holistic and sustainable approach to waste disposal in Kochi, with the potential to inspire similar approaches in other regions facing waste management challenges. Also this study primarily focuses on the environmental and technical facets of site selection, it acknowledges the limitations of not fully addressing social, cultural, political, and financial dimensions. This focus on technical aspects reflects the current state of landfill site selection, which often prioritizes factors like land availability, engineering feasibility, and construction costs. However, this narrow approach has resulted in environmental problems like groundwater contamination and air pollution, as well as social issues like community health risks and property value decline Nevertheless, it aims to lay a strong foundation for future research that can incorporate these critical aspects, ultimately paving the way for a more sustainable and equitable approach to waste management.

14.2 Conceptual Framework

The study's conceptual framework, depicted in Fig. 14.1, begins with identifying the problem of landfill site selection. For background research, various global methods of identifying landfill sites are reviewed to determine common selection parameters. These parameters are then aligned with relevant laws and regulations. The selected parameters are evaluated using Multi-Criteria Decision Analysis (MCDA) with the Analytic Hierarchy Process (AHP). Based on this analysis, guidelines are developed and applied to assess the existing landfill site in Kochi.

14.2.1 Solid Waste Management

Any solid or semi-solid domestic wastes that are disposed of after their initial usage are referred to as "Solid Wastes" in India. A city's solid waste management framework is one of its most important infrastructure elements. The demand for more human

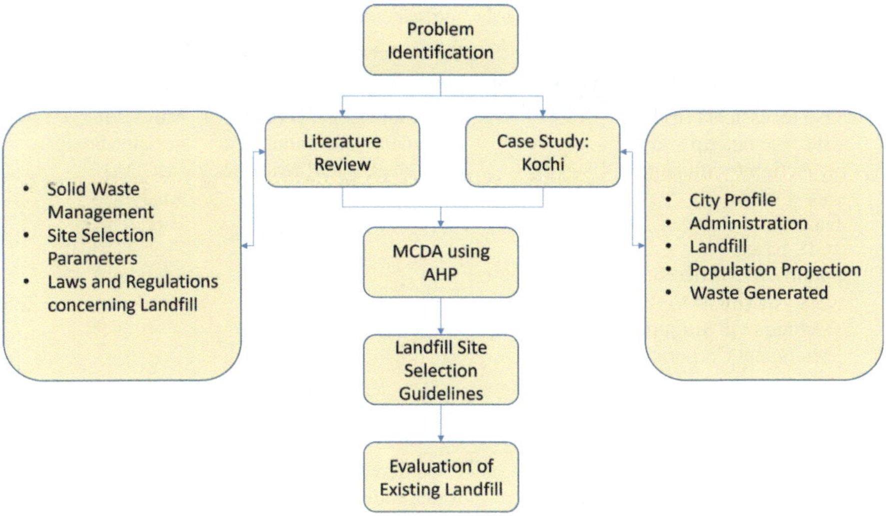

Fig. 14.1 Methodology diagram. *Source* (Author)

capital grows as the city develops, because of increased urbanization. These instances have a significant effect on the management of municipal solid waste (MSW), which is an essential component of any economic development (Paul 2021). India has undergone a massive transformation since 1991, when it opened its doors, as rapid urbanization and population growth have led to changing lifestyles, changing food habits, and changing standards of living. MSW management has been put under considerable stress by all these trends, which were already very stressful (Paul 2021).

According to reports, solid waste management produced over 1.6 billion tons of carbon dioxide equivalent globally, or around 5% of all greenhouse gas (GHG) emissions. Only 8% of waste produced worldwide ends up in sanitary landfills; the remaining 37% is disposed of in landfills. The remaining 33% is disposed of in an unscientific manner in open fields and grounds. Just 19% of the world's waste is recovered through recycling, composting, and reuse (Sohkhlet and Nagargoje 2020).

The Twelfth Schedule 21 of the Indian Constitution states that Local Governments (Urban and Rural) oversee solid waste management. To enhance service delivery, the Kerala government set up organizations at the state level and issued the Solid Waste Management Policy, 2018 (Suchitwa Mission—Govt. of Kerala 2020).

14.2.2 Site Selection Parameters

A comprehensive list of parameters used worldwide for landfill site selection, along with the common methodologies employed is listed in Table 14.1.

Table 14.1 Literature review determining the site selection parameters

Sl. no	Paper title	Year of publication	Parameters	Methodology
1	An assessment of suitable landfill site selection for municipal solid waste management by GIS-based multi-criteria decision analysis (MCDA) technique in Siliguri Municipal Corporation Planning Area, West Bengal, India (Roy et al. 2022)	2022	Elevation, slope, geology, lineament, land value, dist. from river, roads, residence, LULC	Multi-criteria decision analysis (MCDA) Analytical Hierarchical Process (AHP) Saaty's 1 to 9 priority scale
2	Landfill site selection using GIS-based multi-criteria evaluation technique. A case study of Srinagar city, India (Majid and Mir,2021)	2021	Road network, Railway line, water bodies, residential area, built-up area, airport, agricultural field, slope, altitude, size	Multi Criteria Evaluation (MCE) Analytic Hierarchy Process (AHP) Land Suitability Index (LSI) Total Hauling Distance (THD) Cartographic analysis
3	Site selection for hazardous waste using fuzzy logic combined with analytic hierarchy process: A case study in Nahavand, Iran (Najjari and Shayesteh 2019)	2019	Slope, DEM, rainfall, temperature, LULC, dist. from surface water, dist. from faults, vegetation density, vegetation type, dist. from villages, dist. from cites, dist. from road	Fuzzy logic Analytic Hierarchy Process (AHP)
4	Municipal Solid Waste Landfill Sites Selection for Visakhapatnam city under vision 2020 using GIS and AHP (Dara and Reddy 2018)	2018	LULC, geomorphology, geology, structures, soil, drainage, transportation network, watershed, slope	Analytic Hierarchy Process (AHP) Relative Importance Weightage (RIW)
5	Landfill site selection using Simple Additive Weighting (Saw) Method and artificial neural network method; A case study from Lorestan Province, Iran (Mokhtari et al. 2016)	2016	Lithology, dist. from faults, slope, dist. from rivers and drainage system, dist. from wells and springs, dist. from dams, vegetation, dist. from residential area, dist. from roads	Neural network method Simple additive weighting (SAW)

(continued)

Table 14.1 (continued)

Sl. no	Paper title	Year of publication	Parameters	Methodology
6	Landfill site selection for municipal solid wastes in mountainous areas with landslide susceptibility (Eskandari et al. 2016)	2016	Dist. from surface water, distance from springs / wells / qanats, geology and groundwater contamination, slope & aspect, precipitation, LULC, dist. From Source, land cost, temperature, soil depth, road availability, dist. from residential areas, dist. from promenades, wind direction, road type, visibility, dist. from agricultural lands and orchards	Analytical hierarchy process (AHP) Boolean method Sensitivity Analysis
7	Sitting MSW landfills using MCE methodology in GIS environment (Case study: Birjand Plain, Iran) (Motlagh and Sayadi 2015)	2015	Slope, water resources, soil parameters, LULC, fault and protected areas, dist. from road, urban areas, village, airport, historical place, industries	Multi Criteria Evaluation (MCE) Weighted Linear Combination (WLC) Analytical Network Process (ANP) Ordered Weighted Averaging (OWA) Fuzzy Linguistic Quantifier (LQ)
8	Selection of a landfill site for solid waste management: An application of AHP and Spatial Analyst tool (Kumar and Hassan 2013)	2013	Dist. from residential locations, transport connectivity, presence of water bodies, forests, ground water table, and geology	Spatial Analyst Tool Analytic Hierarchy Process (AHP) model Multi-criteria decision-making process

(continued)

Table 14.1 (continued)

Sl. no	Paper title	Year of publication	Parameters	Methodology
9	MSW landfill site selection by combining AHP with GIS for Konya, Turkey (Uyan 2014)	2013	Dist. from residential and industrial areas, dist. from rivers, lakes, wetlands, coastal area and dam, dist. from groundwater wells, dist. from protection areas, dist. from infrastructure, LULC, slope, geology, dist. from roads and railway, elevation, dist. from tourism and historical sites, dist. from airport	Multi-criteria evaluation
10	GIS and the Analytic Hierarchy Process for regional landfill site selection in transitional countries: A case study from Serbia (Zelenović Vasiljević et al. 2012)	2012	Geomorphology, Depth of underground water table, LULC, surface water, protected areas, monuments, aspects, settlements and visibility, land slope, traffic infrastructure, seismology	Analytic Hierarchy Process (AHP) Multi-criteria evaluation

Source Author (2023)

14.2.3 Laws and Regulations Concerning Landfill

India has implemented several MSWM initiatives since 2000, including the Municipal Solid Waste (M&H) Rules, 2000 (Ministry of Environment, 2016), and the Swachh Bharat Mission (SBM) (Ministry of Housing and Urban Affairs, 2017). While existing rules and regulations encompass landfilling of SWM, they are scattered across various sources, such as the one cited below (Tables 14.2, 14.3, 14.4, 14.5, 14.6, 14.7, 14.8 and 14.9).

14.2.4 Kochi City Profile

The State of Kerala's Ernakulam district is home to Cochin, also referred to as Kochi. The Kochi Municipal Corporation (KMC) is in charge of overseeing the city. The Corporation of Kochi currently extends 94.88 square kilometers, comprises 22 health circles and 74 wards, and generates 317 MT of waste per day, which is divided within seven administrative zones (State and Control, 2023). The population density of the city is 6,340 people per square kilometer, which is higher than the 819 people in Kerala, 382 people in India, and the 46 people per square kilometer global average in 2011 (Abhirami, 2021) (Figs. 14.2, 14.3, 14.4 and 14.5).

Table 14.2 Rules and regulations concerning SWM

Year	Rules, policies, schemes, financial plans
1989	The Hazardous Waste (M&H) Rules
1994–1995	MSWM strategy paper by NEERI
	J.S. Bajaj Committee (The High-Powered Committee on Urban Solid Waste Management)
1998	Bio-medical Waste (M&H) Rules
	Supreme Court appointed Barman Committee
2000	MSW (M&H) Rules
	CPHEEO Manual on MSW
2005	Report of the Technology Advisory Group on SWM
	JNNURM (2005–2012)
	UIDSSMT (2005–2012)
2006	Strategy and Action Plan-Use of compost in cities
2007	11th Five-Year Plan (2007–2012)
2008	National Urban Sanitation Policy (NUSP)
	Service Level Benchmarks (SLBs) in MSWM
	Hazardous Waste (Management, Handling & Trans boundary Movement) Rules
	National Mission on Sustainable Habitat (NAPCC)
2010	13th Finance Commission (2010–2015)—Establishing standards for delivery of essential services
2011	Plastic Waste (M&H) Rules
	E-Waste (M&H) Rules
	Draft Bio-medical Waste (M&H) Rules
2014	Swachh Bharat Mission, October 2014
2016	Waste Management Rules, 2016 comprising Solid Waste Management Rules, Plastic Waste Management Rules, Bio-Medical Waste Management Rules, E-Waste Management Rules, Hazardous and Other Wastes (Management and Transboundary Movement) Rules, 2016 and Construction and Demolition Waste Management Rules, 2016
2018	Kerala State Waste Management Policy

Source Eco and Private 2019 Subramony 2020

14.2.5 Administrative

The KMC is split into 21 circles, each with one to five wards, for the purpose of collecting and transporting solid waste (Subramony, 2020). There are 21 sanitation circles in the city, which serve as secondary collection sites for waste after it has been collected initially, using hand carts, wheelbarrows, and tractor trailers. 40% of the waste generated is collected by the city and sent to the treatment plant. In Kochi, officials are aided in the door-to-door collection process by nearly 250 resident

Table 14.3 Cochin municipal corporation SWM

Sl. no	Item	Description
1	Name of municipal body	Cochin corporation
2	Total quantity of solid waste generated tons per day	326 ton
3	Composition of solid waste (Tons per Day)	Wet-226, Dry-100 tons
4	Quantity of waste collected tons per day	317 tons
5	Area considered for collection of waste	98 sq.km
6	Number of sanitary workers	1823
7	Number of trucks	55
8	Number of hand carts	500
9	Approximate quantity of waste at site	100,000 ton
10	Quantity of waste processed	250 ton
11	Quantity of waste disposed by landfilling	99,750 ton

Source Abhirami et al. 2021

Table 14.4 Overview of MSW management facility at Brahmapuram

Item	Particulars	Remarks
Village	Puthencruz	
Taluk	Kunnathunadu	
GP/Mun./Mun. Corp	Vadavukode-Puthencruz	Grama Panchayath
Survey number	32/17	
Area	110 acre	44.5 hectares
Built-up area	17,456 m^2	Compost plant alone 10,000 Sq.m
Capacity	250 TPD	
Facilities	Windrow composting	250 TPD
	Vermicomposting	50 TPD, Not operated
	RDF Facility	100 TPD, Not operated
	Plastic shredding unit	1 TPD, Not functioning at the time of visit
Waste collected from	Kochi corporation	
	Angamaly municipality	
	Aluva municipality	
	Kalamassery municipality	
	Thrikkakkara municipality	
	Thrippunithura municipality	
	Cheranallur GP	
	Vadavukode-Puthencruz GP	
	Kumbalangi	

Source Kochi Corporation (2022)

Table 14.5 Available population data

Year	Population
2011	601,574
2012	628,464
2013	656,116
2014	685,313
2015	715,809
2016	747,662
2017	780,932
2018	815,683
2019	851,980
2020	889,893
2021	929,493

Source Abhirami 2021

Table 14.6 Projected population of Kochi

Year	Projected population		
	Arithmetic increase method	Geometric increase method	Incremental increase method
2031	962,285	970,763	963,698
2041	995,077	1,013,865	999,316
2051	1,027,869	1,058,880	1,036,347

Source Author. 2023

Table 14.7 Kochi's projected daily population growth and total residential waste growth

Year	Kochi		
	Population	Per capita waste generated* (kg)	Total waste (Tons)
2001	595,575	0.53	315.65
2011	601,574	0.53	318.83
2021	929,493	0.62	576.29
2031	970,763	0.62	601.87
2041	1,013,865	0.62	628.60
2051	1,058,880	0.62	656.51

* It is estimated that each person produces roughly 0.62 kg of waste per day (Kumar et al. 2017)

Table 14.8 Pairwise comparison matrix

Less important				Equal importance	More important			
1/9	1/7	1/5	1/3	1	3	5	7	9

* In a pairwise comparison matrix, the intermediate values between the two adjacent judgements are 2, 4, 6, 8

Table 14.9 Pairwise comparison matrix calculation for landfill site selection

AHP	EL	SL	GL	DL	DW	DS	DR	LULC	PA
EL	1	0.50	0.33	0.20	0.20	0.33	0.14	0.50	0.25
SL	2.00	1	2.00	0.50	0.33	3.00	0.20	0.33	0.33
GL	3.00	0.50	1	0.33	0.25	0.50	0.25	0.33	0.50
DL	5.00	2.00	3.00	1	1.00	4.00	0.50	5.00	2.00
DW	5.00	3.00	4.00	1.00	1	3.00	2.00	2.00	3.00
DS	3.00	0.33	2.00	0.25	0.33	1	0.33	0.50	0.33
DR	7.00	5.00	4.00	2.00	0.50	3.00	1	3.00	0.33
LULC	2.00	3.00	3.00	0.20	0.50	2.00	0.33	1	0.20
PA	4.00	3.00	2.00	0.50	0.33	3.00	3.00	5.00	1

* EL—Elevation, SL—Slope, GL—Geology, DL—Distance from lineament, DW—Distance from water body, DS—Distance from road, DR—Distance from residence, LULC—Land use and land cover, PA—Protected area

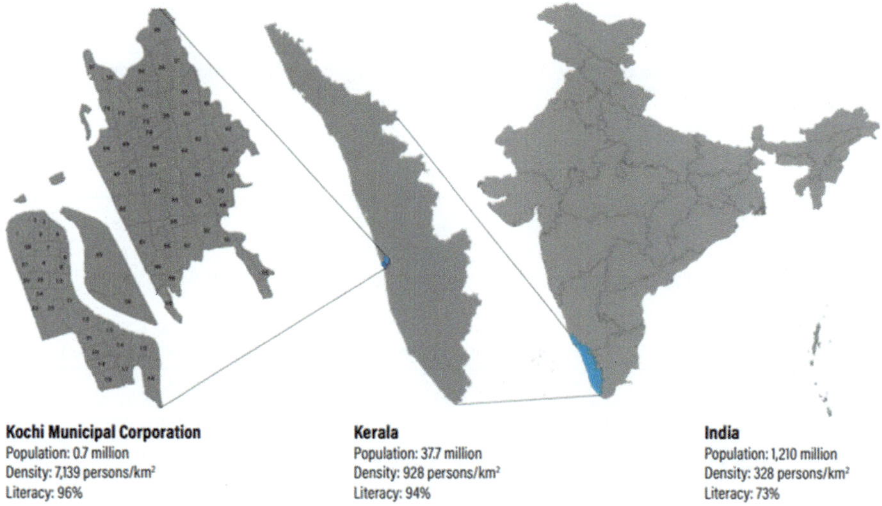

Fig. 14.2 Political map of Kochi municipal corporation. *Source* (DoECC, 2022)

associations, in addition to Kudumbashree self-help workers and municipal staff members. Approximately 136 additional unauthorized sites are utilized for waste disposal in Kochi city and its surrounding areas; a selection of these are displayed below (Eco and Private, 2019).

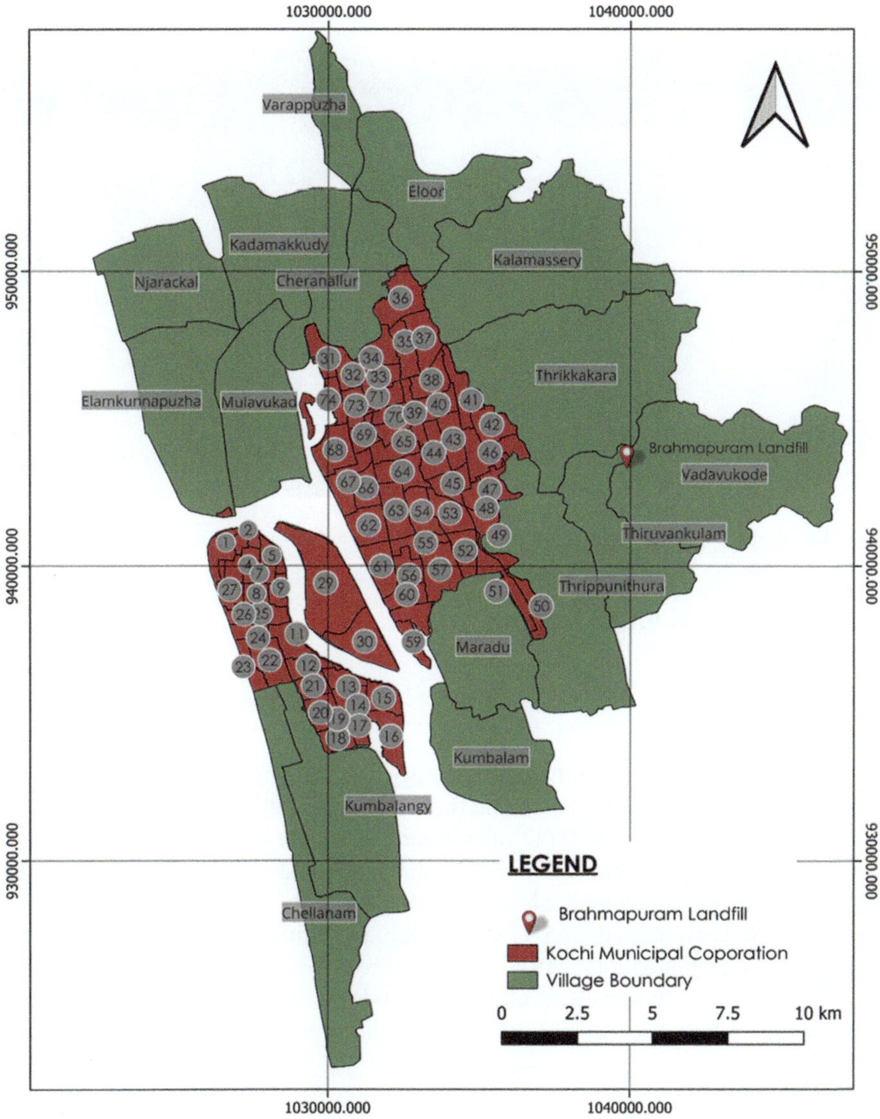

Fig. 14.3 Kochi municipal corporation. *Source* Author, 2023

Fig. 14.4 Brahmapuram landfill. *Source* Global Atlas of Environmental Justice 2021

14.3 Materials and Methods

14.3.1 Landfill

For the management of non-biodegradable waste, Cochin Corporation runs one Resource Recovery Facility (RRF) and three Material Collection Facilities (MCFs). About 2 TPD of the 60–80 TPD of non-biodegradable waste that is collected is moved to the Edappally and Kaloor MCFs that are already in established. Out of the 60–80 TPD of non-biodegradable waste collected, only 20–25 TPD of recyclable plastic materials are sorted. 365 TPD of waste, including biodegradable waste, are dumped at the Brahmapuram MSW Dumpsite from the surrounding three panchayaths (Kumbalangi, Cheranalloor, and Vadavucode-Puthencruz) and five municipalities (Aluva, Angamaly, Kalamassery, Tripunithura, and Thrikkakkara) (State and Control, 2023).

The Brahmapuram solid waste treatment plant, which can process 200 tonnes of waste per day, was opened in 2008 as part of Kochi Corporation's efforts to establish a reliable MSW management system in order to address the rising waste crisis (Siby, 2021). It was constructed following the rehabilitation of the Brahmapuram neighborhood and the acquisition of 33.3 acres of wetland. In its 2018 report, the Central Pollution Control Board highlights the specific section of Kadambrayar between Manckakadavu and Brahmapuram as one of the rivers in India that does not meet water quality criteria (Siby, 2021).

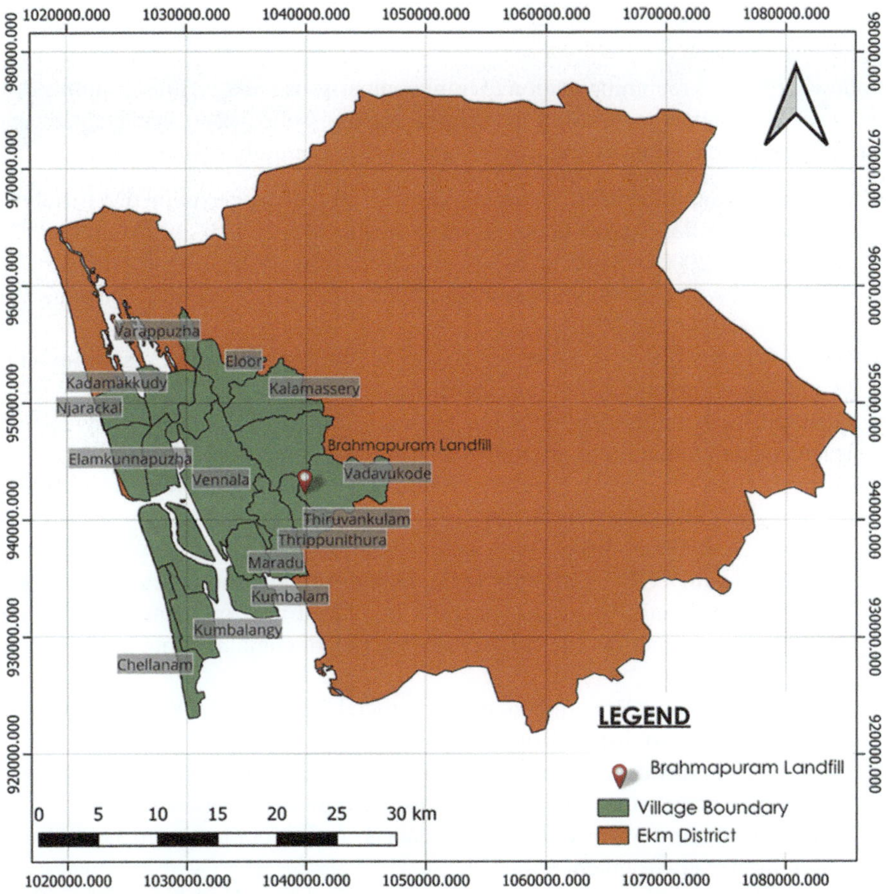

Fig. 14.5 Study area for SWM. *Source* Author. 2023

14.3.2 Population

Three methods of population projection have been used to get the city's population for the 20-year design period.

These techniques are as follows:

1. The method of arithmetic increase
2. The geometric increase technique
3. The method of incremental increase

Three methods are used to estimate population: incremental, geometric, and arithmetic increases. Nonetheless, since the geometric increase method projects a higher value than the other, it is taken into consideration.

14.3.3 Criteria Selection for Suitable Landfill

Following Saaty's recommendation to limit comparisons to seven plus or minus two, the selected parameters were derived from existing solid waste management and landfill regulations in India and a landscape-centric literature review.

- Elevation: Landfill suitability diminishes with increasing elevation due to heightened leachate flow and contamination in steeper slopes (Ali and Ahmad, 2020; Kontos et al. 2005). Flatter areas are preferred for landfill siting to prevent wind dispersal of waste (Abediniangerabi and Kamalirad, 2016; Ali and Ahmad, 2020; Hazarika and Saikia, 2020).
- Slope: Steeper slopes increase the risk of waste dispersal and erosion, making gentler slopes more suitable for landfill sites.
- Geology: Low infiltration rates are favorable for landfill sites as they impede leachate contamination of groundwater.
- Distance from Lineament: Landfills should be sited away from fractured zones to prevent groundwater contamination.
- Distance from Water Body: It is against the law to dispose of solid waste in proximity to any body of water in India, including lakes and rivers (Bhawan and Nagar, 2008). Landfills should generally be placed far from canals, rivers, and other bodies of water in order to decrease the risk of contaminating surface water (CPHEEO, 2016).
- Distance from Road: Proximity to roads is crucial for efficient waste transportation from urban areas to disposal sites.
- Distance from Residence: Landfills should be located away from residential areas to minimize environmental pollution, noise, and unfavorable odor.
- Land use and Land cover (LULC): This data is crucial for identifying landfill sites that minimize environmental impact.
- Protected Areas: Existing laws states setbacks required to be followed for various elements like airport, heritage, eco-sensitive, coastal line, etc.

14.3.4 Assessment of MCDA Technique with AHP Method

The AHP technique, that assigns each distinct criterion an appropriate weight, is the most significant approach for decision-making. Saaty's relative importance scale was used to determine each criterion's weighting value.

14.3.4.1 Pairwise Comparison Matrix Calculation

The Analytic Hierarchy Process (AHP) assigns relative importance weights to each parameter, ranging from 1 (equal importance) to 9 (extreme importance).

The pairwise matrix can be calculated using the following formula, where P_x is the pairwise comparison matrix and x_{nn} is the factors of P_x.

$$P_x = \begin{bmatrix} x_{11} & x_{12} & x_{13} & \cdots & x_{1n} \\ x_{21} & x_{22} & x_{23} & \cdots & x_{2n} \\ \vdots & \vdots & \vdots & \vdots & \vdots \\ \vdots & \vdots & \vdots & \vdots & \vdots \\ x_{n1} & x_{n2} & x_{n3} & \cdots & x_{nn} \end{bmatrix}$$

14.3.4.2 Computing the Normalization of the Weights

The following equation is used to create the normalized pairwise comparison matrix, where NW stands for normalized weight estimation and GM_n for the geometric mean calculation of the nth row in P_x.

$$NW = \left(\frac{GM}{\sum_{n-1}^{N} GM_n} \right)$$

14.3.4.3 Consistency Index (CI)

Saaty's Analytic Hierarchy Process method states that the consistency vector can be found by multiplying P_x by the weight of each criterion and using the consistency index formula, where λ_{max} represents the eigenvalue of the comparison matrix.

$$CI = \frac{(\lambda_{\max} - n)}{n - 1}$$

14.3.4.4 Consistency Ratio (CR)

Saaty's Analytic Hierarchy Process method states that the consistency ratio formula is,

$$CR = \frac{CI}{RI}$$

For nine parameters, the constant value of RI (Table 14.10) is 1.45. Consequently, the consistency ratio value should be less than 0.1 or less than 10% in order to maintain the judgment's consistency.

Number of comparisons = 36
Principal eigen value = 10.114
Eigenvector solution: 6 iterations, delta = 6.6E – 8
Consistency ratio CR = 9.6%

14.4 Results and Discussion

The study used various parameters such distance from water body, distance from lineament, protected areas, distance from residence, LULC, slope, distance from road, geology, elevation which can be used to identify potential landfill site in Kochi Municipal Planning area. The fact that these criteria are regularly and efficiently used to assess whether an area is suitable for landfills makes them a strong case for use. The study resulted in guidelines for selecting suitable landfill sites based on environmental, socio-economic factors, using the Analytic Hierarchy Process (AHP) technique.

14.4.1 Guidelines

14.4.1.1 Environmental Considerations

- Water Resources Protection: Maintain a minimum distance of 200 m from water bodies to prevent contamination.
- Floodplain Avoidance: Exclude areas identified as floodplains within the past 100 years to prevent flooding risks.
- Wetlands Protection: Avoid locating landfills within wetland areas to preserve their ecological functions.
- Critical Habitat Protection: Avoid locating landfills within Critical Habitat Areas (CHA), Sensitive Eco-Fragile Areas (SEFA), or Potentially Unstable Zones (PUZ) to protect sensitive ecosystems.
- Geological Stability: Avoid areas with unstable geological features to ensure structural stability.
- Water Table Depth: In order to prevent groundwater contamination, it is necessary to maintain a minimum distance of 2 m between the base of the clay liner and the highest level of the water table.
- Natural Hazards: Exclude areas prone to natural hazards to protect the landfill from damage and prevent potential environmental risks.
- Subsidence Risks: Avoid areas with subsidence potential to ensure the stability of the landfill site.

Table 14.10 Normalized pairwise comparison matrix for choosing a feasible landfill site

AHP	EL	SL	GL	DL	DW	DS	DR	LULC	PA	Mean
EL	0.03225806	0.0273	0.0155	0.0334	0.045	0.025	0.0181	0.0283	0.03148615	0.02848445
SL	0.06451613	0.0546	0.0938	0.0836	0.0743	0.15	0.0258	0.0187	0.04156171	0.06742522
GL	0.09677419	0.0273	0.0469	0.0552	0.0563	0.025	0.0323	0.0187	0.06297229	0.04681568
DL	0.16129032	0.1091	0.1406	0.1672	0.2252	0.2	0.0645	0.2831	0.25188917	0.17811426
DW	0.16129032	0.1637	0.1875	0.1672	0.2252	0.15	0.2581	0.1133	0.37783375	0.20045373
DS	0.06451613	0.018	0.0938	0.0418	0.0743	0.05	0.0426	0.0283	0.04156171	0.05054104
DR	0.22580645	0.2728	0.1875	0.3344	0.1126	0.15	0.129	0.1699	0.04156171	0.18040475
LULC	0.06451613	0.1637	0.1406	0.0334	0.1126	0.1	0.0426	0.0566	0.02518892	0.08214237
PA	0.12903226	0.1637	0.0938	0.0836	0.0743	0.15	0.3871	0.2831	0.12594458	0.1656185
Sum	1	1	1	1	1	1	1	1	1	1

* EL—Elevation, SL—Slope, GL—Geology, DL—Distance from lineament, DW—Distance from water body, DS—Distance from residence, LULC—Land use and land cover, PA—Protected areas, DR—Distance from road

Table 14.11 Random consistency index value

n	1	2	3	4	5	6	7	8	9	10	11	12	13	14	15
RI	0.00	0.00	0.58	0.90	1.12	1.24	1.32	1.41	1.45	1.49	1.51	1.48	1.56	1.57	1.59

Source Saaty's Random Consistency Index, 1980

- Agricultural and Forest Areas: Avoid locating landfills in agricultural and forest areas to preserve their ecological and economic value and minimize land use conflicts.

14.4.1.2 Socio-Economic Consideration

- Residential Areas: Establish a 200-m buffer zone around the landfill to protect residents from potential environmental impacts.
- Public Park Proximity: Maintain a 200-m buffer zone around public parks to minimize environmental impacts on recreational areas.
- Cultural Heritage Protection: Avoid locating landfills near areas with unique archaeological, historical, paleontological, or religious sites. Maintain a 200-m buffer zone around such areas.
- Highway Proximity: Maintain a 200-m buffer zone between the landfill and the nearest highway to minimize noise and traffic congestion.
- Airport Regulations: Locate the landfill at least 20 km away from airports or helipads. If the site falls within the 10-km to 20-km buffer zone, obtain a No Objection Certificate (NOC).

14.4.2 Discussion

The appropriate weight of the AHP technique is used to identify the factors that need to be taken into account when choosing a landfill site. A matrix of pairwise comparisons has been created to assess the weight of every chosen thematic layer. It came from a consistency ratio that was acceptable ($CR < 0.096$). Table 14.11 displays the calculated weight of each criterion. Of the nine selected parameters, elevation has the lowest suitability weight of 2.7%, while distance from a water body and distance from a lineament have the highest suitability weights of 20.5% and 17.9%, respectively.

14.4.2.1 Brahmapuram Landfill

The Landfill site at Brahmapuram was established in 2008 due to political pressure and without scientific consideration, as a result we analyzed the landfill site using the

Table 14.12 Overview of parameters with AHP weight

Order	Parameter	AHP weight (%)	SD (%)
1	Dist. from water body	20.5	11.4
2	Dist. from lineament	17.9	9.3
3	Protected areas	17.7	13.9
4	Dist. from residence	17.5	8.5
5	LULC	7.8	4.5
6	Slope	6.3	3.3
7	Dist. from road	5.1	1.9
8	Geology	4.5	2.0
9	Elevation	2.7	0.9

Source Computed by the Author

proposed guidelines has been shown in Table 14.12, which indicates that Brahmapuram Landfill site does not fulfil four out of thirteen guidelines. Based on the results, the site is too close to water bodies, and was under subsidence risk which states that it fails to follow the top two ranked parameters of site suitability namely, distance from water body, distance from lineament which has a suitability weight about 20.5% and 17.9%, respectively (Table 14.13).

14.5 Conclusions

This research successfully developed a flexible framework for selecting environmentally and socially responsible landfill sites in Kochi, India. This framework incorporated population projections, waste generation forecasts, and environmental assessments to address the city's municipal solid waste management needs. Utilizing the Analytic Hierarchy Process, the study established comprehensive guidelines for landfill site selection based on key criteria like distance from water body, distance from lineament, protected areas, distance from residence, LULC, slope, distance from road, geology, and elevation.

While the proposed framework offers a robust foundation for landfill site selection, further evaluation, field investigations, and community engagement are recommended before implementation. Nevertheless, this research provides valuable insights for policymakers, urban planners, and the Kochi Municipal Authority as they navigate the relocation of the existing landfill and the future growth of Kochi city.

Table 14.13 Brahmapuram landfill site criteria checklist

Sl. no	Parameters	Distance from landfill		
		Rules and regulations	Existing	Satisfy
1	Water resources protection	200 m	< 200 (30 m)	×
2	Floodplain avoidance	Not be in floodplain	It is a low-lying site, where the water levels rising up to 1 m above ground during monsoon	×
3	Wetland	Not be in wetland	The site is located at wetland	×
4	Critical habitat protection	Not to be in CHA	None (Mangalavanam Bird Sanctuary—10.45 km)	✓
5	Geological stability	Exclude the areas with unstable geological features	None	✓
6	Water table depth	2 m	> 2 m	✓
7	Natural hazards	Not to be in Unstable Zone	No	✓
8	Subsidence risk	Not to be in such areas	Yes	×
9	Habitation	200 m	> 200 (602 m)	✓
10	Public park	200 m	> 200 (2.07 km)	✓
11	Cultural heritage protection	200 m	> 200 (1 km)	✓
12	Highway	200 m	> 200 (>4.7 km)	✓
13	Airport	20,000 m (NOC in case of falling between 10,000 and 20,000)	11.58 km (Kochi Naval airstrip) & 17.32 km (COK)	✓

Source Computed by the Author

References

Abediniangerabi B, Kamalirad S (2016) Landfill sitting using MCDM in Tehran metropolitan. J Urban Environ Eng 10(1):11–24. https://doi.org/10.4090/juee.2016.v10n1.011024

Abhirami KS, Aravind PK, Archa R, Ashik PBHP (2021) Assessment of municipal solid waste management in Satna. Indian J Environ Prot 33(4):321–328. https://www.ijert.org/research/assessment-of-municipal-solid-waste-management-in-kochi-city-IJERTCONV9IS06012.pdf

Abishek Baby K, Elsa Cheriyan S, Professor A, Maria Jacob T (2019) Design of sanitary landfill at brahmapuram solid waste treatment plant 8(6):182–186

Ali SA, Ahmad A (2020) Suitability analysis for municipal landfill site selection using fuzzy analytic hierarchy process and geospatial technique. Environ Earth Sci 79(10). https://doi.org/10.1007/s12665-020-08970-z

Aydi A, Zairi M, Dhia HB (2013) Minimization of environmental risk of landfill site using fuzzy logic, analytical hierarchy process, and weighted linear combination methodology in a geographic information system environment. Environmental Earth Sciences 68(5):1375–1389. https://doi.org/10.1007/s12665-012-1836-3

Bhawan P, Nagar EA (2008) Guidelines and check-list for evaluation of MSW landfills proposals with Information on existing landfills 124:1–46

Chen D, Meng-Chuen B, B. L., Krueger, T., Mishra, A., & Popp, A. (2020) The world ' s growing municipal solid waste : trends and impacts recent citations the world ' s growing municipal solid waste : trends and impacts. Environ Res Lett 15:13

Col L, Raj A (2020) Solid waste management in Kochi : no solution to the age-old problem ? Centre for Public Policy Research, 1–7. https://www.newindianexpress.com/states/tamil-nadu/2016/jun/01/Brahmapuram-dumping-yard-illegal-rules-NGT-South-Bench-935348.html

CPHEEO. (2016). Municipal Solid Waste Management Manual Part II. Ministry of Urban Development, 1–604. http://cpheeo.gov.in/upload/uploadfiles/files/Part2.pdf

Dara PK, Reddy TB (2018) Municipal solid waste landfill sites selection for Visakhapatnam City under Vision 2020 using GIS and AHP. Int J Contemp Res Rev 9(06):20181–20194. https://doi.org/10.15520/ijcrr/2018/9/06/513

DoECC (2022) Kerala State action plan on climate change 2023–2030. https://envt.kerala.gov.in/wp-content/uploads/2022/12/Kerala-State-Action-Plan-on-Climate-Change-2.0.pdf

Eco GJ, Private P (2019) Gj eco power private limited ec for Kochi waste to energy project final EIA / EMP Report Quality Council of India National Accreditation Board for (Issue July). https://environmentclearance.nic.in/DownloadPfdFile.aspx?FileName=yKEIoT749wpqHOTZjQYBcpF1PRCSvq7PiAKyasvbFz6nrsg70SfsnLs3iFkcqamSm3tbGiUpiAChevEDplUwiEhhknXsCKpiq0wvxwqISMMKjnzXfcsdXx0up8fC1s/kjh92ytEXWF2ySSj49Be8Zw==&FilePath=93ZZBm8LWEXfg+HAlQix2fE2t8z

Eskandari M, Homaee M, Falamaki A (2016) Landfill site selection for municipal solid wastes in mountainous areas with landslide susceptibility. Environ Sci Pollut Res 23(12):12423–12434. https://doi.org/10.1007/s11356-016-6459-x

Ministry of Environment, F. and C. C. (2016). Solid Waste Management Rules, 2016_S.O. 1357(E). The Gazette of India, 2(REGD. NO. D. L.-33004/9933004/99), 52–90. http://www.moef.nic.in/sites/default/files/SWM. https://cpcb.nic.in/uploads/MSW/SWM_2016.pdf

Ganesan P (2017) Landfill sites, solid waste management and people's resistance: a study of two municipal corporations in Kerala. Int J Environ Stud 74(6):958–978. https://doi.org/10.1080/00207233.2017.1374076

Gautam S, Brema J, Dhasarathan R (2020) Spatio-temporal estimates of solid waste disposal in an urban city of India: a remote sensing and GIS approach. Environ Technol Innov 18:100650. https://doi.org/10.1016/j.eti.2020.100650

Gbanie SP, Tengbe PB, Momoh JS, Medo J, Kabba VTS (2013) Modelling landfill location using geographic information systems (GIS) and multi-criteria decision analysis (MCDA): case study bo, southern sierra leone. Appl Geogr 36:3–12. https://doi.org/10.1016/j.apgeog.2012.06.013

Gupta S, Mohan K, Prasad R, Gupta S, Kansal A (1998) Solid waste management in India: options and opportunities. Resour Conserv Recycl 24(2):137–154. https://doi.org/10.1016/S0921-3449(98)00033-0

Hazarika R, Saikia A (2020) Landfill site suitability analysis using AHP for solid waste management in the Guwahati Metropolitan Area, India. Arab J Geosci 13(21). https://doi.org/10.1007/s12517-020-06156-2

Justice F, Pillai AVR, Judge F, Secretary TC, Kumar J, Basheer A, Corporation KM, Environmental L, Forum, A (2021) Solid waste disposal site at Kalamassery (vol 2019, Issue January)

Kara C, Doratli N (2012) Application of GIS/AHP in siting sanitary landfill: a case study in Northern Cyprus. Waste Manage Res 30(9):966–980. https://doi.org/10.1177/0734242X12453975

Kaza S, Yao L, Bhada-Tata P (2008) What a waste 2.0. a global snapshot of solid waste management to 2050. https://openknowledge.worldbank.org/entities/publication/d3f9d45e-115f-559b-b14f-28552410e90a

Kontos TD, Komilis DP, Halvadakis CP (2005) Siting MSW landfills with a spatial multiple criteria analysis methodology. Waste Manage 25(8):818–832. https://doi.org/10.1016/j.wasman.2005.04.002

Kumar S, Smith SR, Fowler G, Velis C, Kumar SJ, Arya S, Kumar R, Cheeseman C (2017) Challenges and opportunities associated with waste management in India. Royal Society Open Science 4(3). https://doi.org/10.1098/rsos.160764

Kumar S, Hassan MI (2013) Selection of a landfill site for solid waste management: an application of ahp and spatial analyst tool. J Indian Soc Remote Sens 41(1):45–56. https://doi.org/10.1007/S12524-011-0161-8/METRICS

Lsgd (2019). Brahmapuram-SWM-Reports. https://lsgkerala.gov.in/system/files/2019-02/brahmapuram-SWM-Reports.pdf

Majid M, Mir BA (2021) Landfill site selection using GIS based multi criteria evaluation technique. a case study of Srinagar city, India. Environ ChallS 3. https://doi.org/10.1016/j.envc.2021.100031

Ministry of housing and urban affairs (2017) Swachh bharat mission (Urban). Ministry of housing and urban affairs government of India, 1–68

Ministry of urban development government of India. (2008). National Urban Sanitation Policy. 44. https://www.mohua.gov.in/upload/uploadfiles/files/NUSP_0.pdf

Mokhtari E, Khamehchian M, Montazer G, Nikudel M (2016) Landfill site selection using simple additive weighting (SAW) method and artificial neural network method; a case study from lorestan province, Iran. Int J Geogr Geol 5(10);209–223. https://doi.org/10.18488/journal.10/2016.5.10/10.10.209.223

Motlagh ZK, Sayadi MH (2015) Siting MSW landfills using MCE methodology in GIS environment (Case study: Birjand plain, Iran). Waste Manage 46:322–337. https://doi.org/10.1016/j.wasman.2015.08.013

Najjari A, Shayesteh K (2019) Site selection for hazardous waste using fuzzylogic combined with analytic hierarchy process: a case study in Nahavand, Iran. Avicenna J Environ Health Eng 6(1), 8–15. https://doi.org/10.34172/ajehe.2019.02

Nochian A, Tahir OM, Mualan S, Rui D (2019) Toward sustainable development of a landfill: landfill to landscape or landscape along with landfill? a review. Pertanika J Soc Sci HumIties 27(2):949–962

Paravur N, Municipality T (2013) Central pollution control board zonal office (South), Bengaluru report on municipal solid waste management & processing facility At Kochi, Kerala 1–6

Paul B, Paul D (2021) Comparative analysis of municipal solid waste management in Kochi and Indore. Nat Environ Pollut Technol 20(3):1339–1345. https://doi.org/10.46488/NEPT.2021.V20I03.047

Roy D, Das S, Paul S, Paul S (2022) An assessment of suitable landfill site selection for municipal solid waste management by GIS-based MCDA technique in Siliguri municipal corporation planning area, West Bengal, India. Comput Urban Sci 2(1). https://doi.org/10.1007/s43762-022-00038-x

Siby S, Jayamohan KG (2021) Impact of solid waste effect on ground water, soil, noise and air quality nearer to Brahmapuram solid waste landfill site in Brahmapuram, Kerala, India. An Int J (AMIJ) SingapAn J Sci Res 12(1):32–40. www.sjsronline.com

Saaty RW (1987) The analytic hierarchy process-what it is and how it is used. Mathematical Modelling 9(3–5):161–176. https://doi.org/10.1016/0270-0255(87)90473-8

Suchitwa Mission–Govt. of Kerala. (2020). Introduction and strategic environmental assessment of waste management sector in Kerala. Kerala solid waste management project. https://documents1.worldbank.org/curated/en/788471589794618595/pdf/Environmental-and-Social-Management-Framework-Introduction-and-Environmental-Assessment.pdf

Sohkhlet D, Nagargoje S (2020) Municipal solid waste management: a comparative study between Sydney (Australia) and Pune (India). E3S Web of Conferences 170:1–12. https://doi.org/10.1051/e3sconf/202017004001

State K, Control P (2023) Episodic air pollution at solid waste dumpsite. https://kspcb.kerala.gov.in/assets/uploads/widget/action_plan/Bhramapuram_Report-Final.pdf

Subramony S (2020) Project—IUC : Sustainable and innovative cities and region Sub-project—baseline study for solid waste management in Kochi Deliverable – Final report Acknowledgments. https://www.ieup.eu/assets/knowledge-bank/230220210.4393580016140678 70.pdf
United Nations economic and social commission for Asia and the Pacific. (2002). Chapter 8 Types of wastes. United Nations ESCAP Library, 170–194. http://www.unescap.org/sites/default/files/CH08.PDF
Uyan M (2014) MSW landfill site selection by combining AHP with GIS for Konya. Turkey. Environmental Earth Sciences 71(4):1629–1639. https://doi.org/10.1007/s12665-013-2567-9
Wilson DC, Velis CA (2015) Waste management - still a global challenge in the 21st century: an evidence-based call for action. Waste Manage Res 33(12):1049–1051. https://doi.org/10.1177/0734242X15616055
Yildirim V (2012) Application of raster-based GIS techniques in the siting of landfills in Trabzon Province, Turkey: A case study. Waste Manage Res 30(9):949–960. https://doi.org/10.1177/0734242X12445656
Zelenović Vasiljević T, Srdjević Z, Bajčetić R, Vojinović Miloradov M (2012) GIS and the analytic hierarchy process for regional landfill site selection in transitional countries: a case study from Serbia. Environ Manag 49(2):445–458. https://doi.org/10.1007/s00267-011-9792-3

Chapter 15
Analysing Urban Space Design Principles for the Homeless: Examples of Gezi, Anıt, and Sıracevizler Parks

Gökçen Firdevs Yücel Caymaz and Selva Köroğlu Kurt

Abstract The study aims to investigate the outdoor design requirements for the homeless, who are increasing daily as an important social problem worldwide. For this purpose, the literature review created a checklist including urban space and design parameters deemed necessary to improve the quality of space for the homeless. The data in the literature section were analysed under the titles of Maslow's hierarchy of needs of the homeless, psychosocial needs of users, spatial criteria affecting the quality of the recovery environment, visual aesthetic elements, and functionality of urban furniture. Beyoğlu (Gezi Park), Fatih (Anıt Park), and Şişli (Sıracevizler Park), the three districts with the highest homeless population in Istanbul, were selected as the study area. A total of 21 homeless people living in these parks were surveyed face-to-face. In addition to questionnaires, photographs, video footage, and on-site observations were also used in the evaluations. As a result of the study, it was determined that all three parks generally met the expectations; However, the use of colour in urban furniture, which was considered important in the environments where the homeless lived, was insufficient; Flowering plants were not included in either park except for Anıt Park.

Keywords Homelessness · Urban space · Social exclusion · Parks and homeless · Urban poverty

15.1 Introduction

The homelessness rate is increasing rapidly (Ludden 2023; dos Santos 2005), according to the report prepared by the European Federation of National Organizations Working with the Homeless (Feantsa) and the French Abbe Pierre Foundation

G. F. Yücel Caymaz (✉) · S. Köroğlu Kurt
Istanbul Aydin University, Istanbul, Turkey
e-mail: gokcenfyucel@aydin.edu.tr

S. Köroğlu Kurt
e-mail: selvakoroglu@stu.aydin.edu.tr

L. K. Dashora et al. (eds.), *Urban Sustainability*, Disaster Risk Reduction,
https://doi.org/10.1007/978-981-96-4890-0_15

(Feantsa 2023). The report showed that the highest rates of homelessness were in England at 169%, Germany at 150%, and Ireland at 145%. (Stouhi 2022). Levinson and Marcy (2007) define homeless as 'people without a home who live on the streets and in public places such as parks and riverbanks'. The definition of homelessness varies according to a specific housing situation, a specific minimum standard, the duration of homelessness and lifestyle problems, the use of the welfare system and being part of a certain group of the population, the situation of homelessness, and the possibility of moving or not moving (Richard et al. 2023; Springer 2023). Homelessness is a problem that leads to important situations such as deprivation of the social, psychological, sociological, and cultural benefits that a home provides to an individual, exclusion from socialization, and lack of self-confidence (Sanders and Allard 2021; İlhan and Ergün 2010). Not having a place of one's own, persona non locata, which means almost non-existence as implied by studies on the homeless (Gieryn 2000).

Since the early twentieth century, societies have generally housed the homeless in municipal buildings. At the same time, in the recent past, with the emergence of traditional emergency shelter problems, efforts to provide shelter have become increasingly diversified. According to the National Center on Homelessness and Poverty Law in the United States, many people suddenly become homeless after losing their jobs, rent increases, or home foreclosures (Kingery-Page 2019). Although legal regulations and social service practices exist for the homeless in Turkey and abroad, there are limited sustainable services internationally (Akyıldız 2017). In the United States, hundreds of thousands of people facing poverty, substance abuse, mental illness, and other social concerns are often forced to use public spaces, such as city parks, as temporary homes for sleeping, eating, and living (Rose 2019). Homelessness sometimes manifests as extreme urban poverty even in developed European Union countries (Paraschiv 2013). Homelessness increased rapidly in Japan in the early 1990s, and tent villages began to be set up on riverbanks and in parks in major cities (Cassegård 2014). Low wages, domestic violence, deterioration of mental health, lack of accessible health services, domestic and foreign migration, and systemic racism can also be counted among the causes of homelessness (Spur + Gehl 2024;80 Cities 2023).

One of the main recommendations in Davis (2004) 'Designing for the Homeless' is that public space should function as both a buffer and a transition, accommodating individual desires for privacy. When people face the fact that they have no home, they look for and create new places where they can be at home or feel at home (Jones 2013). Homeless people occupy public spaces due to the lack of space, low-cost housing, and the absence of livable paid employment. Since homeless people do not have their own private spaces, they take shelter in public spaces; and can rest and beg on benches, under bridges, or on sidewalks; Regardless of their shape, scale, and location, they can remain in these areas at night (Wright 1999).

Francis and Hester (1990) in 'The Meaning of Gardens: Idea, Place, and Action', Hester draws attention to new forms of the garden and implies that the garden is also the home of the homeless. This definition may be one of the first theories to include the homeless. Some gardening programmes offer food and temporary shelter for the homeless. Homeless people reorganize urban spaces according to their needs (Bergamaschi et al. 2014). The study results of Koprowska et al. (2020) showed that urban green spaces are important for homeless people (Kirschmann 2023; Speer and Goldfischer 2020) not only in terms of necessity or lack of other options but also more importantly in terms of individual choice and meeting personal needs (Table 15.1). A study by Parker showed that although many homeless people live in urban parks and sidewalks around social service centres, they also frequent places created by and adjacent to transportation infrastructure (Parker 2021).

Homeless individuals find safety in public spaces, such as parks, through interaction with others experiencing the same problem (Speer and Goldfisher 2019; Rennels and Purnell 2017). A park is an open green space with diverse norms, forms, recreational opportunities, and design principles at various scales, including urban, regional, and national. In general terms, it is a type of green space that meets various passive and active recreation needs such as rest, meditation, entertainment, free play, etc., has limited use, shows a flexible form feature, and contains minimum construction and maximum natural elements (American Planning Association 2006; Akdoğan 1982). Park planning aims to organize the interaction between people and the natural environment to create a sustainable and comfortable living space for individuals (Table 15.2) (Jodidio 2022; Low et al. 2005).

Table 15.1 Share of recorded homeless locations found within the different categories of urban green spaces (Koprowska et al. 2020)

Urban green spaces	With a 5 m buffer (%)	No buffer
Parks	2.3	2.1%
Forests	0.3	0.3%
Allotment gardens	0.9	0.9%
Cemeteries	0.2	0.1%
Large and medium-sized informal UGS (according to the database of topographic objects)	6.8	–
Small pieces of informal UGS (according to the European settlement map)	25.5	–

Table 15.2 Areas with open space used for homeless people and the facilities within them (Spur + Gehl 2024, Kingery-Page 2019)

Eddie Maestas Park, Denver, CO	A place where people can eat, sleep, socialize, get counselling, help finding a job, and access health care
Oppenheimer Park, Vancouver, British Columbia	Open visibility sites, providing services to the homeless rather than focusing on isolating them from the area
The Living Room Project, Seattle, WA	Emergency shelters are necessarily focused on overnight stays

The results of a study conducted by the National Recreation and Park Association (NRPA) in 2017 emphasized the importance of providing homeless people with space to support personal hygiene (toilets, showers, etc.), shelter in emergencies, access to computers/phones, and connecting people to social services (Colman and Barrett 2022). In his Hierarchy of Needs, American psychologist Abraham Maslow (1943) explains that an individual can only move from one level or 'stage' to another by meeting each of the previous needs. Physiological needs, which have partial goals, cease to be active determinants or regulators of behaviour when they are chronically met (Maslow 1954). He mentions food, water, shelter, sleep, clothing, and sex as the most basic needs. These are followed by safety, love/belonging, dignity, and finally self-actualization (Table 15.3) (Maslow 1943; Walsh 2011). Among the psychosocial needs that will contribute to the design of spaces (parks) for the homeless, the issues of 'Belonging, Privacy, and Supportive atmosphere' will also be important (Ross 2011).

Homeless people can be found throughout the park, including areas considered off-limits to park users; they may coexist with residents and visitors in the park (Low et al. 2005). In a 2019 survey conducted by the Guadalupe River Park Conservancy, 43% of respondents expressed concerns about homeless people in the open-ended response section of the survey (Spur + Gehl 2024). To fully benefit from the many benefits that parks and public spaces can provide to community health and well-being, communities must embrace inclusive design and management. The installation and maintenance of public restrooms will provide adequate space for all park users to maintain privacy and meet their basic needs (Safe Routes Partnership 2021). Park spaces should be designed to accommodate the street furniture needs of homeless people.

Attention should be paid to colour and lighting design in areas used by homeless people. Folkets Park in Copenhagen uses a dim lighting model that not only allows certain areas to be illuminated but also provides dimly lit areas for homeless people to sleep comfortably at night without fear of theft (Spur + Gehl 2024). Increased visibility, achieved through improved lighting and brush clearance, will enhance the

Table 15.3 Studies that relate Maslow's Hierarchy of Needs to reflect the lives and situations of homeless people (Arnett 2014)

La Gory and Pipkin (1981)	Safety, love and belonging, and self-esteem must be met before reaching self-actualization, lack of shelter
Burt and Cohen (1989)	Food
Fischer (1992); North, Smith, and Spitznagel (1994)	Safety
Solarz and Bogat (1990)	Love and belonging
Heather Mosher (2010)	Survival, psychological, power, and structural
Anthony Warnes and Maureen Crane (2000)	Offering services
Warnes and Crane (2000)	Hygiene and better nutrition

safety of all park users, including vulnerable individuals experiencing homelessness (Safer Routes Partnership 2022). Colours have varying effects on spatial perception at different times of day due to light. Since light colours reflect light, they create a bright and soft impact in the morning and evening hours, while they may appear pale due to strong light at noon. On the contrary, dark colours absorb light, making them appear vivid and attractive at noon but less interesting in the morning and evening (Holtzschue 2012). Open toilets, water fountains, hand washing, and sanitation facilities will create a quality environment for the homeless, pregnant women, young children, older adults, and others throughout the year. The presence of moving or stagnant water elements will increase the comfort of the environment with features such as recreational use, physiological effects, and adding liveliness and movement to the environment (Feyizoğlu 2008).

With the loss of physical space comes a serious loss of community and social life, and these losses should be considered in solutions to homelessness (Taşçı et al. 2024). The problem of homelessness has been addressed in scientific studies with its psychological (Somerville 2013; Phelan et al. 1997; Goodman et al. 1991) and sociological dimensions (Rose 2014; Kohut and Patterson 2022; Grainger 2021; Christian et al. 2011; Akyıldız 2017; Shlay and Rossi 1992; Shinn and Weitzman 1990), and there are very few studies on architectural design approaches of physical spaces for the use of the homeless. This study will address a gap by emphasizing design details to ensure that people living on the streets can live at an acceptable level. Urban parks provide shelter for the homeless and people who live on the streets (Abel 2019; Mitchell 1992). Public parks serve as a social bond for homeless individuals where they live with their friends who experience the same difficulties. A previous study conducted by Reitzes et al. (2011) at Robert W. Woodruff Park in Atlanta, Georgia, found that homeless individuals may seek out resources, including service organizations, in the park.

Davis (2006, 2017) mentions the lack of public toilets, fences, gates, and 'homeless-proof' benches among the designs in public spaces that discriminate against and exclude homeless people. The increasingly hostile design of public spaces has created parks that exclude homeless people (French 2023). As many authors worldwide (Wandalowski 2021; Hu 2019) have stated, hostile designs such as spikes, granite stones, fences, or benches with partitions to prevent lying down, which also aim to deter homeless people from using them, should be avoided (Jensen 2018). It is important to ensure that people who have to live on the street in some way can live at an acceptable level. In this context, this study investigates which design parameters should be taken into consideration to increase the outdoor living comfort of homeless individuals living in constantly increasing open spaces all over the world, under the titles of Maslow's hierarchy of needs, spatial criteria affecting the needs of users and the quality of the recovery environment, and furniture usage standards in parking areas, respectively.

15.2 Methodology

15.2.1 Research Areas

According to a study titled 'Those Spending the Winter on the Streets of Istanbul: Where Are the Homeless in the City and Politics'? conducted on one of Turkey's least talked about problems, the homeless, the three districts with the highest number of homeless people in Istanbul are Beyoğlu, Fatih, and Şişli. The reason why the homeless, whose number is estimated to exceed a thousand in the city, live in these districts is that they feel safe due to the crowds at all hours of the day. According to a report prepared by the Istanbul Zoning Agency in consultation with homeless people, tradesmen, and authorities, the areas most frequently used by homeless people to spend the night are Gezi Park in Beyoğlu, Anıt Park in Fatih, and Sıracevizler Park in Şişli (Sayın 2023). For this reason, these three parks were selected as sample areas in the study.

Gezi Park (also known as Taksim Gezi) is a city park in the Beyoğlu district of Istanbul. It is located between Cumhuriyet, Asker Ocağı, and Mete streets, northeast of Taksim Square. Anıt Park (also known as Saraçhane Park) is located in the Fatih district, just in front of the Bozdoğan Aqueduct, and features a monumental statue depicting Fatih Sultan Mehmet conquering Istanbul by riding through an arch on his horse, surrounded by his teachers. The renovation of Sıracevizler Park in Şişli was completed in 2021 and includes a playground for pets, a basketball court, and a new children's playground (Table 15.4).

15.2.2 Research Methodology

In the study, an attempt was made to develop a checklist to enhance the quality of life for homeless individuals in urban areas. The first part of the checklist was organized according to the survey results conducted with the homeless. A face-to-face survey was conducted with 21 homeless people in the selected sample areas. Of the 12 homeless individuals interviewed at Gezi Park, eight reported staying in the park permanently, while the remaining four mentioned moving from place to place. In Anıt Park, 8 of the interviewees stated that they always stay in this park. In Sıracevizler Park, three people said that they do not stay in the park continuously. In the results of the survey, it was determined that the issues that the homeless people find missing in their environment are the need for value (temporary space use—homeless people who change their places), belonging (permanent space use—homeless people who live permanently), meeting the need for water, and having garbage bins. The other parts of the checklist included the safety and physical needs of the homeless; belonging, privacy, and supportive environment criteria; and the functionality and appearance of furniture use criteria, which are considered important within the scope of the literature study. Positive answers to the examples evaluated in line with the

Table 15.4 Characteristics of research areas (Beyoğlu, Fatih, and Şişli municipality archives)

	Gezi park	Anit park	Siracevizler park
Construction date	1943	1985	1980
Locations	Beyoğlu-İstanbul	Fatih-İstanbul	Şişli-İstanbul
Coordinates	41°02′18″K 28°59′13″D	41°0′56"N28°57′15"E	41°02′39″K 28°59′23″D
Area	38,000 m^2	16,220 m^2	9000 m^2
Area usage			
Shade and sun areas			
Use of water element			
Garbage bin			
Benches			

(continued)

Table 15.4 (continued)

	Gezi park	Anit park	Siracevizler park
Sculptures			
Lighting equipments			

*All the photos were taken by the authors

checklist are marked with a tick (✓), while negative results are expressed with a cross (x).

15.3 Results

According to the study, it was found that the criteria based on survey data were satisfied in Gezi and Anıt Parks (Table 15.5). However, Sıracevizler Park did not meet the needs for value, belonging, and water, including the need for a fountain in the park. In the literature-based criteria evaluation, it was observed that all three parks fulfilled security needs such as cameras and police, as well as physical needs like food and care services. Under the heading of belonging, which includes the psychosocial needs of the users, it was determined that only Sıracevizler Park did not provide special areas where users could experience a sense of belonging; in terms of privacy, special sections that allow individual use in common areas in Gezi Park were not provided, and the use of colour was insufficient in terms of supportive atmosphere in all three parks. It was determined that there was no problem with the functionality of the furniture and that there was no use of flowers in Gezi Park in terms of visual aesthetic use (Table 15.5).

Table 15.5 Evaluation of park design parameters for the homeless (improved by the authors)

				1	2	3
Criteria based on survey results	Value need (temporary space utilization—homeless people in displacement)			✓	✓	X
	Belonging (permanent space use—homeless people living permanently)			✓	✓	X
	Meeting the need for water (existence of a fountain in the park)			✓	✓	X
	Having garbage bins			✓	✓	✓
Criteria determined by literature study	**In the Context of Maslow's Hierarchy**		Security needs (cameras, presence of security personnel)	✓	✓	✓
			Physical needs (Nutrition-Food Service)	✓	✓	✓
	Psychosocial needs of users and spatial criteria affecting the quality of the healing environment	Belonging	Providing spatial elements that support independent movement and orientation in the use of space (Presence of signboards)	✓	✓	✓
			Providing options that allow for personalization/customization of the space (The layout of the benches, the concept of shadow-sun)	✓	✓	✓
			Providing flexibility options to change usage areas according to personal preferences (Portability of Urban Furniture)	✓	✓	✓
			Providing special spaces where you can feel a sense of belonging (Presence of areas that may be suitable for the use of materials such as tents and sheds)	✓	✓	X
		Privacy	Providing special sections for individual use in common areas (benches located in pocket areas)	X	✓	✓

(continued)

Table 15.5 (continued)

				1	2	3
			Providing spatial arrangements that allow personal separation in multiple use in sleeping areas (If the dimensions of the benches are in different sizes)	✓	✓	✓
		Supportive environment atmosphere	Providing an attractive effect outside the perception of an institutional place in the environment (being in an area accessible to every person)	✓	✓	✓
			Ensuring that a positive effect is created in the space by considering the effects of colours (use of colour in urban furniture)	X	X	X
			Providing a pleasing effect by allowing sufficient natural light into the space (no dense trees to block the sunlight)	✓	✓	✓
			Providing relaxing, positive effects thanks to warm light close to daylight thanks to artificial lighting (presence of lighting elements)	✓	✓	✓
	Furniture usage in the park area	Functionality of park furniture	No separatist elements on the benches	✓	✓	✓
			Sufficiency due to Material Use (Material not affected by external conditions such as rain etc.)	✓	✓	✓
			Availability of ergonomic length on benches	✓	✓	✓
			Having a backrest on the benches	✓	✓	✓

(continued)

Table 15.5 (continued)

				1	2	3
		Visual aesthetic elements	Existence of ornamental pools	✓	✓	X
			Availability of flower plants	X	✓	X
			The existence of monumental sculptures in the park	✓	✓	X

1. Gezi Parkı, 2. Anıt Parkı, 3. Sıracevizliler Parkı

15.4 Discussion

Designing supportive living environments for homeless people is important. Koprowska et al. (2020) determined in their interviews with homeless people that urban green spaces are seen as a place to sleep, a necessity or choice to avoid staying in a shelter, a place to socialize and find resources, as well as a place to find silence, rest and enjoy nature. Addo's (2018) study shows that homeless individuals often visit the park seeking a sense of belonging and acceptance.

In the initial survey results section of the first part of the evaluation, it was determined that the criteria were met in Gezi and Anıt Parks. However, the need for water, including elements such as value, a sense of belonging, and the presence of a fountain in the park, was not met in Sıracevizler Park. It is important to have well-maintained public toilets and fountains in parks (French 2023). Garbage bins should be placed in areas where they are needed to avoid obstructing pedestrian passage and wheelchair access (Türkoğlu 1991). In all three parks, proper placement and a sufficient number of garbage bins are satisfactory.

In all three parks, it was observed that security needs such as cameras, and police officers and physical needs such as food and nursing home services were met. Under the heading of belonging, which includes the psychosocial needs of the users, it was found that only Sıracevizler Park did not provide special areas that could make users feel a sense of belonging; in terms of privacy, Gezi Park did not have special sections that allow individual use in common areas, and in terms of supportive atmosphere, the use of colour was insufficient in all three parks.

The use of colour in parks is important for user comfort. The use of colour should be increased in all three parks. Different colours will create positive psychological and sociological factors in people. Colours can have effects such as warm-cold, stagnant-moving, constructive-destructive, or repulsive-attractive (Çetinbaş 2012). While warm colours evoke excitement and happiness, cool colours have a more calming and soothing effect (Korkut et al. 2010). Considering the colour harmonies in all three parks, it would be good to use colours in water, soil, walls, floor surfaces, and furniture (Altınçekiç 2000).

Nielsen (2024) mentions in his study that homeless individuals find the urban furniture in public spaces to be insufficient. Ensuring that homeless park users have access to bathrooms and fountains, electrical outlets, grill areas, seating, and shelters in parks throughout the year will reduce the likelihood of other park users being disturbed by activities such as littering (Pitas et al. 2024). No significant problems were observed in terms of the quantity and quality of furniture in all three parks. Many homeless people are afraid to sit among people they do not know; therefore, they prefer to sit on the periphery of the area (Jost 2009). Additional seating can be added to the perimeters of the parks in all three areas. Homeless people using the parks can be offered flexible options that allow them to change according to personal preferences.

While it is emphasized that urban furniture should be made of wood in the standards in the parameter of functionality (adequacy due to material use), users stated that they prefer it to be made of material that dries quickly after external conditions such as rain. Seating units should be resistant to corrosive and damaging effects such as natural events (wind, rain, sun, snow). They should be positioned so as not to be affected by vehicle and pedestrian traffic and should be placed in areas that people can easily reach (Türkoğlu 1991). Climate, socio-economic, and geographical factors influence material use. In urban seating elements, in addition to the selection of the carrier construction from durable materials in terms of durability, the selection of the seating part from materials that do not absorb water, such as wood, and plastic, are easy to clean and have low conductivity provides diversity in design (Akyol 2006). Seating in public spaces can include armrests in the middle of the benches, sloping or curved benches, studded surfaces, leaning bars, and rough stones instead of grass in open areas (French 2023; Stahl 2013). It is pleasing that the benches in all three parks do not have elements that would prevent the homeless. As a result of the study, one of the important results of this study is that although they were told that the possible material to be used in park design should be wood, they requested that the benches to be placed in the park be made of a material that they can wipe and dry themselves even if they get wet (Since they are less likely to experience rain while sleeping) so that they can spread their blankets over them and sleep. While sculptures are used in Gezi Park and Anıt Park, they are not used in Sıracevizler Park. Sculptures have functions such as integrating society and transmitting culture. In addition, sculptures are uniquely created by artists and become individual and social symbols within the space and the city. Sculptures are signs that can be perceived from different angles and are generally used in city squares, parks, and places where awareness of expression is desired (Öztürk Kurtaslan 2005).

Lighting elements should emphasize urban identity and contribute to the reading of urban identity (Boduroğlu 2001). Tholl (2017) stated in his study that homeless people feel insecure because strong night lights in parks can cause homeless people to be subject to mugging and attacks. Users did not express a negative opinion about the light level in the parks that were the subject of the study. Area lighting can be used to increase the quality of space use and suppress the panopticon effect in public spaces. The choice of artificial lighting is important for achieving the desired ambiance in a space. Adding flower beds to Gezi and Sıracevizler parks will increase the comfort of use. Trees in parks provide shade and help to provide a pleasant and calming atmosphere for the homeless. Flower beds can be used for aesthetics and naturalness and to delimit the area, especially in places with insufficient green space, hard ground, and ground that is not suitable for growing plants (Durmuş 2008; Anılsın 2001).

Given the hostile social and physical landscapes that homeless people face in all three parks, park managers can use public education to address host community members' preconceptions about homelessness. For example, signage and written communications should emphasize that public space is shared and open to everyone. The organization of space in parks can impact its usability for homeless individuals. Area lighting, park edges, and separate areas can be used to create dynamic public spaces that meet the needs of homeless people (French 2023).

The interaction between the urban environment and homelessness supports the need for a paradigm shift in the attitudes and evaluation approaches to services and shelters for individuals experiencing homelessness (Stahl 2013). Homeless people generally live in urban public spaces. Public space does not belong to any particular social group (Bergamaschi, Francesconi, 1996). Individuals experiencing homelessness and homeowners are increasingly experiencing conflict over the use of public spaces, leading to efforts to regulate how individuals experiencing homelessness use public spaces (Addo 2018). In many parts of the world, the number of homeless individuals is increasing rapidly, with many struggling to survive in public spaces. These spaces were not designed as homes, and homeowners feel that the presence of homeless residents disrupts public spaces, making them inhospitable and even unsafe. Nevertheless, many homeless people are forced to live in public spaces such as parks, plazas, and squares (Huttenhoff 2021). At this stage, it will be important to design the physical space quality of public open spaces and parks that can offer a lot of usage opportunities, taking into account the use of homeless people.

15.5 Conclusion

Parks can provide equal use opportunities for everyone with inclusive design practices. Including homeless individuals in the planning and design processes, and considering their needs and wishes can help them secure their spatial rights in public parks (French 2023). At this stage, parks and other public spaces play a vital role in providing daily comfort for the homeless. Therefore, parks should be designed considering the needs of the homeless. As a result of the study, it can be said that current park design approaches generally meet the expectations of the homeless. Evaluation of the three parks where the field study was carried out suggests that the use of colour in urban furniture, floor coverings, and three-dimensional structures throughout the park should be increased. It was also determined that wood, which is a comfortable material in furniture design, is not preferred by the homeless because it dries slowly when exposed to rain.

References

Abel D (2019) Turning a playground into a home: the impact of homelessness on public parks. Doctoral dissertation, California State University, Northridge

Addo R (2018) Homeless individuals' social construction of a park: A symbolic interactionist perspective. Qual Rep 23(9):2063–2074

Akdoğan G (1982) Bitki Kompozisyonu Yüksek Lisans Ders Notları. Yıldız Teknik Üniversitesi Fen Bilimleri Enstitüsü Peyzaj Planlama Programı

Akyildiz Y (2017) Dünyada Ve Türkiye'de Evsizlik Sorunu Ve Çeşitli Uygulamalar. EUL J Soc Sci/LAÜ Sosyal Bilimler Dergisi 8(1)

Akyol E (2006) Kent mobilyaları tasarım ve kullanım süreci. Master Dissertation, Istanbul Technical University

Altınçekiç H (2000) Peyzaj mimarlığında renk ve önemi. J Faculty For Istanbul Univ 50(2):59–78. https://doi.org/10.17099/jffiu.65143

American Planning Association (2006) Planning and urban design standards. John Wiley & Sons

Anilsin F (2001) Ürün Kullanıcı İlişkileri Açısından İstanbul'da Kent Mobilyaları, 1.Uluslararası Kent Mobilyaları Sempozyumu, İstanbul Büyükşehir Belediyesi, Istanbul, 9–11 May 2001

Arnett JE (2014) Expanding the public realm: inclusive design guidelines for integrating homeless and non-homeless users. Doctoral dissertation, University of Georgia

Bergamaschi M, Castrignanò M, Rubertis PD (2014) The homeless and public space: urban policy and exclusion in Bologna. Revue Interv Écon Pap Polit Econ (51). https://doi.org/10.4000/interventionseconomiques.2441

Bergamaschi M, Francesconi C (1996) Urban homelessness: the negotiation of public spaces. Vis Sociol 11(2):35–44. https://doi.org/10.1080/14725869608583764

Boduroğlu Ş (2001) Kentsel dış mekanların aydınlatılmasının kentsel tasarım ilkeleri açısından incelenmesi. Master Dissertation, Mimar Sinan University

Cassegård C (2014) Space, art and homelessness. In: Youth movements, trauma and alternative space in contemporary Japan. Brill, pp 117–179

Cheng MA (2007) Finding home: making a place for the homeless in the urban landscape. Doctoral dissertation, Massachusetts Institute of Technology

Christian J, Clapham D, Abrams D (2011) Exploring homeless people's use of outreach services: applying a social psychological perspective. Hous Stud 26(5):681–699

Colman A, Barrett A (2022) Solutions to homelessness—the role of parks and recreation. https://www.nrpa.org/blog/solutions-to-homelessness-the-role-of-parks-and-recreation/

Çetinbaş Ş (2012) İç mekan tasarımında insan ve renk ilişkisi: Konut örneği. Master Dissertation, Karabük University

Davis M (2006) City of quartz: excavating the future in Los Angeles. Verso Books

Davis M (2017) Fortress Los Angeles: the militarization of urban space. In: Cultural criminology. Routledge, pp 287–314

Davis S (2004) Designing for the homeless: architecture that works. University of California Press, Berkeley, CA

dos Santos MCL (2005) Design waste and homelessness. Des Philos Pap 3(3):155–165. https://doi.org/10.2752/144871305X13966254124635

Durmuş Ç (2008) İstanbul kent mobilyaları üzerine bir araştırma. Master Dissertation, Yıldız Technical University

Feantsa (2023) A worrying increase in homelessness. https://www.feantsa.org/en/press-release/2023/09/05/?bcParent=27

Feyizoğlu S (2008) Kent mobilyaların kentsel mekanlarda yeri ve oluşum süreçlerinin irdelenerek sınıflandırılması. Master Dissertation, Mimar Sinan University

Francis M, Hester RT (eds) (1990) The meaning of gardens: idea, place, and action. MIT Press

French M (2023) Inclusive park design for people of all housing statuses: tools for restoring unhoused individuals' rights in public parks. The Ralph and Goldy Lewis Center for Regional Policy Studies, UCLA. https://escholarship.org/uc/item/1f1619sk

Gieryn TF (2000) A space for place in sociology. Ann Rev Sociol 26:463–496. http://www.jstor.org/stable/223453

Goodman LA, Saxe L, Harvey M (1991) Homelessness as psychological trauma: broadening perspectives. Am Psychol 46(11):1219

Grainger GL (2021) Punishment, support, or discipline? Taking stock of recent debates about homeless governance in neoliberal cities. Sociol Compass 15(8):e12909

Holtzschue L (2012) Understanding color: an introduction for designers. John Wiley & Sons

Hu W (2019) 'Hostile architecture': how public spaces keep the public out. The New York Times. https://www.nytimes.com/2019/11/08/nyregion/hostile-architecture-nyc.html

Huttenhoff M (2021) Coexistence in public space: engagement tools for creating shared spaces in places with homelessness. SPUR. https://www.spur.org/sites/default/files/2021-01/spur_gehl_coexistence_in_public_space.pdf

İlhan ÖGN, Ergün YDDA (2010) Evsizler ve Toplum Sağlığı. Sosyal Politika Çalışmaları Dergisi 20(20):79–90. https://doi.org/10.21560/spcd.08722

Jensen OB (2018) Dark design: mobility injustice materialized. In: Mobilities, mobility justice and social justice, pp 116–128. Routledge. eBook ISBN9780429434587

Jodidio P (2022) A sustainable future: urban parks & gardens. Prestel Publishing

Jones S (2013) Mean streets. A report on the criminalisation of homelessness in Europe. https://www.housingrightswatch.org/sites/default/files/Mean%20Streets%20-%20Full.pdf

Jost D (2009) Homeless haven. Landsc Archit 99(11):68–83

Kingery-Page K (2019) Designing for public space inclusive of unhoused people. https://thefield.asla.org/2019/03/07/designing-for-public-space-inclusive-of-unhoused-people/

Kirschmann A (2023) Homelessness in urban green spaces: an analysis of cultural ecosystem services through a lens of environmental justice in Munich, Germany

Kohut C, Patterson M (2022) Being homeless at the "end" of homelessness navigating the symbolic and social boundaries of housing first. Can Rev Sociol/revue Canadienne De Sociologie 59(1):59–75

Koprowska K, Kronenberg J, Kuźma IB, Łaszkiewicz E (2020) Condemned to green? Accessibility and attractiveness of urban green spaces to people experiencing homelessness. Geoforum 113:1–13. https://doi.org/10.1016/j.geoforum.2020.04.017

Korkut A, Şişman EE, Özyavuz M (2010) Peyzaj Mimarlığı. Verda Yayıncılık, Kayseri

Levinson D, Marcy R (2007) Homelessness handbook. Berkshire Publising Group, America

Low S, Taplin D, Scheld S (2005) Rethinking urban parks: public space and cultural diversity. University of Texas Press.

Ludden J (2023) Homelessness in the U.S. hit a record high last year as pandemic aid ran out

Maslow AH (1954) Personality and motivation. Harper and Brothers, New York, NY

Maslow AH (1943) A theory of human motivation. Psychol Rev 50(4):370–396. https://doi.org/10.1037/h0054346

Mitchell D (1992) Iconography and locational conflict from the underside: free speech, people's park, and the politics of homelessness in Berkeley, California. Polit Geogr 11:152–169

Nielsen CH (2024) Typologization of exclusionary design: An exploration of design interventions excluding unhoused people from urban public spaces. Des Stud 93:101264

Öztürk Kurtaslan B (2005) Açık Alanlarda Heykel-Çevre İlişkisi ve Tasarımı. Erciyes Üniversitesi Sosyal Bilimler Enstitüsü Dergisi 1(18):193–222

Paraschiv M (2013) Urban characteristics and homelessness in Bucharest. Urban Arhit Constr 4(2):27–34

Parker C (2021) Homelessness in the public landscape: a typology of informal infrastructure. Landsc J 40(1):49–66. https://doi.org/10.3368/lj.40.1.49

Phelan J, Link BG, Moore RE, Stueve A (1997) The stigma of homelessness: the impact of the label "homeless" on attitudes toward poor persons. Soc Psychol Q:323–337

Pitas N, Rose J, Mullenbach L, Russell Z (2024) Unhoused users in parks and public greenspace: an intergroup conflict approach. Landsc Urban Plan 249:105113. https://doi.org/10.1016/j.landurbplan.2024.105113

Reitzes DC, Crimmins TJ, Yarbrough J, Parker J (2011) Social support and social network ties among the homeless in a downtown Atlanta park. J Community Psychol 39(3):274–291. https://doi.org/10.1002/jcop.20431
Rennels T, Purnell D (2017) Accomplishing place in public space: autoethnographic accounts of homelessness. J Contemp Ethnogr 46:490–513
Richard L, Nisenbaum R, Brown M, Liu M, Pedersen C, Jenkinson JI, Hwang SW et al (2023) Toronto, Kanada'da evsiz insanlar arasında SARS-CoV-2 enfeksiyonunun görülme sıklığı. JAMA Netw Open 6(3):e232774–e232774
Rose J (2014) Ontologies of socioenvironmental justice: homelessness and the production of social natures. J Leis Res 46(3):252–271. https://doi.org/10.1080/00222216.2014.11950325
Rose J (2019) Unsheltered homelessness in urban parks: perspectives on environment, health, and justice in Salt Lake City, Utah. Environ Justice 12(1):12–16
Ross K (2011) Bud clark commons/holst architecture. ArchDaily. https://www.archdaily.com/189376/bud-clark-commons-holst-architecture. Accessed 25 May 2022
80 Cities (2023) Addressing homelessness in parks: an inclusive practices guide. https://www.880cities.org/wp-content/uploads/2021/09/Homelessness_in_Parks_AARP.pdf. Accessed 10 July 2024
Safe Routes Partnership (2021) Working with unhoused people in parks: steps to take in your community. https://www.saferoutespartnership.org/resources/fact-sheet/working-unhoused-people-parks
Sanders B, Allard M (2021) 'Barely breaking even': the experiences and impact of in-work homelessness across Britain. Crisis, London. https://www.crisis.org.uk/media/246566/barely-breaking-even-the-experiences-and-impact-of-in-work-homelessness-across-britain-2021.pdf
Sayın BC (2023) İstanbul'da büyüyen kriz gizli evsizler. https://gazeteoksijen.com/turkiye/istanbulda-buyuyen-kriz-gizli-evsizler-169594
Shinn M, Weitzman BC (1990) Research on homelessness: an introduction. J Soc Issues 46(4):1–11
Shlay AB, Rossi PH (1992) Social science research and contemporary studies of homelessness. Ann Rev Sociol 18(1):129–160
Somerville P (2013) Understanding homelessness. Hous Theory Soc 30(4):384–415
Speer J, Goldfischer E (2020) The city is not innocent: Homelessness and the value of urban parks. Cap Nat Social 31(3):24–41. https://doi.org/10.1080/10455752.2019.1640756
Springer S (2023) Global homelessness: neoliberalism, violence, and precarious urban futures: neoliberalism, violence, and precarious urban futures. The Palgrave encyclopedia of urban and regional futures. Springer International Publishing, Cham, pp 636–639
Spur+Gehl (2024) Coexistence in public space engagement tools for creating shared spaces in places with homelessness. https://www.spur.org/sites/default/files/2021-01/spur_gehl_coexistence_in_public_space.pdf
Stahl JC (2013) Interactions between the urban environment and "the homelessness": observations and responses. https://trace.tennessee.edu/utk_gradthes/2459
Stouhi D (2022) Exclusionary architecture: how design interventions in public spaces are dismissing the homeless. https://www.archdaily.com/983864/exclusionary-architecture-how-design-interventions-in-public-spaces-are-contributing-to-homelessness
Taşçı MN, Hatipoğlu Şahin B, Tereci A (2024) Shelter design for the homeless in the context of quality of life and social sustainability. In: Dakam Conference
Tholl S (2017) In Copenhagen, a "people's park" design includes dark corners. https://nextcity.org/features/copenhagen-park-design-includes-dark-corners
Türkoğlu G (1991) Vandalizmve Kent Mobilyası. Kamu Mekanları Tasarımı Kent Mobilyaları Sempozyumu IMSÜ Mimarlık Fakültesi Yayınları
Walsh PR (2011) Creating a "values" chain for sustainable development in developing nations: where Maslow meets porter. Environ Dev Sustain 13(4):789–805. https://doi.org/10.1007/S10668-011-9291-Y
Wandalowski Z (2021) Hiding homelessness through hostile architecture only hurts the community. https://dbknews.com/2021/10/19/hostile-architecture-homeless-city-planning-community/

Wright T (1999) New urban spaces and cultural representations: social imagineries, social-physical space, and homelessness. In: Hutchison R (ed) Constructions of urban space. Research in urban sociology, vol 5. Emerald Group Publishing Limited, Leeds, pp 23–57. https://doi.org/10.1016/S1047-0042(00)80020-3

GPSR Compliance

The European Union's (EU) General Product Safety Regulation (GPSR) is a set of rules that requires consumer products to be safe and our obligations to ensure this.

If you have any concerns about our products, you can contact us on ProductSafety@springernature.com

In case Publisher is established outside the EU, the EU authorized representative is:

Springer Nature Customer Service Center GmbH
Europaplatz 3
69115 Heidelberg, Germany

Batch number: 08719677

Printed by Printforce, the Netherlands